The Theory of *Hunyuan* Whole Entity

Foundation of *Zhineng Qigong* Science

Dr. Pang Ming

Copyright © 2014 Zhang Yuhong

All rights reserved. Copyright is under Universal Copyright and other international conventions. No part of this book may be reproduced or utilized in any form or by any means, electronic or mechanical, including photocopying, recording, or by any information storage and retrieval system, without permission in writing.

ISBN: 978-0-9864388-1-3

ZQ Educational Corporation is honored to assist in publishing this English translation of the Textbook Series of *Zhineng Qigong* Book II by Dr. Pang Ming.

ZQ Educational Corporation, Louisville, Kentucky, USA

Inquiries: sales@zqec.biz

Translator Zhang Yuhong with Dr. Pang Ming

Dr. Pang Ming reading in his study

Dr. Pang Ming writing in his study

Dr. Pang Ming with the translator on Nov. 5, 1998

The Theory of Hunyuan Whole Entity by Professor Pang Is the Guidance for Human Beings to Cultivate *Qigong* and Enhance Physical and Mental Health

Dedication to the English Version of *The Theory of Hunyuan Whole Entity*

Mr. Hao Huaimu

Ms. Zhang Yuhong, as a Master of Arts majoring in English who graduated from Tsinghua University, has eventually finished the translation of *The Theory of Hunyuan Whole Entity* written by Mr. Pang Ming, the most famous theorist, practitioner and educator of *qigong* culture in China today. Ms. Zhang has worked assiduously on the translation for seven years in her spare time, with successive revision of it for twelve times. The English version of *The Theory of Hunyuan Whole Entity*, being anticipated and awaited by many readers for so many years will be soon published.

As a person fascinated by traditional Chinese *qigong* culture and a previous administrator and promoter of *qigong* cause in China, I feel exhilarated on hearing the news of the upcoming publication of the book and write this dedication on the invitation of the translator to express my happiness and extend my congratulations.

The translation and publication of the English version of *The Theory of Hunyuan Whole Entity* give great convenience to and are good news worthy of celebration for readers of English, both in the east and west and readers all over the world, who intend to deeply understand and practice Chinese *qigong* and want to get rid of disease, tap intelligence and prolong life. It is also a bond and bridge to enhance communication of thought and culture and to reinforce friendship between people in the east and the west hemisphere.

The reason why I am so dedicated to recommending *Zhineng Qigong* created by Mr. Pang Ming to the world for people to learn is that I have been in close contact with Teacher Pang Ming for 20 years and has firsthand experience in *Zhineng Qigong* practice. I keenly feel that Teacher Pang Ming is the best human being in this world, that *Zhineng Qigong* he created is the highest level *qigong* method in the history of *qigong* development in China. Sometimes when I search into my heart of hearts and try to find an appropriate word to describe Teacher Pang Ming, I feel that using such words as "good", "very good" and "the best" can only express, to some extent, the character of Mr. Pang Ming as a true teacher and his contribution to human beings. As far as human science is concerned, or called life science of human beings, or called *qigong* science as a particular subject, merely in this realm, as far as practical problems solved in the two aspects of theory and practice are concerned, and according to contributions he made to this world, I have not seen talents surpassed Mr. Pang Ming.

Works on *Zhineng Qigong* written by Teacher Pang Ming, if put one upon another, which have been issued by publishing houses and are nearly twenty volumes in over seven million Chinese characters so far, plus additional materials and lectures having not been officially published, can reach the height of a human being. Although Teacher Pang Ming's works are named with *Zhineng Qigong*, these works in fact cover all aspects of traditional Chinese culture. You can find answers to all professional questions concerning *qigong* in his books. Therefore, I once suggested to Teacher Pang one day in the autumn of 1998 that, if he omitted the two Chinese characters of "*zhi neng*" (which means intelligence and ability or energy in Chinese), "Your works can surely be the textbooks for universal use for all *qigong* schools in China". Teacher Pang spoke his mind to me, "It is great to have a variety of *qigong* schools, and this is the demand of the development of our time." As time goes on, I read gradually more about Teacher Pang's works and understand his works more deeply, finding that Teacher Pang illustrates *Zhineng Qigong* in the grand background of entire traditional culture of China. In essence, Chinese *qigong* was born and has developed in the great background of traditional Chinese culture with a history of over 5000 years. The extensiveness and profundity of *qigong* originates from the extensiveness and profundity of traditional Chinese culture.

When talking about religion, people naturally think of Buddhism, Daoism (not Daoism as a school of thought, but the Daoism as a religion), Islam, including Confucianism though it is not a religion, and also Christianity, Catholics and so on, even some old and new heresies under criticism. On the one hand, for people not knowing much about religion, they do not know what is actually talked about in these religions and what exact relationship between them and *qigong* is. On the other hand, people do not know the depth and the advantage of each of them. It is just like breathing a sigh on seeing the vastness of oceans, or watching the fire on the other side of the bank—too far to catch a clear glimpse, being surrounded by the mist of not knowing which way to follow. However, readers will find that Teacher Pang gives answers to these questions.

He expounds the great amount of knowledge concisely and explains the profound truth with ease, making it easy for readers to understand.

Qigong of China is also closely related to martial arts, *taiji*, traditional Chinese medicine and so on that are associated with traditional Chinese culture. Teacher Pang gives very clear analysis about them. He is against narrow-minded nationalism, believing that all cultures created by human beings and beneficial to human lives should be respected, inherited and developed. Therefore, he not only masters traditional Chinese medicine, knows profoundly about the essence of Chinese martial arts and *taiji*, but is also a master of western medicine. He has reached a very high level of medical treatment in both traditional Chinese medicine and western medicine.

In addition to this, Teacher Pang pays very close attention to human beings' development of modern science and technology. He advocates that *qigong* science should make efforts to create conditions to do research related to systems theory, informationism, cybernetics, synergetics, dissipative structure theory and mutationism, make them an organic whole and try to rid ourselves of the biased period of modern human beings. What a great accommodating attitude!

To say that Teacher Pang Ming learns knowledge of both the ancient and the modern, of both the east and the west is not an exaggeration. To say that he has well digested and completely mastered the essence of the traditional culture of the Chinese nation and the achievements in the development of human beings' modern science and technology is also perfectly justified and most appropriate. Such laudatory evaluation of Teacher Pang Ming is already beyond a great many learners, scholars and experts. Yet, the description above has neglected a more outstanding and significant feature of Teacher Pang Ming, namely, his creative spirit, or his great creativity in culture. Personally, I think Teacher Pang Ming is highly creative in naming his *qigong* method as *Zhi Neng* which means intelligence and energy (or ability). This distinctly helps guide people to gradually advance from familiar intelligence state into the state of super intelligence. It helps end the life style with the mind distracted and limited by senses from the sensory organs and helps realize the species attribute of human beings that is free and self awakened.

It is a brand new creation that Teacher Pang names the theory of his *qigong* method "The Theory of *Hunyuan* Whole Entity" in order to have the theory agree with the practice method in spirit. No matter who the practitioner is, as long as the practitioner realizes the "*hunyuan* whole entity" in himself or herself (with *hunyuan* meaning merged into oneness with the universe), this practitioner will have achieved the fundamental condition to advance into the super intelligence state. In the theory of consciousness, Teacher Pang Ming has created many new concepts and new vocabulary, for example, "the Theory of Mind Oneness Entity", and this is firstly established in thousands of years of history of *qigong* development, with very profound impact.

Why Zhang Yuhong, the translator of *The Theory of Hunyuan Whole Entity*, chooses this book first in only more than 200,000 Chinese characters from the Chinese originals of more than ten books of *Zhineng Qigong* Textbook Series, which is for academic school students majoring in *qigong*, as the beginning step to introduce to English readers? This question caused a great interest in me to write this dedication. Thinking again and again about this, I come to realize that the translator Zhang Yuhong, as the re-creator of *The Theory of Hunyuan Whole Entity*, is also an admirer and a follower of Teacher Pang Ming and a practitioner of *Zhineng Qigong* created by Teacher Pang. She respects Teacher Pang Ming and has learned, perceived and grasped to a certain extent the system of the entirety of *Zhineng Qigong*. She has relatively deeply realized the significant position of this book in the system of the entirety of *Zhineng Qigong* and its function as an outline and overview in the panorama of this brand new science. It can well be said that reading this book first and to grasp its spiritual essence is a shortcut for learning *Zhineng Qigong* practice method and the entire Chinese *qigong* culture, for exploring the mystery of life and for realizing the goal of good health and long life. This is also the reason why I entitle this dedication "*The Theory of Hunyuan Whole Entity* by Professor Pang Is the Guidance for Human Beings to Cultivate *Qigong* and Enhance Physical and Mental Health."

Hao Huaimu

Former Minister of *Qigong* Department in Martial Arts Administrative Center

of State Sport General Administration of China

Beijing

Aug. 18, 2014

The Translator's Preface

Teacher Zhang Yuhong

The idea to translate Professor Pang Ming's books into English first occurred to me in 1994, when I was practicing *Zhineng Qigong* in *Huaxia Zhineng Qigong* Recuperation and Training Center in Qinhuangdao, a seaside city in China. I then went on with my master's degree studies in linguistics and applied linguistics in Tsinghua University and began earnestly wishing that I could translate the books written by Prof. Pang. In 2000, I was fortunate to become a student learning and practicing *Zhineng Qigong* in the *Qigong* Teacher Training Class, a two-year academic training program in *Huaxia Zhineng Qigong* Training Center affiliated to Hebei Sports College at that time. The dream of translation, however, had to be temporarily put aside because of busy work after my graduation. Since I have more time now, the translation of one of Prof. Pang's works, *The Theory of Hunyuan Whole Entity—Foundation of Zhineng Qigong Science* (originally published in Chinese by International Cultural Publishing House in China), is eventually accomplished.

During the translation of this book, I studied from Prof. Pang Ming and consulted him on whatever questions I had while reading his book. This deepens my understanding of the Theory of *Hunyuan* Whole Entity: This theory is a statement of the existential form of a special substance that has not been recognized by modern science yet and is called by Prof. Pang Ming the Hypothesis of Three Levels of Substances. This hypothesis and the term of three level substances have been mentioned several times in this book, and what is expounded in this book is unexceptionally the statement of this hypothesis from different stratums and aspects. In order to make it easy for readers to have a clear understanding of it, the Theory of Three Level Substances or the Hypothesis of Three Levels of Substances is now briefly summarized as follows:

According to the Theory of Three Level Substances, the universe is constituted of substances at three levels. They are: Firstly, the level of concrete substance. This refers to the level from atoms to macro-realm concrete substances, which is the form of objective existence that takes mass as its unique characteristic while energy and information reside in mass. Secondly, the level of substance in the form of energy field, such as electric, magnetic and gravitational fields. It is the form of objective existence that takes energy as its unique characteristic with mass (in the hidden state) and information residing in energy. The substances of the above-mentioned two levels have been profoundly studied in modern science, yet the ultimate depth has not been reached so there are still a lot of mysteries in the universe that baffle modern science. This is because there is a third level of objective existence in the universe, namely, the level of *hunyuan* whole entity substance (which is briefly called *hunyuanqi*). This is a level of substance that has not been recognized by human beings as a whole. It is an objective existence that takes time-and-space information as its unique characteristic with both mass and energy in the hidden state (However, this substance can transmute or generate into different kinds of energy under certain circumstances. At first sight, this hypothesis is seemingly similar to the potential *quan xi* energy field, the zero point energy field or vacuum energy at the subquantum level, the idea of which is established by Ervin Laszlo, the famous developer of systems philosophy. Yet they are essentially different.). It exists in an accommodating state with the two levels of substances mentioned above. Substances at this level include: a. Primordial *hunyuanqi* which is the most fundamental material state in the universe and all existences in the universe evolve from it. When all existences form in the universe, it becomes their background and driving force. b. *Hunyuanqi* of all existences. All existences in the universe have their own *hunyuanqi*. The substantial entities of all existences gather into form with it and it permeates in and around all existences. It is the noumenon or the thing-in-itself that makes features of all existences manifest. c. *Yi Yuan Ti*'s *hunyuanqi* (the *hunyuanqi* of the Mind Oneness Entity). It is the material foundation, i.e., noumenon, of the activity of human consciousness. It has not only the ability to reflect and remember existences but also the function of reflecting *Yi Yuan Ti* and other existences at choice. It has the initiative feature and the feature of making choice compared with the primordial *hunyuanqi* and the *hunyuanqi* of all existences.

According to the Theory of Three Level Substances, the three levels of substances can mutually transmute. The relationship between them is *qi* gathering into energy and energy gathering into form and concrete substance dispersing into energy and energy dispersing into *qi*. This process can naturally go on and can also be carried out by applying super intelligence techniques to turn *hunyuanqi* into different forms of energy. It can be carried out by mobilizing *hunyuanqi* to interact with all existences as well and make them change according to human requirements and plans. Super intelligence techniques can be acquired through specialized training. Features of the third level substance can be proven by means of experiments.

China has long been a country rich in the mystery of life cultivation and wisdom of great philosophy. *Zhineng Qigong* is a great inheritance and development of the essence of traditional Chinese culture, meanwhile well targeted at the great demand of our time for Good Health, Good Environment, and Harmonious Relationships between humans and nature among people all over the world. This target is realized by deep recognition of the nature of the world, of human life and of human relationship. *The Theory of Hunyuan Whole Entity*, which is about the Theory of Three Level Substances or the Hypothesis of Three Levels of Substances, is a profound work giving revelations not only to *qigong* practitioners who are actively improving their physical and mental health level, to researchers, scientists who are keen in knowing more about the truth of the universe and human life, but also to all people who are interested in the essence of Chinese culture. The translator is conscientiously presenting this great wisdom inherited from Chinese ancestors and developed with new creativity by Grand Master Pang Ming to the world.

It should be noted that many terms in this book are creatively new and have never been used before, and the translation of them was not very easy. Although I have tried my best, I might not have fully expressed all the author's meanings. However, with further study and improvement, I will have deeper understandings.

Zhang Yuhong

Mid-Autumn, 2011

A Brief Introduction

This book as the second textbook in the textbook series was written according to the needs of teaching academic school students majoring in *Zhineng Qigong*. It is the theoretical foundation upon which *Zhineng Qigong* Science is established. This textbook illustrates essentially the theory of the concept of *hunyuan* whole entity, which includes the theory of *hunyuan*, the theory of whole entity, human *hunyuanqi*, the theory of consciousness, the theory of *Dao De* (morality), the theory of the enhancement of life and the concept of *hunyuan* medical treatment. The theory of *hunyuan* whole entity is the outlook and the methodology in which super intelligence interacts with the objective world. It is the outcome of the combination of the classical concept of life as a whole entity with modern times as well as the crystallization of mass *qigong* practice. Although many links in it still need further enrichment and improvement, it is already a remarkably well-structured theoretical system.

The Author's Preface

Grand Master Pang Ming presides over the wedding ceremony of Master Liu Zhijun
and his wife, Xue Xiuchun, at Huaxia Zhineng Traing Center

For over a decade, epoch-defining changes have taken place in the cause of *qigong* development in China and an unprecedented promising prospect has been unfolded in *qigong* history. This can be seen in the following facts:

a. A great multitude of tens of millions of *qigong* practitioners have come into being, which changes traditional *qigong* of "the sages and the men of virtue" into *qigong* for the masses;
b. Achievements obtained in the research into miraculous *qigong* phenomena with technology in modern science have provided convincing evidence for the objective substantiality of *qi*;
c. Different levels of effects in *qigong* practice (such as getting rid of disease and prolonging life, enhancing and beautifying the human heart and body and tapping human potential) have been preliminarily yet comprehensively displayed;
d. *Qigong* has developed from the stage of internal *qi* (the stage of practicing *qigong* to build up one's own body) into the stage of external *qi*. It has developed from the level of "doing well in one's own life cultivation" as practiced in the past into a higher level of benefiting the world as well;
e. Traditional *qigong* theories and methods have been appropriately reformed in different *qigong* schools in order to meet the demand of the times.

Zhineng Qigong was born in the vigorous campaign for the reformation of traditional *qigong*. Its birth further facilitates the promising development of *qigong*. *Zhineng Qigong* has inherited the essence of a variety of traditional *qigong*s, including Confucianist *qigong*, Buddhist *qigong*, Daoist *qigong*, medical *qigong*, martial arts *qigong* and folk *qigong* and has assimilated the

i

reasonable quintessence of the whole entity concept of "heaven and human in one" in ancient Chinese civilization. It has also absorbed achievements from modern science, modern medicine and philosophy as well. It has established an integrated and well-organized theoretical system and has compiled hierarchical methods easy to learn and practice. It has also created a brand new approach to teach learners efficiently on a large scale in their *qigong* practice—organizing *qi* field. Moreover, it combines *qigong* with socialist spiritual civilization and integrates *qigong* into communist ideology. All these developments result in entirely new look in *qigong* and enable traditional *qigong* to develop into *Zhineng Qigong* Science.

Zhineng Qigong is now developing all over China like a raging fire. This is only the prelude to the mighty live drama of *Zhineng Qigong* Science, however. In order to welcome the arrival of the high tide of *qigong* science, more far-reaching and extensive mass *qigong* practice activities are needed since *qigong* science is a typical practical science. Yet this alone is not enough as *qigong* science is also a science of high rationality. It not only has its own special methods and methodology but also its unique foundation of methodology—and this is the fundamental difference between *qigong* and other modern sciences. In view of this, in order to ensure that *qigong* will develop along the road of *qigong* science, in order that mass *qigong* practice activities can break away from the confinement of one method and one technique and enter the realm of *qigong* science, the circles of *qigong* must improve their theoretical knowledge in *qigong* as well as their level of cultivation in *qigong* science.

In order to meet *Zhineng Qigong* practitioners' needs in their different levels of *qigong* practice, I wrote this *Zhineng Qigong* textbook series after the publication of *Zhineng Motional Qigong Methods for Elementary Learners* and *Concise Zhineng Qigongology* (intermediate level textbook). There are altogether nine volumes in this textbook series which include the fundamentals, techniques and applications of *Zhineng Qigong* Science: *A Survey of Zhineng Qigong Science*, *The Theory of Hunyuan Whole Entity—Foundation of Zhineng Qigong Science*, *Essence of Zhineng Qigong Science*, *Practice Methods of Zhineng Qigong Science*, *Technology of Zhineng Qigong Science—Super Intelligence*, *A Comprehensive Statement of Traditional Qigong Knowledge*, *Qigong and Human Culture*, *A Brief History of Qigong Development in China* and *Modern Scientific Research in Qigong*.

This textbook series was written according to the needs in teaching *Zhineng Qigong* academic school students. In *A Survey of Zhineng Qigong Science*, the main subject is the connotation and denotation of *qigong*, *qigong* science and *Zhineng Qigong* Science. The methods, methodology, special features of the foundation of the methodology in *qigong* science as well as their relationship with modern science are illustrated in this book. The position and significance of *qigong* science in human culture (particularly in the aspect of epistemology) have also been explored. This is a programmatic work which everyone who engages themselves in *Zhineng*

Qigong Science should carefully and conscientiously study and comprehend.

Essence of Zhineng Qigong Science is where the essence of *Zhineng Qigong* aggregates. In it, mysteries in *Zhineng Qigong* and traditional *qigong* are analyzed from the different aspects of consciousness, cultivation, adjusting breath and body postures. It contains both inheritance and development of traditional *qigong*. In this textbook, not only the common features among the many *qigong* schools in China, each of which has its own strong points, have been illustrated, but the secret techniques in a lot of *qigong* schools have also been revealed. What should be particularly noted is that, in the chapter entitled "Applying Consciousness" in this book, the formation and the laws of the motion of consciousness as well as the effect of the motion of consciousness upon the human body's life activities have been comprehensively and profoundly elaborated. This is the brand new content of *qigong* science, which has sublated the idealistic and god-orientated notion of consciousness in classical *qigong* and has established the dialectical materialistic concept of consciousness. It provides readers with the text to digest, the mental path to follow, the theories to go by and the methods to apply, demonstrating in words the changes of consciousness once so difficult to describe and only reliant on the learners' intuitive ability to comprehend. It is significantly instructive not only to *qigong* practice, external *qi* transmission, consciousness perception (telepathy, telekinesis) and consciousness transportation (telekinesis, teleportation) at the present stage, but also to guiding *qigong* practice at higher levels afterwards.

In *Essence of Zhineng Qigong Science*, in accordance with *qigong* classics and with support from *qigong* practice, the nature of illusion in the process of *qigong* practice and the nature of struggle between true fact and illusion in the process of *qigong* practice are elaborated. This not only helps eradicate superstition in *qigong*, but also points out the route to advance to higher levels. This is the critically important literature that people who intend to make profound research into *Zhineng Qigong* by personal practice must carefully and assiduously study so as to pass this critical juncture in their *qigong* practice.

The two books of *A Survey of Zhineng Qigong Science* and *Essence of Zhineng Qigong Science* in fact belong in the domain of the overview of *Zhineng Qigong* Science (and of *qigong* science as well). They state the laws of the science of *qigong* in the overall sense and are the common laws and shared content of all *qigong* schools. They are the quintessence of *qigong* as a branch of learning and the life-blood that runs through all schools of *qigong*. Bearing this in mind, if we take a look at each *qigong* school and method vying with each other for glamor, we come to know that they are in fact one family without distinction of being high or low among them. With this insight into *qigong* practice, advantageous advance and progress will be inevitable and irresistible.

The textbook of *The Theory of Hunyuan Whole Entity—Foundation of Zhineng Qigong Science* essentially illustrates the theory of the concept of *hunyuan* whole entity, which includes the theory of *hunyuan*, the theory of whole entity, human *hunyuanqi*, the theory of consciousness, the theory of *Dao De* (morality), the theory of the enhancement of life and the concept of *hunyuan* medical treatment. The theory of *hunyuan* whole entity is the theoretical foundation upon which *Zhineng Qigong* Science is established and is the outlook and methodology in which super intelligence interacts with the objective world. It is the outcome of the combination of the classical concept of life as a whole entity with modern times as well as the crystallization of mass *qigong* practice. Although many links in it still need further enrichment and improvement, it is already a remarkably well-structured theoretical system.

Practice Methods of Zhineng Qigong Science makes a detailed introduction into the theories and the methods of the first, second and third series of *Zhineng Qigong* practice methods. With vivid illustrations and revealing text, it is both a good teacher and a helping friend for *Zhineng Qigong* practitioners.

In *Technology of Zhineng Qigong Science—Super Intelligence*, the theories and the practice methods of the technology of super intelligence in *Zhineng Qigong* Science are comprehensively introduced. It elaborates upon at the theoretical altitude such subjects as detecting disease with *qigong* (including *qigong* penetrating sight, consciousness perception, and diagnosing disease away from patients) and treating disease with *qigong* (treating disease with external *qi*, with the doctor in one place while the patient is in another place, and by organizing *qi* field, etc.). It also reveals with perfect candor a variety of practical methods of how to acquire super intelligence and abilities efficiently and introduces all the secret techniques of super intelligence and abilities in practical use. Even without teachers, so long as practitioners assiduously practice according to these methods, they can quickly grasp them through self-instruction.

A Comprehensive Statement of Traditional Qigong Knowledge briefly introduces many kinds of traditional Mahayana *qigong* methods. What is covered in this book is mostly handed down to the author of this book from his teachers. By studying this book, readers can not only know something about the gist of traditional *qigong*, but can also more clearly understand this inside information that *Zhineng Qigong* Science has literally assiduously researched and well mastered the mysteries of a great variety of *qigong* schools and methods.

A Brief History of Qigong Development in China emphatically illustrates that the development of *qigong* in China has gone through a spiral form of development of negation of negation, during the process of which *qigong* developed first from simplicity to complexity and then from complexity to simplicity at a higher level. As for the two textbooks of *Qigong and Human Culture* and *Modern Scientific Research in Qigong*, the position of *qigong* in past, present and future civilizations of human beings as well as its rarely-known yet highly significant functions are comprehensively investigated in the entire background of both ancient and modern human civilizations.

It should be pointed out that although the author of this textbook series has made great efforts to express the many profound theoretical aspects of *qigong* in simple language, either by drawing support from facts in everyday life or by using existent scientific terms, there are still many terms and theories that are not easy for people to understand. This is because *qigong* science is a brand-new science. There is neither rational knowledge of *qigong* science nor perceptive knowledge or practical experience of *qigong* in people's common sense. In actual fact, any new and epoch-making scientific theory is at the beginning hard to understand. Isn't it so? When Albert Einstein was explaining his theory of relativity at the beginning of the 20th century, how many people could truly understand? Yet, it indeed wasn't very long before the theory of relativity became extremely important in advanced physics.

Another example is Percy Bridgman, a Nobel Prize winner in modern physics who, when giving an academic report on thermodynamics of irreversible processes in Brussel in 1946, was not only not understood by his audience but was also strongly opposed by them. A famous thermodynamicist even interrogated him sarcastically, saying that it was an absolute waste of time conducting research into such a something that would "pass in a twinkling of the eye". Decades later, the theories of irreversibility, nonlineality and instability today have not only become grand projects in which people invest great interest, but have immersed or are immersing into the field of multidisciplinary subjects. The science of *qigong*, though not as strange or new to the Chinese people as the theories mentioned above, is after all a completely brand-new field for modern science. Therefore, it is entirely understandable how incompatible people might feel when *qigong* is considered as science. Yet we believe that, with the development of *qigong*, people will gradually recognize the truth of the theories and viewpoints of *qigong* science in their *qigong* practice.

Lastly, another point that should be made clear to readers is that all the theories and the practice methods of *Zhineng Qigong* are newly created based on what the author has been sincerely studying and resolutely practicing. Although it has been investigated by millions of people in their actual practice, the science of *Zhineng Qigong* is after all something newly born that requires further enrichment. In this sense, this textbook series of *Zhineng Qigong* Science will undoubtedly have its imperfect points. I sincerely hope that friends in the *qigong* circles will kindly give criticisms and corrections so as to enable this textbook series to better exercise its functions in improving people's level of mental and physical health and to make its due contribution to human civilization.

Pang Ming

May, 1992

Intentions for Human Living

健康长寿

和谐自控

自由自觉

美满平等

Living in health with a long life

Being harmonious with self-control

Free with true-self awareness

Enjoying perfect beauty and equality

CONTENTS

Chapter II The Theory of *Hunyuan*   **44**

Chapter III The Theory of Whole Entity   **76**

Chapter IV Human *Hunyuanqi*   **124**

Chapter V The Theory of Consciousness 176

Chapter VI The Theory of Morality (*Dao De*)   **264**

Chapter VII The Theory of the Enhancement of Life and the Concept of *Hunyuan* Medical Treatment   319

混元靈通

Chinese Calligraphy by Dr. Pang Ming

Chapter I Introduction

The theory of *hunyuan* whole entity is a unity of outlook, methodology, epistemology and ontology of *Zhineng Qigong* Science. It is gradually established by absorbing the reasonable essence of the theory of *qi*, the theory of *qi* transformation and the theory of concept of human-and-heaven whole entity from traditional *qigong* and traditional Chinese medicine, by drawing support from the achievements of both modern science (including medical science, biology and the new and old three theories[1]) and philosophy, and through the practice and investigation of *qigong* (or super intelligence[2]) under the guidance of dialectic materialism and historical materialism. It is the foundation upon which the science of *Zhineng Qigong* is established. The theory of *hunyuan* whole entity seems to be a further development of holism in modern systems science yet differs in essence from it. What kind of theory is it then?

Section I Synopsis of the Theory of *Hunyuan* Whole Entity

I. What Is the Theory of *Hunyuan* Whole Entity

The theory of *hunyuan* whole entity is a theory that expounds the formation of the whole entity of existences, the special entity and features of the whole entity as well as how to recognize and use the features of the whole entity. The content of it not only differs from the common-sense concept of entirety, but also essentially differs from the concept of wholeness in systems science.

The concept of entirety in people's common sense is formed with human consciousness that processes the partial attributes of objective existences obtained with different human sensory organs and abstracts them with thinking. In other words, people's grasp of the entirety feature is not directly through cognition with human sensory organs, but indirectly through thinking. It reflects, to a certain extent, the holistic essential attribute of objective existences, but it is merely the entirety of the sum of partial features.

In modern systems science, the world is regarded as a unified whole made up of multiple-

Note:

1. The new and old three theories: Systems theory, cybernetics, informationism; dissipative structure theory, mutationism, synergetics.

2. Super intelligence and abilities (or super intelligence for abbreviation) refer to supersensory awareness and capabilities in receiving and sending information, in thinking and bodily functions, such as telekinesis, teleportation, telepathy, clairvoyance and clairaudience. These are the intelligence and capabilities tapped and manifested when *qigong* practitioners have reached a certain level in their *qigong* practice. Some people are naturally born with these abilities shown in their daily life.

stratum substances and the content of this multiple-stratum world is enumerated. For example, quarks, particles, atoms, and molecules in the micro material world and the earth, the solar system, galaxies, clusters of galaxies and island universes in the macro material world. In the biological world, there are macromolecules, subcellular fractions, cells, tissues, organs, systems, individuals and species. This is what is called system in modern science, with each level of existence constituted of components from the level below. All this is recognized and illustrated from the perspective of the structure of substances. Although it has also been pointed out in systems science that a certain structure will display a certain function, it has merely realized that structure is the foundation to bring about function in the system and that function is the external expression of structure. It has not realized and cannot realize the special entity and nature of the whole entity in which the structure and the function of the object are unified in oneness. Therefore, wholeness stated in modern systems science is just part of the attribute of the whole entity of existences. It is still not the entire true whole entity of existences itself.

The whole entity illustrated in the theory of *hunyuan* whole entity is the whole entity itself as well as its special attribute formed with the *hunhua* (literally meaning merge and change, here referring to mergence and combination) of the form and structure of an existence with its functions. This whole entity itself and its special attribute can be directly grasped through super intelligence. For people without super intelligence, this special entity and its attribute are hard to understand, as people with familiar intelligence[1] do not have this kind of sensory and perceptual experience yet. In order to make it easy to understand, this whole entity will be explained with specific things at different levels as forced analogies.

i. The *Hunyuan* Whole Entity of Plants

Although in a plant there are such differences as root, stem, leaf and flower, the plant itself is a *hunyuan* (merged-into-oneness) whole entity. This whole entity is not the sum entity of the root, the stem, the leaves and the flowers of the plant (in a dead plant, though its shape or body is still intact, it already has no vitality), but the unified entity in which the form and the structure of the plant merge into oneness with its metabolic functions. This *hunyuan* whole entity is a special state that is "shapeless and formless" and cannot be sensed or perceived by people with familiar intelligence. It fills both the inside and outside of the entire body of the plant and contains all the information about the entire features of the plant. The trail of this state of the plant can also be observed by people with familiar intelligence through certain means. The method is: Look at the sun in the early morning for five minutes and then look at the trees around. You will see some-

Note:

Familiar intelligence and abilities (which is called familiar intelligence for abbreviation) refer to a) the awareness acquired with the familiar modes of perceptions, such as seeing, hearing, smelling, tasting, touching and conception, and b) the abilities of the normal visual sense, aural sense, olfactory sense, gustative sense, tactile sense and making use of concepts.

thing fog-like around the canopy of the trees, and this is the rough reflection of the *hunyuan* whole entity of trees. The biological light observed with modern science and technology is but the expression of part of the features of the *hunyuan* whole entity. When people's super intelligence reaches a certain level, they can sense the biological light and cognize the whole entity feature with it. Some readers might ask: How can the root and the stem and the leaves and the flowers of a plant that are so different from each other be unified in a *hunyuan* whole entity? The answer to this question will be comprehensively expounded in the following chapters on *hunyuanqi*. Here I would just invite you to think about this—a tiny little plant seed can grow into a plant with deep roots and exuberant foliage. Doesn't this demonstrate the fact that the root, the stem, the leaves and the flowers of a plant are unified in the seed of the plant?

ii. The *Hunyuan* Whole Entity of Various Machines

This whole entity is not the structure of the machine itself, nor is it the sum of each of its components and their functions. The *hunyuan* whole entity forms when the value of the machine is realized as the machine operates under normal conditions with the structure of the machine driven by a certain power and the function of the structure of the machine displayed. Namely, the whole entity is the functional state in which the structure and the function of the machine and the power that drives the machine are unified. This kind of unified whole entity is a merged and combined whole entity in the oneness state that cannot be divided. Although people with familiar intelligence cannot completely sense and cognize it, they can probably come to understand certain special features about it. For example, a normal-intelligence person who wants to check why a machine stops has to check the components one by one according to the structure of the machine before this person can recognize whether there is anything wrong with the machine or not and also the position where the stoppage takes place. But an experienced mechanic can recognize the normality or the abnormality of the whole of the machine according to its overall performance, such as its vibration, sound or the feeling of some special touches and can even locate the position of the stoppage from these aspects. It is true that this kind of judgment made according to experience is still not the cognition of the *hunyuan* whole entity, but the understanding of a partial manifestation of it. A person with super intelligence, however, can directly perceive the entire content in the state of this whole entity.

From the two forced analogies mentioned above, it can be seen that the *hunyuan* whole entity of existences is a kind of special substantial-entity state perceived with super intelligence. It is a unified whole entity formed with the form and the function of existences merged and combined (*hunhua*). What the theory of *hunyuan* whole entity expounds is the substantial-entity characteristics of the *hunyuan* whole entity of existences as well as the laws of this substantial entity's *hunyuan* (merging-into-oneness) change.

II. Content of the Theory of *Hunyuan* Whole Entity

The theory of *hunyuan* whole entity can be applied to different existences at all levels in the universe and is very extensive in content. For the convenience of explanation, the theory of *hunyuan* whole entity is to be stated in six chapters in this book from the perspective of *qigong* science, i.e., the theory of *hunyuan*, the theory of whole entity, human *hunyuanqi*, the theory of consciousness, the theory of morality (*Dao De*), the theory of the enhancement of life and the concept of *hunyuan* medical treatment. What is provided here is merely an abstract of its content.

i. The Theory of *Hunyuan*

The theory of *hunyuan* essentially expounds the content characteristics of the *hunyuan* whole entity of existences. It is believed in the theory of *hunyuan* that existences in the universe are the display of *hunyuanqi* in different forms, which includes:

i) The Concept of *Hunyuan* Substance: Everything is made up of not merely one material factor but is a *hunyuan* whole entity (*hunyuanqi*) consists of at least two kinds of material factors that are merged and combined. There are different levels of *hunyuanqi* in the universe and the *hunyuan* particle (*hunyuanzi*) is the most fundamental *hunyuanqi* in it. Then there are primordial *hunyuanqi*, *hunyuanqi* of all existences (including inorganic matter and organic matter), human *hunyuanqi*, and finally the *hunyuanqi* that brings about consciousness activities—*Yi Yuan Ti* (Mind Oneness Entity).

ii) The Concept of *Hunyuan* Motion (or Change): Everything is in *hunyuan* (mergence-into-oneness) change. The mergence and change (*hunhua*) between each existence and its external environment is going on at all times.

iii) The Concept of *Hunyuan* Time and Space: Both time and space can show a merging and accommodating state, e.g., the past and the future of time can merge into an instant at present and extended space can merge into a limited space.

ii. The Theory of Whole Entity

In the theory of whole entity, what is essentially expounded is the characteristics of the *hunyuan* whole entity of existences in form, which includes the concept of universe whole entity, the concept of human-and-universe whole entity and the concept of human body whole entity.

iii. Human *Hunyuanqi*

Human *Hunyuanqi* mainly elaborates upon the entity and attribute of human body *hunyuanqi*, the laws of its generation and change as well as its dialectical relationship with human life motion.

iv. The Theory of Consciousness

In the theory of consciousness, it is pointed out that consciousness activity is a special mode of material motion. It is the content and process of the activity of *Yi Yuan Ti* (Mind Oneness Entity) which is the special form of manifestation of human body *hunyuanqi*; it is the fundamental expression of human nature. The content and laws of consciousness activity are then further expanded upon, together with its guidance position in life activities.

v. The Theory of Morality (*Dao De*)

In the theory of morality (*Dao De*), the significant position of *Dao De* as content of consciousness activity in all consciousness activities is stated.

vi. The Theory of the Enhancement of Life and the Concept of *Hunyuan* Medical Treatment

Exposition of human physiology, pathology, medical treatment as well as how human beings fantastically progress into more advanced levels is given in light of the theory of *hunyuan* whole entity.

III. Fundamental Differences between the Theory of *Hunyuan* Whole Entity and the Ancient and Modern Holistic Theories
i. Difference from the Holistic Concept in Modern Science

The theory of the concept of *hunyuan* whole entity is a materialistic monistic concept of whole entity:

i) According to the theory of *hunyuan* whole entity, every existence has its own *hunyuan* substantial entity and this *hunyuan* substantial entity can be cognized with super intelligence. Holistic structure and function stated in modern systems science are the partial attribute of different aspects of the *hunyuan* whole entity.

ii) Spiritual activity is itself the form of expression of the activity of human body *hunyuanqi* at a higher level. It is both the content of activity in the subjective world of humans and an objective existence that can be cognized with super intelligence. This is what modern science and philosophy find hard to accept.

iii) Although there is the method of analysis (the whole entity of substance is constituted of its parts) and the method of synthesis (the whole entity of substance determines its parts) in the cognitive process according to modern scientific methodology, in the foundation of their methodology is both the separateness and antagonism of spirit and substance. In the concept of *hunyuan* whole entity, however, it is monism of the unity of spirit and substance. The concept of

5

hunyuan whole entity is a concept of whole entity in which subject and object are unified.

iv) The theory of *hunyuan* whole entity is established upon the basis or the level of super intelligence. Although this theory can give extremely great inspirations to modern science, it is mainly applied in giving guidance to tap and apply human super intelligence. This forms an opposite yet complementary wing to the holistic theory in modern systems science which is hard to be applied to the depth of research into the realm of life and consciousness.

ii. Essential Difference from the Holistic Concept in Traditional *Qigong*

i) The theory of *hunyuan* whole entity has illustrated that *jing* (body, form, essence), *qi* and *shen* (spirit) as stated in traditional *qigong* are different forms of expression of *hunyuanqi*. This solved the problem in traditional *qigong* that *jing*, *qi* and *shen* are antagonistic. It gives "refining *jing* into *qi*" and "refining *qi* into *shen*" stated in ancient *qigong* an internal association and reason and solves the difficult point in the transformation between *jing*, *qi* and *shen*.

ii) The theory of *hunyuan* whole entity has endowed mental activity with the material feature of *hunyuanqi*. It has dissolved the myth of the concept of *shen* (spirit) in traditional *qigong* and enables *shen* to enter into the realm of human life and living, which, therefore, blocks the secluded tunnel to religion or theology where *qigong* might escape.

iii) The theory of *hunyuan* whole entity reveals the progressing order in which all levels of substances continuously evolve with humanity as its summit. Yet humans are not perfect and need to further improve and advance into higher levels. Practicing *qigong* is the means to accomplish this task. This breaks the enclosed circle in the holistic concept in traditional *qigong* which states that the natural world is from *Dao* to human and that *qigong* practice is from human back to *Dao*. According to the theory of *hunyuan* whole entity, *qigong* practice can elevate humanity to a new and upward spiral.

iv) It is clearly pointed out in the theory of *hunyuan* whole entity that change in existences is not simple change, but a *hunhua* (merge and change) process of the internal *qi* and external *qi*, and this is the power that facilitates development and change. This theory provides a theoretical foundation for the multiplicity of existences in the universe, thereby eliminating the factor of gods creating all in traditional holistic theories.

IV. The Theory of *Hunyuan* Whole Entity—a Unity of Outlook, Ontology, Epistemology and Methodology

According to the theory of *hunyuan* whole entity, the universe is a *hunyuan* whole entity, a substantial entity formed with the mergence and combination of substantial matter that has form and insubstantial matter that has no form, and nothing in the universe is an exception. For example, the organism is the *hunyuan* whole entity formed with the mergence and combination of its structure and its metabolic function; inorganic substance is the *hunyuan* whole entity formed with the reaction and mergence and combination of its material structure and its positive and negative energy entities (e.g., positive and negative electricity); quantum is the *hunyuan* whole entity formed with the mergence and combination of its features of vibration and particle; energy field is the *hunyuan* whole entity formed with its energy and its structure of time and space; the most fundamental level of substance, *hunyuan* particle (*hunyuanzi*), is the *hunyuan* whole entity formed with the mergence and combination of its structure of time and its structure of space. And so on and so forth. This *hunyuan* substantial entity is the overall worldview, namely, outlook, of *Zhineng Qigong* Science. Meanwhile, as the *hunyuan* whole entity itself is the substantial entity of all existences in the world, it also belongs in the domain of ontology. It is believed in the theory of *hunyuan* whole entity that this substantial entity, though it cannot be cognized with familiar intelligence, can be cognized with super intelligence. Therefore, it unveils the mystery of agnosticism in Immanuel Kant's ontology and enables it to step on a scientific track.

In the theory of *hunyuan* whole entity, it is expounded that super intelligence is also the innate intelligence of humans. Drastically different though it is from familiar intelligence, it follows the fundamental principle of practice-knowledge-practice again-further knowledge in epistemology whether it is in cognizing or in acting on objective existences and it can be tested and examined by the practice of familiar intelligence. For example, if someone is cognized with super intelligence as suffering from a disease, familiar intelligence can be applied to check whether the diagnosis is correct or not; when a patient's disease (a tumor, for instance) is dispelled in an instant with super intelligence, the means of familiar intelligence can also be used to check the reliability of the result. Although familiar intelligence means cannot explain the mechanism of super intelligence, the result of super intelligence acting upon the objective world can be confirmed with familiar intelligence. Therefore, the two methods of familiar intelligence and super intelligence in perceiving and reforming the objective world are both innate human functions. Moreover, both these methods follow the fundamental principle—the principle of practice in epistemology in cognizing objective existences. In view of this, methodology and epistemology are also combined in the theory of *hunyuan* whole entity.

Section II Theoretical Origins of the Theory of *Hunyuan* Whole Entity

Although the theory of *Hunyuan* Whole Entity is a brand new theory, in traditional *qigong* theories (including traditional Chinese medicine theories) and in the theory of systems science in

7

modern science, there is also a wealth of knowledge about the holistic concept. They provide a lot of materials for the Theory of *Hunyuan* Whole Entity and are its theoretical origins. A brief introduction of these origins from the aspect of traditional *qigong*, including the concept of holism and the theory of *qi* in traditional *qigong* theories, will be given as follows:

I. The Classical Concept of Holism in Which Humans and the Universe Are Harmoniously in Oneness
i. Materialistic World Outlook

The universe (equivalent to the two characters of *Yu Zhou* in the Chinese language) has been stated a long time ago in *Shi Zi*, "Above, below and the four directions (front, back, left and right) are called *Yu*, and that which is since time immemorial is called *Zhou*." It is also said in *Zhuang Zi*, "What exists yet has nowhere is *Yu*. What has length (or process) yet has no beginning and end is *Zhou*." In *Ling Xian*, it is further pointed out, "*Yu*'s borderline is unlimited. *Zhou*'s edge has no end." From this it can be seen that the two characters *Yu* and *Zhou* are actually the entire space and time as well as all existences accommodated in it. This is identical with the world outlook in materialism today.

It should be pointed out here that "heaven and earth" or literally "sky and land" (*tian di*) are usually used in ancient Chinese classic works to describe nature. Yet the connotations of *tian di* (heaven and earth, or sky and land) and *yu zhou* (the universe) are not the same. "Heaven and earth" refers mainly to the sun, the moon, stars, mountains, rivers, and the vast land, all of which are part of the universe. *Tian di* has its own process of changing into form and ruins (Please refer to *Ling Xian* for its formation and *Lie Zi* for its ruins).

i)   The Materiality of the Universe

Ancient *qigong* masters believe that there are different levels of matter in the universe which can exist independently and transform into and permeate into each other. According to broad classification, there are the two categories of *QI* (器, things with form) level which has form and the *Dao* level which is formless. This is what is said in *Xi Ci* in *Yi Jing* (*The Book of Changes*), "That which is above the level of form is called *Dao*. That which is below the level of form is called *QI*." If we classify in detail, five levels can be found as what is stated in *Dao De Jing* (*The Scripture of Dao and Its Functions*), "*Dao* generates one (oneness state). One generates two (*yin* and *yang*). Two generates three (*qi* images). Three generates all existences." In *Xi Ci* in *Yi Jing*, a classic work of Confucianism, it is said, "In *yi* there is *taiji* (Confucianists of late generations say that *wuji* generates *taiji*), which gives birth to two *qi* aspects (*yin yang*). The two *qi* aspects give birth to four *qi* images, which give birth to the eight trigrams." Here *Dao* and *yi* (or *wuji*) are at the same level; one and *taiji* are at the same level; two and two *qi* aspects are at the same level; three and four *qi* images are at the same level. "Three generate all existences" and "Four *qi* images give birth to the eight trigrams" are at the same level. These levels will be stated in detail as follows:

a. *Dao*

The level of *Dao* is the root of the universe. Although it is something formless, ancient Chinese masters have given descriptions of its entity, attribute and functions. In Confucianism, the description of *Dao* is relatively simple. For example, "silent and still (entity)"; "If it is communicated, it gets completely connected (attribute)"; "giving new life force and being inexhaustible (function)". The statement of *Dao* in Daoism is more profound. For example, in *Dao De Jing*, it is said, "*Dao* as a substance is something not as distinct and clear as what sense organs can perceive. 'Indistinct and not clear (so-called)', it has its own (*qi*) image. 'Not clear and indistinct', it has its own state of somethingness. Far-reaching and profoundly deep, it has quintessence. The quintessence is rather real. It has honesty and integrity." In *Zhuang Zi*, "*Dao* has great love and integrity in it, yet it is in inaction and has no form... It is the origin and root itself and has existed since the remotest antiquity before the existence of heaven and earth...It is earlier than *taiji* yet it does not consider itself as being aloft. It comes before heaven and earth and this is not considered as being long. It is older than the remotest ages and this is not considered as being aged."

In the passage above written by Lao Zi, the description of the state of *Dao* as a substance was given. In the passage from *Zhuang Zi*, the state of *Dao* as being surpassing time and space was described. If we combine the two descriptions, the entity and the state of *Dao* can be clearly understood. When practitioners have reached a certain level in *Dao* cultivation, this entity and state can be perceived. As to what is stated in *Dao De Jing*: "There exists something in a merging state born before heaven and earth. Silent and infinitely vast, independent and unaltered, it is motional all around without an end", and "It is clearly oneness and seemingly exists", what has been stated here is the description of *Dao*'s entity and attribute. While "*Dao* is abundant and never seems to be over sufficient in use. Profound and deep, it seems to be the ancestor of all existences." "It can be regarded as the mother of the universe." "*Dao* generates one." "*Dao* is what is immersed in all existences."...These statements are about the functions of *Dao*. That is, all existences come from *Dao* and are dependent on *Dao*.

In sum, *Dao* is a kind of objective, long-lasting substance that has no birth and death. It does not have beginning or end in time. Nor does it have borderline or end in space. It is everywhere in the universe and can generate the substance level of oneness, which further generates all existences in the universe. *Dao* penetrates through every level of substance in the universe.

b. *Taiji* (One, Oneness)

The substance level of *taiji* refers to *yuan qi* (primordial *qi*). "*Yin* and *yang* have not divided and *qi* is merged in oneness" mentioned in later generation refers to this. *Taiji* is also called *taiyi* (greatest oneness) (In *Lv Shi Chun Qiu* is recorded "From *taiyi* come the two *qi* aspects"). What kind of substance is *taiji* in the universe then? It is said in *Zhou Yi Zheng Yi*, "*Taiji* is the primordial *qi* merged in oneness before the division of heaven and earth." In *Gongyang Zhuan*, "*Yuan* is *qi*. It begins from having no form. It divides when *qi* images appear. It gives birth to heaven and earth and is their beginning." That is to say, *taiji* is an endless mass of primordial *qi* in greatest harmony. Its entity image is a state of being "'indistinct' and natural" (*Tai Ping Jing*), with not any differentiation in it. Its function is to generate and to give life force to all existences. It is said in *Tai Ping Jing*, "Existences come from primordial *qi*. Primordial *qi* accommodates heaven, earth and all directions. Heaven and earth as well as all things in the universe come from primordial *qi*". From this it can be known that heaven and earth and all other existences are all generated from *yuan qi*. *Yuan qi* can give life force to all existences. As to *yuan qi*'s attribute, some people think that *yuan qi* flows between heaven and earth and never comes to an end. In fact, as far as *yuan qi* is concerned, there is actually no motion as there is not relative stillness in it. (Note: Please do not equate *yuan qi* in the human body to *yuan qi* as the level of substance in the universe.)

Another point that should be mentioned is that the reason why *yuan qi* can generate all existences lies in that it abides by the laws of *Dao*. *Yuan qi* and *Dao* seem to be the same at first sight, yet they are not completely the same. *Yuan qi* develops from *Dao*, and its entity and attribute are more concrete compared with Dao. It is the fundamental driving force of any specific thing.

c. Two (Two *Qi* Aspects)

The material level of "two" refers to *yin qi* and *yang qi*. They can evolve from *yuan qi* and differ from the general concept of *yin* and *yang* as usually called. *Yin* and *yang* commonly referred to are the two contrastive attributes of things having forms. *Yin* and *yang* talked about here refer to the substantial or material state of formless *qi*. They differ from the *taiji* state of greatly harmonious and undivided *qi*. The *qi* with the feature of *yang* is "motional and prevalent", while the *qi* with the feature of *yin* is "still and condensing". The two of them not only depend on each other but also transform into each other. Here being motional and still refers to two motion styles with two different characteristics. When the motions of *yin qi and yang qi* reach a certain limit, they can generate the material level of "three".

d. Three (Four *Qi* Images)

With respect to three in "two generating into three", there have always been different explanations. The explanations can be categorized mainly into two types: a) Three refers to all existences, and three the term itself refers to all kinds of *QI*s (things that have forms). b) Three refers to a special material level. It is the step when formless *Dao* and *qi* (气) transform into *QI* (器). We incline to agree with the latter.

According to Lao Zi, "All existences are in contact with *yin* and embrace *yang*. Their *qi*s blend to be harmonious unity." This can tell us that three refers to *yin qi*, *yang qi* and *he qi* (blended *qi*) that is formed with *yin qi* and *yang qi*.

On the other hand, in *Xi Ci* in *Yi Jing*, "The two *qi* aspects generate four *qi* images." "Two" become "four", and this seems to differ from Lao Zi's statement of two generating three. In fact, they are not different. The four images refer to *tai yin*, *tai yang*, *shao yin* and *shao yang*. From the perspective of *qi*, *tai yin* and *tai yang* are respectively *yin qi* and *yang qi*, while *shao yin* and *shao yang* are the combined state of *yin qi* and *yang qi*. Though they are called four images, they actually refer to three kinds of *qi*. Therefore, "three" and four *qi* images still belong in the category of "being formless above the level of things having forms". They are substances of the same level.

The three substance levels mentioned above all belong in the category of *Dao* compared with *QI*.

e. Three Generate All Existences

The material level of all existences refers specifically to the world of *QI* which has form, *qi* and quintessence (*zhi*, referring to information). All substance with form (atoms, molecules and tangible substances) belongs to this category. This level of substance results from the motion of four *qi* images. "The pure clear *yang* aggregates into heaven (sky) while the dense *yin* condenses into earth (land)". And based on this, new motion begins. That is, "The motional and the still attract each other; the above and the below border on each other; *yin* and *yang* interact and transform and changes accordingly come." Furthermore, "It is *qi* in the sky. It is form on the land. Form and *qi* attract each other, and all existences manifest." (*Tian Yuan Ji Da Lun* in *Su Wen*) When the substance level of all existences has formed, the generating and changing process of automatic multiplication of things begins. This is what is said in *Xi Ci* in *Yi Jing*, "Heaven and earth are in great mist of *qi*; all existences change decently into form; man and woman carry out sexual coupling; all existences are generated." People are familiar with the world of *QI* that has forms, so further statement about it is omitted here.

The five levels stated above, though they generate into the next level one after another, are the material states inherent in the universe. Among the five substance levels, the four formless material states not only accommodate each other, but also penetrate the world of *QI*.

ii) The Motion and Change of Substance in the Universe
a. Transformation between Somethingness (Being in Form) and Nothingness (Being Formless)

Transformation between the state of having form and the state of having no form is actually about changes between the two levels of formless *Dao* and "in-form" *QI*, which include the generating process of changing from "nothingness" into "somethingness" and the returning process of changing from "somethingness" into "nothingness". Ancient Chinese masters have given more descriptions of the former than they have given about the latter. Lao Zi has put forward that "Having (somethingness) is born of having not (nothingness)." Zhuang Zi has also said "All come from nothingness." Such statements seem to be absurd yet are actually scientific. This "having not" or nothingness state cannot be regarded as having actually nothing or being entirely empty. It is in fact a kind of special state compared with "having" or somethingness (the state of having form, *qi* and quintessence). For example, it is said in *Dao De Jing*,

"You look and yet cannot see it and it is called remote, you listen yet cannot hear it and it is called rare, you touch yet cannot feel it and it is called subtle. The three cannot be further questioned, thus they merge into oneness. It is neither bright above nor dark below. Being endless and beggaring description, it returns into nothingness. This is called the form of no form, the image of nothingness. It is called *huang hu* (literally meaning seemingly obscure state)."

Therefore in later generations, there is also such saying as all existences come from *huang hu* (the seemingly obscure state that can be sensed with a person's whole being instead of merely with the eyes, ears, nose, tongue or body), which can be said to be the explanation of nothingness.

Moreover, there also exists a process in "Having (somethingness) is born of having not (nothingness)". It is said in *Zhi Le* in *Zhuang Zi*,

"To inspect its beginning, there is actually no life; not only no life, but also no form; not only no form, but also no *qi*. Being in the midst of vast nothingness, it changes into *qi*, and *qi* changes into form, and form changes into life."

The key reason why "nothingness" can change into "somethingness" lies in the condensing and expanding of *qi*. Zhang Zai in the Song Dynasty said that "*Qi* is originally void and clearly onenss having no form. Once being connected, it starts to have life and condenses to manifest image." (*Zheng Meng Qian Kun*) That is to say, having form and having no form are but the different external manifestations of *qi*, namely, things having form are the condensed state of *qi*, while the image of having no form is the dispersed state of *qi*. Humans as the most intelligent of all existences are no exceptions. Wang Chong in the Han dynasty has pointed out that "*Qi* condenses into humans." In *Tai Ping Jing*, it is also said, "Humans are born of *qi* which is in the mergence state. *Qi* generates *jing* (form, essence); *jing* generates *shen* (spirit); and *shen* generates brightness and enlightenment..."

In addition, *QI* or things with forms that change to a certain degree can also return to nothingness or "having no form". "Nothingness generates somethingness" and "somethingness generates nothingness" are the inevitable laws of the change of *qi*. Zhang Zai said,

"In the utmost void, there cannot be no *qi*. *Qi* cannot refuse to condense into all existences. All existences cannot stop dispersing into the utmost void." And "*Yin qi* and *yang qi* are widely different when being in expansion (or dispersed) and humans cannot catch a glimpse of them. They are in mergence when combined and humans do not see anything special in them. Form gathers into things, and things decay and return to oneness (or the original state)." (*Zheng Meng Qian Kun*)

It should be pointed out here that "nothingness generates somethingness" and "somethingness generates nothingness" demonstrate the mutual transformational relationship among different levels of substances. The world of *QI* having forms we are in contact with and the specific things in this world are all experiencing this process.

The statement of "somethingness and nothingness mutually generate" seems to be easy to understand now. This is because a) People with super intelligence have perceived the existence of a special substance that instrument in scientific experiments cannot detect; b) The insubstantial or intangible material - the energy (field), can transform into material having concrete forms, and this has been proven in science. What is more thought-provoking is that the statement of matter is the condensed form of energy put forward by Albert Einstein that is widely acknowledged in the circle of science and the statement of "*qi* condenses into form" seem to come from the same origin. As to how to gather *qi* into something that has form, this is just what is considered as the mystery and miracle of *qigong* science.

b. Changes of the World of *QI* (Things with Forms)

It is considered by the ancient Chinese that heaven and earth as well as all existences continuously change. The content of change includes:

a) Transformation between All Existences

In the process of growth and decay, all existences between heaven and earth not only affect each other, but also transform into each other. All existences are born from heaven and earth or nature and finally return to nature. The cycle like this incessantly goes on without an end. Jia Yi has talked about this in his *Peng Niao Fu*:

"All existences change without having a rest. They change, going with the passing of time and then back to where they come from. Form and *qi* continue to change with each other and alternate to become different. It goes infinitely on and how can I describe it? Myriads of changes never come to an end."

With all these changes taking place under the control of Mother Nature, it is just like what is said in *Xi Ci* in *Yi Jing*:

"With image formed in the sky and form coming into being on land, changes take place. So hardness and softness massage each other, and the eight trigrams interact with each other. There is the drumming of thunder and the nourishment of wind and rain. The sun and the moon navigate in space and winter and summer alternate. The *Dao* of *kun* turns into man and the *Dao* of *qian* turns into woman. *Qian* knows the great beginning, and *kun* develops into being."

All existences, under the effect of the inherent regularity of Mother Nature, form the internal consistency and create conditions for the mutual influence among heaven, earth and all other existences and for the transformation between form and *qi* as well as their continuation. The reason why *qigong* masters enhance their life force by means of changes in heaven, earth and all other existences lies in this. The world of *QI* is complex, and in order to grasp the varied laws in this world of *QI*, ancient Chinese have established the mathematical mode of eight trigrams which reveal the laws of change among all existences. The correctness of the eight trigrams is being gradually proven in modern science.

b) Change in All Existences

This change includes the change of life and death in all existences and also of renovation in all existences during the period of relative balance. These two kinds of change are interrelated. In their balanced and calm state, motion and change inside things do not stop. When the motion reaches a certain extent, balance will be broken and change of life and death takes place. It is just like what is said in *Liu Wei Zhi Da Lun* in *Su Wen* (*Plain Questions* of *The Yellow Emperor's Classic of Internal Medicine*):

"Qi Bo said, 'Success and failure rely on each other and are born of motion. When motion occurs and does not stop, change appears'. The Yellow Emperor said: 'Does this have duration?' Qi Bo said, 'When there is no life and death, it's the duration of stillness.' The Yellow Emperor asked, 'No life and death?' Qi Bo said, 'When the going out and the coming in of *qi* stop, the magical mechanism (*shen ji*) comes to an end. When the upward and downward motions (referring to the spiral motion) stop, *qi* is isolated and endangered. Therefore, without the going out and the coming in of *qi*, there will be no birth, growth, maturity, aging or death. Without the upward and downward motions, there will be no living, growing, changing, harvesting and storing. Therefore, the going out and the coming in of *qi* and the upward and downward motions are in all *QI*s or existences. Thus the *QI* is the "universe" of life and growth. When a *QI* falls apart, its balanced motions of *qi* are separated, and its life and growth cease. So there is nothing that has not the coming in and going out of *qi*, nothing that has not the upward and downward motions of *qi*. There are small and big changes and there are long and short durations.'"

The statement has not only talked about the dialectic relationship between internal motion and external relative stillness in substance, but has also pointed out that the duration of external relative stillness depends on the process of change in substance, that is, big change has longer duration and small change has shorter duration.

ii. The Concept of Holism in the Human Body
i) Heart-and-Body (Mind-and-Form) Holistic Concept of the Human Body

Ancient Chinese people regarded the human spirit and the human body as a whole entity and investigated human life activity with it. According to the ancient Chinese, the human being is a unity of *jing* (body, form, essence), *qi* and *shen* (mind, spirit). In *Yuan Dao Xun* in *Huai Nan Zi*, "Form is the shelter of life. *Qi* is what fills life. Spirit is the controller of life." That is to say, body (*jing*) is the foundation of human life activity; mind is the commander of human life activity, and *qi* is the special substance of human life activity. It permeates the entire human body and combines body and mind into a unity. In this unity, the ancient Chinese especially emphasized the commanding effect of "mind". In *Liu Jie Zang Xiang Lun* in *Su Wen*, it is said "Heart (referring to human spirit) is the root of life and the change of mind." In *Guan Zi*, "Heart to the human body is the throne to the emperor." In *Yuan Dao Xun* in *Huai Nan Zi*, further illustration was provided:

"Heart is the master of the five internal organs. It is used to control the four limbs, to make *qi* and blood flow, to ride in the realm of human relationship and to come in and go out of the gate of the natural world. Therefore, if a person has not attained his or her own heart and intends to administrate *qi* under heaven, it is just like wishing to adjust the bell and the drum while having no ears and loving articles while having no eyes. A person like this is not qualified to do it."

Moreover, the activity in the heart (human mind) is also related to change and safety of life of the human body. In *Ling Lan Mi Dian* in *Su Wen*, it is said:

"Heart is a commander as a king. Spirit and enlightenment come from it. ... When the master is wise, his followers are peaceful, and using this in life cultivation, a person can enjoy a long life. When the master is not wise in mind, the twelve organs (referring to the internal organs[1] and internal hollow organs[2]) are in danger. Dao is blocked and the human body suffers harm. Using this in life cultivation is misfortune."

This classic paragraph talks incisively about the important position of heart as the commander of consciousness and spirit in human life activity. In *qigong*, the function of human spirit is particularly emphasized. For example, in *Ben Zang Pian* in *Ling Shu* (*Miraculous Pivot of The Yellow Emperor's Classic of Internal Medicine*), it is said:

"Mind and will are used in controlling human spirit, in keeping soul and courage collected, in adapting oneself to the change in temperature like coldness and heat, and in harmonizing such

Note:

The internal organs also called five *zang* organs refer to the liver, the heart, with pericardium included, the spleen, the lungs and the kidneys. The internal hollow organs also called six *fu* organs refer to the gallbladder, the small intestine, the large intestine, the stomach, the bladder and the triple energizers.

emotions as joy and anger. When one is harmonious in the mind, one is concentrated in spirit. *Hun* and *po* (human soul and courage) are not distracted. Regret and anger are nowhere. The five internal organs do not suffer."

It is pointed out in *Zhuang Zi*, "When embracing one's own spirit and being calm, one's body will be self-adjusted, straight and balanced." It is said in *Da Xue* (*The Great Learning*), "If you want to cultivate your body, be upright in your heart (cultivate your mind). If you want to be upright in your heart, keep yourself entirely in honesty." It is said in *Nei Ye* in *Guan Zi*,

"When heart is in integrity inside, body is in integrity outside. When a person is like this, the person does not meet disaster in nature or any harm in human relationship. And this person is called a sage. When someone can really be upright and calm, he or she has fine texture in skin, being sensitive in the ears and the eyes, healthy in tendons and strong in bones. This person can sustain great sphere (heaven, sky), step on great square (earth, land), reflect great clarity (*Dao*) and see great brightness (enlightenment)."

In *qigong* in later generations, the adjustment of the whole entity of *jing*, *qi* and *shen* are emphasized.

ii)   The Concept of Holism in the Human Body (Form)

The human body is an organic whole entity. In traditional Chinese medicine, it is believed that each part of the human body—the four limbs outward, the five internal organs and the six internal hollow organs inward, is related to each other and controls each other, forming an organic whole entity that centers on the internal organs and internal hollow organs. In this whole entity, the internal organs are organically related, so are the internal hollow organs; the internal organs and internal hollow organs are organically related, and the internal organs, internal hollow organs are also organically related to each part of the human body.

a.   It is a Mutually Reinforcing-and-Restricting Relationship between the Five Internal Organs

According to the theory of the five elements, the five internal organs are associated with the five elements respectively. The liver belongs to wood, the heart belongs to fire, the spleen belongs to earth, the lungs belong to metal and the kidneys belong to water. The five internal organs generate each other and constrain each other according to the relationship of the five elements.

Mutual Generation:

Wood -generates- Fire -generates- Earth -generates- Metal -generates- Water -generates- Wood

Mutual Restriction:

Wood -restricts- Earth - restricts - Water - restricts - Fire - restricts - Metal - restricts - Wood

Because of the mutually reinforcing relationship between the five internal organs, life *qi* of the five organs are inexhaustible; meanwhile, as they can constrain each other, there won't be the tendency of growing alone without control. Hence balance between the five internal organs is guaranteed. This is what is called "interpromotion and interrestraint" in traditional Chinese medicine.

b. It is a Relationship of *Yin-Yang* and Exterior-Interior Association between the Internal Organs and Internal Hollow Organs

The internal organs belong to *yin* and are the interior. The internal hollow organs belong to *yang* and are the exterior. An internal organ is accompanied with an internal hollow organ. The liver is accompanied with the gallbladder, the heart is accompanied with the small intestine, the spleen with the stomach, the lungs with the large intestine, the kidneys with the bladder, and the pericardium with the triple energizers.

c. The Organic Connection between the Five Internal Organs and Each Part of the Human Body

In *Huang Di Nei Jing*, not only is each part of the human body associated with the five internal organs, but physiology and pathology of the human body as well as the diverse and complex phenomena in Mother Nature are also associated with them. The inductive method based on *yin yang* and the five elements is applied in the expression method. The content of connection stated seems to be somewhat forced and farfetched, yet practice in traditional Chinese medicine and in *qigong* has proven that the statement agrees with reality in this world. The complicated system stated in *Huang Di Nei Jing* is simplified as Table of *Wu Zang Pang Tong* (Table of Five Internal Organs and Their Universal Connections). (See page 18)

The first part of the table is about the relationship inside the human body, and the latter part is about the holistic organic connections between humans and Mother Nature. Detailed content will be stated in The Holistic Concept of Heaven and Humans (iii) in this section.

Why can each part of the human body and the five internal organs become an organic whole entity? According to ancient Chinese, this results from smooth connection of *jing luo* or main and collateral (*qi*) channels (including twelve *jings* or regular meridians, fifteen *luos* or largest collaterals, and the eight extra meridians) in the human body. And the continuance of the life activity of this organic whole entity depends on the function of *qi* moving through each part of the human body.

It needs to be pointed out here that the heart-and-body whole entity and the human-body

Table of Five Internal Organs and Their Universal Connections

(Reflecting the Human-and-Heaven Whole Entity)

Five *Zang* Internal Organs	Liver	Heart	Spleen	Lungs	Kidneys
Six *Fu* Internal Organs	Gallbladder	Small Intestine	Stomach	Large Intestine	Bladder
Five Aspects in the Human Body	Tendons	Blood Vessels	Flesh	Skin	Bones
Five Emotions	Anger	Joy	Thinking	Worry	Fear
Five Spiritual Aspects	Soul	Spirit	Mental Activity	Daring Vigor	Aspiration
Five Apertures	Eyes	Tongue	Mouth	Nose	Ears
Five Sounds	*Jiao*	*Zhi*	*Gong*	*Shang*	*Yu*
Five Aspects Controlled by Internal Organs	Color	Smell	Taste	Voice	Body Fluid
Five Colors	Dark Green	Red	Yellow	White	Black
Five Tastes	Sour	Bitter	Sweet	Hot	Salty
Five Body Fluids	Tear	Sweat	Liquid in the Mouth	Nasal Mucus	Saliva
Five Voices	Shout	Laugh	Sing	Cry	Chant
Five Luxuriances	Hands	Color in the Face	Lips	Body Hair	Hair
Five Elements	Wood	Fire	Earth	Metal	Water
Five Forms	Straightness	Pointedness	Squareness	Thickness	Roundness
Five Normalities (Humanity)	Kindness	Respect	Integrity	Righteousness	Wisdom
Five Directions	East	South	Middle	West	North
Five Crops	Sesame	Wheat	Glutinous Millet	Rice	Bean
Five Fruits	Plum	Apricot	Date	Peach	Chestnut
Five Domestic Animals	Chicken	Sheep	Cattle	Dog	Pig
Five Normalities (Climate)	Wind	Heat	Humidity	Driness	Cold
Five Changes	Generation	Growth	Change	Gathering in	Storage

whole entity complement and realize each other and unite with each other. Traditional Chinese doctors have in mind the whole entity with internal organs and internal hollow organs as the kernel, whereas they also regard *jing*, *qi* and *shen* as the most valuable treasure. Masters of life cultivation emphasize *jing*, *qi* and *shen*, while they never ignore the internal organs, the internal hollow organs or the human body. It is just that the emphases in the two groups are not the same.

iii) Part of the Human Body Reflects the Whole Entity

Each part made up of the human body can reflect the situation of a person's entire life movement, and this is another manifestation of human-body whole entity. Since the *qi* that maintains human body life activity moves through the entire human body, in a certain extent, the information of all the human body can be shown in a particular part of the human body. For example, the condition of the internal organs and the internal hollow organs can be reflected from the color of the tongue (including coating on the tongue). It can also be reflected in the five rings and eight octants (The five rings: the upper and lower eyelids are called the ring of flesh which belongs to the spleen. The inner and outer corners of eyes are called the ring of pulse which belongs to the heart. The white of the eye is the ring of *qi* which belongs to the lungs. The black part of the eye is the ring of wind which belongs to the liver. The pupil is the ring of water which belongs to the kidneys.). As to the three regions and their nine subdivisions of *cunkou* pulse (radial artery at the wrist), they are associated with the internal organs and internal hollow organs respectively. These are used in traditional Chinese medicine to diagnose disease, and people are already familiar with this. There are more examples as follows:

a. Face Reflecting the Whole Entity of the Human Body

A statement about the face reflecting the whole entity of the human body is found in *Wu Se Pian* in *Ling Shu* and the facial acupoints as well as what they reflect are shown in Fig.1. (See page 20) Is this true in reality? In Long Hua Hotel in Shanghai, experiments of acupuncture point anesthesia have been conducted according to the positions of the acupoints. Doctors had performed 1251 operations from 1976 to 1977 and the rate of successful anesthesia reached 96 percent. This has sufficiently proven that the statement given by ancient Chinese is correct. Concrete positions of such acupoints are shown in the diagram of Face Reflecting the Whole Entity of the Human Body. (Fig. 1)

b. *Chi Fu* (Elbow-to-Wrist Skin) Reflecting the Whole Entity of the Human Body

A statement about *Chi Fu* reflecting the whole entity of the human body is found in *Mai Yao Jing Wei Lun* in *Su Wen* and the content is shown in Fig. 2. (See page 20)

c. Reflection of the Entire Human Body in *Jing Luo* (Main and Collateral *Qi* Channels) and in *Jing Xue* (Acupoints Distributed along *Jing Luo*)

The fact that the twelve regular meridians in the human body reflect the condition of internal organs and the internal hollow organs has been stated in *Huang Di Nei Jing* a very long time ago, and it will not be stated in detail here. It should be known that each *jing mai* or regular meridian is not only related to the change in the internal organ or the internal hollow organ that the *jing mai* is associated with, but it also contains the information of the five internal organs in the entire human body. In the acupoints of each regular meridian, there are the acupoints of *jing* (third tone, first rising then falling tone, in the Chinese language), *xing*, *shu* (In *yang jing* or the *yang* meridian, there is also the acupoint of *yuan* which is the same as the acupoint of *shu* in nature), *jing* (first tone, flat tone) and *he*, which are together called the Five *Shu* Points. The Five *Shu*

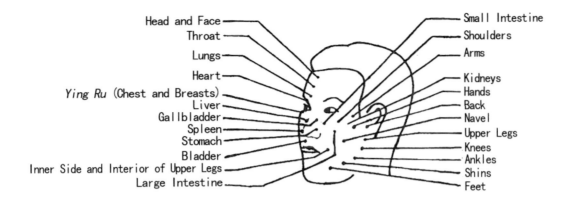

Fig. 1 Face Reflecting the Whole Entity of the Human Body

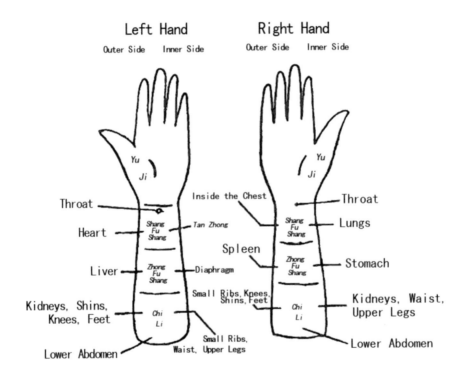

Fig. 2 *Chi Fu* Reflecting the Whole Entity of the Human Body

Points belong to the five internal organs respectively. In the six *yin* meridians, the acupoint of *jing* (third tone) belongs to the liver (wood), the acupoint of *xing* belongs to the heart (fire), the acupoint of *shu* belongs to the spleen (earth), the acupoint of *jing* (first tone) belongs to the lungs (metal), and the acupoint of *he* belongs to the kidneys (water). In the six *yang* meridians, the acupoint of *jing* (third tone) belongs to the lungs (metal), the acupoint of *xing* belongs to the kidneys (water), the acupoint of *shu* belong to the liver (wood), the acupoint of *jing* (first tone) belongs to the heart (fire), and the acupoint of *he* belongs to the spleen (earth). The Five *Shu* Points are the important acupoints for treating disease and adjusting the internal organs through *jing luo*—the main and collateral meridians or *qi* channels. The interrelationship between the internal organs and the internal hollow organs can also be known from them.

Moreover, in each acupoint, there are different kinds of information about skin (lung), flesh (spleen), blood vessels (heart), tendons (liver) and bones (kidney).

In sum, each part of the human body contains the information of the entire human body.

iii. The Holistic Concept of Heaven and Humans

Ancient Chinese have established the theory of "Heaven and Humans in Oneness" by means of long-time observation of human body life activity as well as its relationship with Mother Nature. This theory consists of the two aspects of "Connection of Heaven and Humans" and "Similarity between Heaven and Humans".

Heaven and Humans Are Connected

What is called Connection of Heaven and Humans? Mr. Gao Shizong said in the explanatory note in *Sheng Qi Tong Tian Lun* in *Su Wen*, "Lively *qi* being connected to heaven means that the *qi* of *yin yang* and of the five elements in the human body is inexhaustible and connected to heaven." Heaven here is a general term of heaven and earth or sky and land which refers to nature or the entire universe in this context. The theory of Heaven-and-Human Connection can be found in the classic treatises as early as the Zhou dynasty and the Qin dynasty. It is said in *Bing Neng Lun* in *Su Wen*, "*Shang Jing* is a book about (human) *qi* being connected to heaven above." It is a pity that this book was long lost. There are also a lot of viewpoints about the connection of heaven and humans stated hither and thither in *Huang Di Nei Jing*. For example, it is said in *Sheng Qi Tong Tian Lun* in *Su Wen*, "Ever since the ancient times, the root of life is the connection with heaven. It is based on *yin* and *yang*, between heaven and earth, and in the accommodation of the six directions. Human *qi* of the whole human body (with three energizers included), of the nine points (the eyes, the ears, the nostrils, the mouth, the anus, the external genitalia), of the five internal organs, and of the twelve major joints (shoulders, elbows, wrists, hips, knees and ankles) are all connected to heaven. It is said in *Yin Yang Ying Xiang Da Lun* in *Su Wen*, "The *qi* of heaven is connected to the lungs. The *qi* of food and drink is connected to the throat (and then sent to the stomach). The *qi* of wind is connected to the liver. The *qi* of thunder is connected to the heart. The *qi* of crops is connected to the spleen. The *qi* of rain is connected to the kidneys." It is said in *Ben Zang Pian* in *Ling Shu*, "The five internal organs are correspondent to heaven and earth, in agreement with *yin* and *yang*, connected to the four seasons and they adapt to changes in the five elements." In sum, human life activity and heaven, earth,

Mother Nature are mutually connected. This will be stated respectively as follows:

i)   Heaven, Earth and Humans Come from One *Qi*

As it has been previously stated, according to ancient Chinese belief, humans together with heaven, earth and Mother Nature are born of one *qi*. It is said in *Qi Jiao Bian Da Lun* in *Su Wen*:

"It is said in *Shang Jing*, '*Dao* knows astronomy above, geography below and human relationship between heaven and earth.' The Yellow Emperor asked, 'What does that mean?' Qi Bo said, 'For *qi*, it's 'location' that makes the difference. When it is in the sky, it is astronomy. When it is on land, it is geography. When it is between sky and land, it is human relationship.'"

That is to say, heaven, earth and humans result from one *qi* distributed in different realms, and this can be known and grasped. For this, it is stated even more clearly in *Tai Ping Jing*, "Heaven, earth and humans are originally one same *yuan qi* (primordial *qi*). They are divided into three bodies and each has its or their own beginnings." And "*Yuan qi* has the three names of *tai yang* (utmost *yang*), *tai yin* (utmost *yin*) and *zhong he* (Golden Mean and harmony) and, equivalently, there are the three forms of heaven, earth and humans." That is to say, humans as well as heaven and earth are part of Mother Nature and they enjoy its nourishment together. Meanwhile, if humans and all other existences are regarded as subject, heaven and earth become the surroundings for humans and things. This is what is said in *Su Wen*, "Heaven and earth are the above and below of all existences." "Earth is something that conveys all existences." Since humans and all other existences together with heaven and earth share the essence *qi* of Mother Nature, the situation of essence *qi* circulating incessantly in the three parts (humans, all other existences, heaven and earth) is formed. In *Yin Fu Jing*, this is further elaborated and it is pointed out that humans steal the *qi* of heaven and earth, all other existences steal the *qi* of humans, and heaven and earth steal the *qi* of all existences. And the *qi* circulates like this without an end, hence the inexhaustible growth and change in nature. The aim of ancient *qigong* masters is to steal the essence of heaven and earth and keep their own essence.

ii)   Humans and All Other Existences Are Closely Connected

Although humans and all other existences differ widely, they are the product of one *qi* of heaven and earth. With regard to all existences, it is said in *Zhi Zhen Yao Da Lun* in *Su Wen*, "What is from heaven is the *qi* of heaven. What is from earth is the *qi* of earth. With the combination of the *qi* of heaven and earth, six steps are divided (the six steps involve the change of wind, fire, humidity, heat, dryness and coldness) and all existences come into being." With regard to humans, it is said in *Bao Ming Quan Xing Pian* in *Su Wen*, "Humans are born on the earth while human life is connected with the above. Heaven and earth combine their *qi* and human life begins." It is also said in *Ben Shen Pian* in *Ling Shu*, "What is bestowed in me from heaven is morality (*De*). What is bestowed in me from earth is *qi*. Morality mobilizes, *qi* contacts (*Dao* of heaven and *Dao* of earth combine and *yin* and *yang* merge) and life manifests." While it is further pointed out in *Liu Wei Zhi Da Lun* in *Su Wen* that humans and all other existences share the same *qi* and that "In the midst of above and below and in the interaction of *qi* is where humans live. Therefore, it is said 'Above the mechanism of heaven, *qi* of heaven commands. Below the mechanism of heaven, *qi* of earth commands. In the combination of heaven-and-earth *qi*, human *qi* follows and all existences originate.'" Of course, humans and all other existences are not the same. In *Bao Ming Quan Xing Lun* in *Su Wen*, it is said "Among what is accommodated by heaven and conveyed by earth, none are nobler than humans." In *Wang Zhi* in *Xun Zi*, the reason why humans are so noble is further illustrated, "Water and fire have *qi* yet not life. Grasses and trees have life yet not senses. Beasts have senses yet not integrity. Humans have *qi*, life, senses as well as integrity. Therefore, humans are the noblest under heaven."

Although human *qi* has its own special motional ways, the basic forms—upward motion, downward motion, going-out motion, coming-in motion, closing motion and opening motion, are identical with and connected with the motional ways in all other existences. Therefore, it is said in *Zhuang Zi*, "Creatures contact with each other with their breath." Such going-out and coming-in motions are between heaven and earth and connect humans with all other existences. The statements given by *qigong* masters in later generations, such as "When humans breathe in, heaven and earth breathe in with them; when humans breathe out, heaven and earth breathe out with them" and "Heaven and humans integrate in oneness", seem to be exaggerations, yet in fact the statement is reasonable in a certain sense.

iii) Heaven, Humans and All Other Existences Are a Whole Entity

Humans live between heaven and earth and among all other existences and form an organic whole entity connected with heaven, with the five internal organs as the nucleus. It is said in *Yin Yang Ying Xiang Da Lun* in *Su Wen*:

"The Yellow Emperor said, 'I heard that sages in ancient times stated and stressed the human form (body), listed and differentiated (form and function of) the internal organs and internal hollow organs, knew the distribution of main and collateral meridians, connected with the six directions, and let (*qi* in) each *jing* flow along its route, each acupoint of *qi* have its name, … 'space' in muscles together with joints all have their beginnings, directions of the flow of *qi* in relevant positions have their orders, the four seasons and *yin* and *yang* have their regularities, and the correspondence between inside and outside have the interior and exterior features. Are these all true?'

"Qi Bo said, 'In the east begins the wind. The wind generates wood. The wood generates sourness. Sour taste reinforces the liver. The liver reinforces tendons. Tendons reinforce the heart. The liver is related to the eyes... The spirit (of human) is wind in the sky, wood on land, tendons in the (human) body, and the liver among the internal organs. It is related to green in regard of color, mi (3) in music, exhaling in sound, grasping in (diagnosing) pathological change, the eyes as far as human apertures are concerned, sourness in taste, and anger in emotion. Anger harms the liver. Sadness conquers anger. Wind harms tendons, and dryness conquers wind. Sourness harms tendons and peppery taste conquers sourness.

"In the south generates heat. Heat generates fire. Fire generates bitter taste. Bitter taste reinforces the heart. The heart reinforces blood. Blood reinforces the spleen. The heart is related to the tongue. It is heat in the sky, fire on land, blood vessels in the human body, the heart among the internal organs. It is related to red in regard with color, sol (5) in music, laughter in sound, worry in pathological change, the tongue as far as human apertures are concerned, bitterness in taste, and joy in feeling. Hilarity harms the heart. Fear conquers hilarity. Heat harms qi, and coldness conquers heat. Bitter taste harms qi and saltiness conquers bitter taste.

"In the middle is born humidity. Humidity generates earth. Earth generates sweet taste. Sweet taste reinforces the spleen. The spleen reinforces flesh. Flesh reinforces the lungs. The spleen is related to the mouth. It is humidity in the sky, land on earth, flesh in the human body, the spleen among the internal organs. It is related to yellow in color, do (1) in music, song in sound, omitting in pathological change, the mouth as far as human apertures are concerned, sweetness in taste, and thinking in mental motion. (Excessive) thinking harms the spleen. Anger conquers thinking. Humidity harms flesh, and wind conquers humidity. Sweet taste harms flesh and sour taste conquers sweetness.

"In the west is born dryness. Dryness generates metal. Metal generates peppery taste. Peppery taste reinforces the lungs. The lungs reinforce skin and hair. Skin and hair reinforce the kidneys. The lungs are related to the nose. It is dryness in sky, metal on land, skin and hair in the human body, and the lungs among the internal organs, white in color, re (2) in music, crying in sound, coughing in pathological change, the nose as far as human apertures are concerned, hot in taste, and worry in emotion. Worry harms the lungs. Joy conquers worry. Heat harms skin and hair, and coldness conquers heat. Hot taste harms skin and hair and bitter taste conquers hot taste.

"In the north is born coldness. Coldness generates water. Water generates saltiness. Saltiness reinforces the kidneys. The kidneys reinforce marrow. Marrow reinforces the liver. The kidneys are related to the ears. It is coldness in sky, water on land, bones in the human body, the kidneys among the internal organs, black in color, la (6) in music, moaning in sound, fear in pathological change, the ears as far as human apertures are concerned, saltiness in taste, and fear in emotion. Fear harms kidneys. Thinking conquers fear. Coldness harms blood, and dryness conquers coldness. Saltiness harms blood and sweet taste conquers saltiness."

The statement here seems to be forced and farfetched. Yet it is full of precious experiences of ancient Chinese people. It has always been cherished and emphasized by traditional Chinese medicine doctors and *qigong* masters (For details, please refer to Table of *Wu Zang Pang Tong* that reflects the human-and-heaven whole entity).

In sum, heaven and earth, *yin* and *yang*, the four seasons and all existences all have an effect upon human life activities. Ancient Chinese life cultivation masters sought to attain their own balance in this grand system. First of all, it is to follow the laws and make use of various conditions to serve themselves. Therefore, it is said in *Hui Qi Tiao Shen Da Lun* in *Su Wen*:

"The four seasons together with *yin* and *yang* are the root and foundation of all existences. So sages nourish *yang* in spring and summer and nourish *yin* in autumn and winter in order to adhere to the root and to flow with all existences at the gate of growth. Doing adverse things against what the root and foundation demand cuts the root and harms the truth."

Heaven and Humans are Similar

The theory of heaven and humans being similar is a further development of the theory of Heaven and Humans Coming from One *Qi* which belongs to the theory that humans and heaven are connected. It is believed in the theory of similarity between heaven and humans that heaven and earth are one *yin yang* and one *taiji*, and the human body is also one *yin yang* and one *taiji*. Therefore, Dong Zhongshu said, "The human body is like heaven (meaning universe here)." This is what is stated in later generations, "The human body is actually one little *tian di* (referring to universe). There are a lot of statements about this theory in *Huang Di Nei Jing*. For example, in *Jin Kui Zhen Yan Lun* in *Su Wen*, it is said:

"From before dawn to midday, it is *yang* of heaven, the *yang* of *yang*. From midday to dusk, it is *yang* of heaven, the *yin* of *yang*. From night to when roosters crow, it is *yin* of heaven, the *yin* of *yin*. From when roosters crow to the time (3:00am and 5:00am) before dawn, it is *yin* of heaven, the *yang* of *yin*... So the back is *yang*, and the *yang* of *yang* is the heart. The back is *yang*, and the *yin* of *yang* is the lungs. The stomach is *yin*, and *yin* of *yin* is the kidneys. The stomach is *yin*, and the *yang* of *yin* is the liver. The stomach is *yin*, and the utmost *yin* of *yin* is the spleen. All this is about the correspondence of *yin* and *yang*, the exterior and the interior, the outside and the inside, and the female and the male, hence to correspond to the *yin* and *yang* in heaven (the universe)."

It is said in *Yin Yang Li He Lun* in *Su Wen* in *Huang Di Nei Jing*:

"I heard that heaven is *yang* and earth is *yin*, that the sun is *yang* and the moon is *yin*. Lunar months of 30 days and of 29 days add up to 360 days as one year, and in humans there exists such correspondence."

It is also stated in *Yin Yang Ying Xiang Da Lun* in the same book:

"Heaven and earth are considered as *yin* and *yang*. The sweat of *yang* is named as the rain of sky and land. The *qi* of *yang* is named as the wind of sky and land. The *qi* of anger is like thunder. The *qi* with 'heat' or 'fire' move upward..."

This theory is further developed in classic works of *qigong* in later generations. Not only the four seasons, the five elements, the twelve two-hour periods (into which a day is traditionally divided in China) and the degrees in "rotation and revolution" (symbolizing universe) and so on are used as marks in the human body in *qigong* practice or as terms in *qigong* practice, but heaven and earth, the sun, the moon and the stars and so on all became terms of different parts of the human body. Content as this already belongs to specialized theories of *qigong* and will not be further elaborated here.

In the classic concept of life whole entity, human life is considered as an organic whole integrated with the universe as one entity, with *jing* (body, form, essence), *qi* and *shen* (mind, spirit) combined as a trinity, and with the five internal organs as the kernel or nucleus. The practice for thousands of years in traditional Chinese medicine and *qigong* has proven that this theory is correct. It should be acknowledged that, in the grand system of humans and universe as one whole entity, ancient Chinese merely saw the panorama of it. They still could not reveal all its inherent laws and concrete details.

II. The Theory of *Qi*
The Theory of *Qi* is the kernel and essence in the theories of traditional *qigong* and traditional Chinese medicine. It includes the two aspects of *qi* and *qi* transformation in content.

i. *Qi*
The earliest statement of *qi* is found in the book *Guan Zi*. *Qi* (referring to quintessence *qi*) can generate sky, land, humans and all other existences. It can move around between heaven and earth. The most statement of *qi* is found in *Huang Di Nei Jing*. It has not only comprehensively illustrated heaven and earth, (12 in the 24) fortnightly periods and the change in *yin* and *yang* outside the human body, but has also found the law of "the five elements (wood, fire, earth, metal and water) and six *qi*s (wind, fire, heat, humidity, dryness and coldness)", and made it a particular branch of learning. It has also given detailed and clear descriptions of various kinds of *qi* in the human body. There are *zhen qi* (true *qi*), *ying qi* (nourishing *qi*), *wei qi* (guarding *qi*), *zong qi* (ancestral *qi*), *zheng qi* (upright *qi*), *jing luo qi* (*qi* in the main and collateral meridians), *zang fu qi* (*qi* in the internal organs and internal hollow organs) and so on. The description of guarding *qi* is very similar to some characteristics of human body *hunyuanqi*. For example:

"Guarding *qi* is the strengthful *qi* in water and grains. Bold, swift and smooth in motion, it is unable to enter blood vessels, so it flows in the skin and muscles, immerses among *huangmo*, the thin membranes between the five internal organs, and is distributed in the chest and the stomach." "Guarding *qi* is used in warming up flesh, filling up the skin, nourishing the grain of skin and the texture of the subcutaneous flesh, and carrying out the opening and closing motions (of *qi*)." "The clearer is called nourishing *qi* (*ying qi*) and the less clear is called guarding *qi* (*wei qi*). *Ying qi* is in the blood vessels and guarding *qi* is outside them." "Nourishing *qi* is from the middle energizer[1] and guarding *qi* is from the lower energizer[2]."

In addition, there is also a statement about "*qi* street" in this classic work of traditional Chinese medicine. It is pointed out in this book that "Chest *qi* has *qi* street; stomach *qi* has *qi* street; head *qi* has *qi* street and shin *qi* has *qi* street." It also pointed out that "The four *qi* streets are the shortcut for *qi*." This is very similar to human body *hunyuanqi* in the category of human *hunyuanqi*. In *Nan Jing*, the concept of human body *hunyuanqi* is put forward for the first time. *Tai Ping Jing* in the East Han dynasty further illustrated the primordial *qi* (*yuan qi*) of heaven, earth and Mother Nature. It pointed out that "Heaven, earth and humans are originally from one same primordial *qi*". It is not until the Song dynasty, in *Yun Ji Qi Qian* by Zhang Junfang, that the writing of the treatise *Yuan Qi Lun* was accomplished with very rich content. Let me try stating it briefly.

In *Yuan Qi Lun*, *yuan qi* is considered as *Dao* and the root and foundation of heaven, earth and all existences. It is stated in the book, "What is *Dao* exactly? *Dao* is actually *yuan qi*;" and "Nature is actually oneness. Great *Dao* is actually oneness. *Yuan qi* is also actually oneness. Oneness is truly the most original pure *yang qi*." Since the author of the book regarded *yuan qi* and *Dao* as the same, he thought that *yuan qi* is the most fundamental to bring about heaven, earth and all existences. He pointed out that "*Yuan qi* has no formal name yet it brings about things that have names. *Yuan qi* accommodates all and generates existences." "*Yuan qi*, being firstly clear, rises and generates sky. *Yuan qi*, being secondly not so clear, falls and turns into earth." Then, "Humans and things all come into being because of oneness *qi*." And humans come into being because of the *yuan qi* of heaven and earth. It is said in the book, "Humans are endowed with life from primordial *qi* which helps form human spirit and body. Humans receive oneness *qi* that helps form human bodily fluid and essence." *Yuan qi* is very important in the human body. It concerns the change and safety of the human being. In the book, it is said, "Human primordial *qi* is exquisite in its naturalness and calmness and embraces a clearly void and profoundly magical entity... The root of life is exactly primordial *qi*." *Nan Jing* is also

Note:

Three energizers (*san jiao*) of the upper energizer, the middle energizer and the lower energizer make up one *fu* of the six internal hollow organs. They include the portions of both the chest and the stomach, with all the internal organs and internal hollow organs included. Three energizers divide the human body into three parts. The portion above the diaphragm is called the upper energizer, including the heart and the lungs. The portion below the diaphragm to the navel is called the middle energizer, including the spleen, the stomach, the liver and the gallbladder. The portion below the navel is called the lower energizer, including the kidneys, the large intestine, the small intestine and the bladder.

quoted in this book:

"*Yuan qi* is the source for generating *qi* in humans. It is the motional *qi* between the kidneys. It is the foundation of the five *zang* internal organs and the six *fu* internal hollow organs, the root of the twelve regular meridians, the gate of breath, the source of three energizers, with another name called God Guarding against Evils." "When *yuan qi* is abundant in the human body, marrow condenses into bones and intestines turn into tendons. Because the purely true essence (*zhen jing*), primordial spirit (*yuan shen*) and primordial *qi* (*yuan qi*) do not leave the human body, the human being can enjoy long life."

Human *yuan shen* has to be nourished with *yuan qi*. It is pointed out in the book, "Human *yuan qi* is the savory food of human spirit and soul." "Human spirit cannot be magical without *yuan qi*. *Dao* cannot generate without *yuan qi*." It is just because of this, the cultivation of *Dao* must involve the cultivation of *yuan qi*. It is pointed out, "Learning of *Dao* is called learning of internal cultivation. Learning of internal cultivation is what is related to one's own body and own heart and is called three *dan tian*s, three *yuan qi*s... When *yuan qi* is cultivated to an extremely high level, *yuan qi* is stored inside and there is no manifestation of *yuan qi* going out and coming in at all." It is further pointed out:

"With calmness in mind and harmony in *qi*, *yuan qi* comes accordingly. When *yuan qi* automatically comes, the five internal organs are well connected and nourished. With the five internal organs well connected and nourished, blood in all the blood vessels and *qi* in all meridians flow well. When blood in all blood vessels and *qi* in all meridians flow well, body fluid corresponds to what is required from above in an enhanced state, and a person is not desirous of food and drink. *Dao* in three *dan tians* is both well cultivated and accomplished, and a person, with sufficient *qi* in the body and the internal organs, is young in countenance and can live a long life."

In sum, *Yuan Qi Lun* has given a brilliant exposition of the content, the attribute and the function of *yuan qi* in the human body, of how to cultivate *yuan qi* as well as of the changes after *yuan qi* cultivation. It needs to be pointed out that *yuan qi* is considered the same as *Dao*, and this is one point that needs development.

ii.　*Qi* Transformation

"*Qi* transformation" stated here is a special term of traditional *qi* monism (Note that it is not the process of liquid changing into gas in physics). It refers generally to the fact that changes in all existences result from the function of *qi*. It refers to that *yuan qi* in heaven and earth generates and nourishes all existences and that it also brings about the various changes of *yuan qi* inside one existence (including human beings). With regard to the statement of *qi* transformation, the mostly detailed illustration is in *Huang Di Nei Jing*. In this book, it is believed that the void entity of *Dao* of the universe is the foundation for *yuan qi* to generate all existences. It is through the motion and transformation of the five elements and the six *qi*s that the generation and change of all existences are carried out. This is what is said in *Tian Yuan Ji Da Lun* in *Su Wen*:

"The greatest void is boundless and deep. It is the foundation to form oneness. All existences rely on it to begin. The five elements distribute throughout it. The distribution of *qi* is magical, and it controls *kun*, the root and source of all existence on land. The nine stars (The Big Dipper and *You Bi*, *Zuo Fu*) are high above so bright. The seven luminaries (the Sun, the Moon, Venus, Jupiter, Mercury, Mars, and Saturn) move in their own orbit. There is *yin*. There is *yang*. There exists something soft and something hard. The hidden and the obvious are all appropriately located. Coldness and hotness expand and spread. With generation and change all around, all existences manifest."

The driving force to bring about change in all existences is that "The motional and the still attract each other; the above and the below border on each other; *yin* and *yang* mix and transform and changes accordingly come." For this, it is said in *Yuan Qi Lun*:

"So *yuan qi* divides. Hardness and softness begin to separate. *Yin* and *yang* interact. The clear and the obscure depart. Sky is formed outside. Land is formed inside.

"Sky comes into being because of *yang*... It brings about form since there is something merged in oneness (referring to *Dao*). It covers and nourishes all with the arch of heaven. It makes life start on account of motion.

"Land comes into being because of *yin*... The nameless (referring to *Dao*) brings about essence. Laws and regularities convey life. Stillness lets life come to an end. And this is called 'Sky is accomplished in formation and land is levelly formed.'"

"Motion is used to bring about change. Stillness (referring to calmness) is used to contain and change. Abundant *qi* helps in interaction and intercourse. Time raises all existences... In the sky, there appear images. On the land, there appear landscapes. There are nine positions in the sky. There are nine areas on the land. The sky has three *chen*s (the sun, the moon and stars). The land has mountains and rivers. There are images to imitate and landscapes to measure. Though they differ in all possible ways, they are actually connected. Generating mutually in nature cannot be calculated."

In this quoted paragraph, it is stated that *yuan qi* in the universe first changes into sky, land, the sun, the moon and the stars and existences on the earth also gradually generate into being afterward. There is another paragraph which concisely illustrates that *yuan qi* generates into *yin* and *yang*, into the five elements and all existences.

"The five *qi*s have not formed, the three *cai*s (heaven, earth and human) have not divided, the two *yi*s (*yin* and *yang*) have not established. This is called *hundun*, *hunyuan* and also *yuan kuai ru luan* (literally meaning original mass of *qi* like an egg) or *wu qi hun yi* (five *qi*s merging into oneness). Then oneness divides its *yuan qi*, which develops into five *qi*s (wood, fire, earth, metal, water). The *qi* has images, which are called *qi* images."

That is to say, the process of *yuan qi* changing into all existences is through changing into five *qi*s, and then various *qi* images in Mother Nature manifest. In *Huang Di Nei Jing*, the five *qi*s are further systematically called five *yun*s (referring to wood, fire, earth, metal and water) and six *qi*s (referring to wind, fire, hotness, humidity, dryness and coldness) to carry out *qi* transformation (*Jun Huo* or "Monarch fire" in them does not carry out *qi* transformation). The *qi*-transformation effect of *yuan qi* upon all existences is carried out through five *yun*s and six *qi*s. As to how this process of *qi* transformation is accomplished, in *Huang Di Nei Jing*, it is only pointed out that this depends on the effect of the "mechanism of spirit" inside all existences and the *qi* (or *qi* establishment) in the external environment, namely, "That which is rooted in the Golden Mean (balanced spiritual state without any feelings, such as gladness, anger, sadness or delight) is called mechanism of spirit…That which is rooted outside it is called *qi* establishment." And "If the coming in and the going out of *qi* stop, the mechanism of spirit vanishes. If the upward motion and the downward motion of *qi* cease, the *qi* is isolated and endangered." That is to say, *qi* transformation in all existences is accomplished through the upward and downward motions (the spiral motion) and going-out and coming-in motions of the *qi* in them. As to how *qi* transforms in all existences, it is still impossible to be clearly stated at the time of ancient Chinese.

As to *qi* transformation inside all existences, a lot of statements have been given in *Huang Di Nei Jing* about the *qi* transformation in humans. There are relatively specific descriptions about the generation, motion, function and change of the *qi* inside the human body. This is an important component part in basic traditional Chinese medical theories and we will not go into detail about it here. What we want to point out is that each and every tissue, organ and also the metabolic process needs the participation of *qi* transformation. For example, it is said in *Lan Tian Mi Dian Lun* in *Su Wen*, "The bladder is an organ like a minister of state and body fluid is stored there. The fluid in it can be released with *qi* transformation." Here an example of one internal hollow organ is given to show that realization of functions of the twelve internal organs and the internal hollow organs (with pericardium included) needs the involvement of *qi* to carry out *qi* transformation. The relationship between metabolism and *qi* transformation in the human body is rather profoundly and comprehensively stated in *Huang Di Nei Jing*. It is said in *Yin Yang Ying Xiang Da Lun* in *Su Wen*:

"*Yang* is *qi*. *Yin* is taste. Taste returns to form. Form returns to *qi*. *Qi* returns to quintessence.

Quintessence returns to change (meaning human spirit here). Quintessence eats *qi*. Form eats taste. Change (human spirit) generates quintessence. *Qi* generates form… Quintessence changes into *qi*."

There are two levels of meaning in this classic paragraph: One is the assimilation process (inside the human body) of the nutritious substance that the human body takes in from the outside world, i.e., taste→ form→*qi*→quintessence→change (human spirit). The other is the dialectic relationship of generation and change between the five steps. That is:

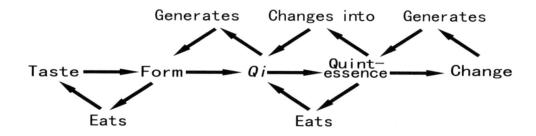

From this it can be seen that all the metabolic process depends on the generation and the participation of *qi* to complete *qi* transformation (*qi* to change or digest). Meanwhile, the *qi* that life activity in the human body needs generates in the metabolic process (changing into *qi*). Thus the processes of *qi* to change and changing into *qi* are the fundamental content in human body's life activity (If we go into detail, the three words of *qi* in the Chinese quotation differ in what they actually refer.).

iii. *Hunyuanqi*

In traditional *qigong* theories, the system of *hunyuanqi* theories has not formed. *Hunyuanqi* is talked about hither and thither in different classic works and the terms referring to *hunyuanqi* are not the same. There are such terms as *hunyuan* (mergence into oneness), *hundun* (mergence into a chaos state), *hunlun* (mergence state), *kongdong* (void hole), *hunyuanqi*, *hunyuan* oneness *qi*, and prenatal *hunyuan* oneness *qi*. Moreover, connotations of each term are not all the same. Generally speaking, *hunyuanqi*, *hunyuan* oneness *qi* (terms with *qi*) and so on refer mainly to human *hunyuanqi*, while *hunyuan* (mergence into oneness), *hundun* (mergence into a chaos state), *hunlun* (mergence state) (terms without *qi*) refer mainly to the beginning of the universe. In *Yun Ji Qi Qian*, *hunyuan*, *hundun*, *kongdong* and *hundong* (mergence hole) are illustrated respectively:

"*Hunyuan* (mergence into oneness) records before *hundun* (mergence into a chaos state) and it is the beginning of primordial *qi*. When primordial *qi* has not formed, what is in the silent void? The purest quintessence attracts and the true oneness is born. *Yuan qi* is in motion and heaven and earth are established. The creator of the universe (referring to *Dao*, to *yuan qi*) exerts functions and all existences benefit. As for *hundun*, there is only void in it and only nothingness outside it. Being vast and mighty, it is impossible to name it."

From here it can be seen that *hunyuan* refers to the prenatal *yuan qi* of the universe and is reasonably at the same level of *Dao*. It is even more clearly stated in *Zhi Yan Zong*:

"The *qi* of *hunyuan* makes somethingness out of nothingness. Somethingness is called *taiji* that generates two aspects (*yin* and *yang*). With the two aspects being divided, the four images come clearly into being. With the change of *yin* and *yang*, all existences are born."

As to *hundun*, it is said in *Yun Ji Qi Qian*:

"When the two aspects of *yin* and *yang* have not divided, it is called *hong yuan ming xing*. Being a great grand mass of mergence *qi* like an egg, it is named as *hundun*. In the mixture of heaven and earth, there is no light, no image, no voice, no sound and no ancestor. It is profoundly deep and infinitely boundless. In it there is essence. The essence is rather real. Accommodating where there is no outside, in the crystal clear void state, one *qi* is born in deepest and remotest background."

Hundun stated in this quoted paragraph seems to refer to *yuan qi* in the universe. As to this, it is more clearly stated in *Yuan Qi Lun*, "The five *qi*s have not formed. The sun, the moon and humans have not divided. The two aspects of *yin* and *yang* have not established. This is called *hundun* or *hunyuan*." As for *kongdong*, it is said, "*Yuan qi* is distributed in extensiveness and beyond profundity and infinity. It is born of *kongdong* (void hole)." And "The nine *qi*s come before the utmost void state and is hidden in the void hole. There is no light, no image, no form, no name, no color, no four seasons, no voice and no sound." This empty void hole (*kong dong*) seems to be the origin of *yuan qi* of the universe and it is very similar to the black hole in modern astronomy. As for *hunlun*, it is said in *Lie Zi*:

"The earliest beginning is the start of *qi*. The earliest start is the start of form. The greatest simplicity is the start of quality (referring to information). When form, *qi* and quality (information) have come into being yet have not divided, it is called *hunlun* (entirely merged state)."

Hunlun talked about here refers to the chaos oneness state when the two *qi*s of *yin* and *yang* are generating all existences before all the existences are formed. For this, it is directly stated in *Xing Ming Gui Zhi*, "When *Yin* and *yang* are divided, heaven and earth establish themselves and humans come into being. This is called two (*yin* and *yang*) generating into three (three *qi* images of *yin qi*, *yang qi* and the *qi* with *yin* and *yang* combined) and is called *hunyuan* (merged into oneness)." From this it can be seen that *hunyuan* and the chaos state of two generating three with all in a merged chaos entity are also the same. This gives me a great revelation to put forward the stratum feature of *hunyuanqi*.

In *qigong* classic works, the most detailed statement about human body *hunyuanqi* is found in *Bao Hun Yuan Xian Shu*:

It is said in *Bao Hun Yuan Xian Shu* in *Xian Shu Mi Ku*:

"What is *hunyuan*? It is *jing*, *qi* and *shen* before one's birth. Bai Ziqing said, 'The *jing* is not what is released when sexual contact is carried out. It is saliva in Jade Emperor's mouth (Jade Emperor refers to human spirit). The *qi* is not human breath, but the mist-like substance of utmost simplicity. The *shen* is not the level of human spirit in thinking, but the deepest level human spirit that is as great as Yuan Shi Tian Zun (The great Daoist figure earlier than Lao Zi.). This is the *yuan* (primordial) *jing*, *yuan* (primordial) *qi* and *yuan* (primordial) *shen*. To call *jing*, *qi* and *shen* before one's birth with *yuan* (primordial) is because they are the thing-in-itself or original being. Before humans have their bodies, they first have this original thing. After a person gets this original thing, form comes into being from formlessness, and quintessence come into being from where there seems to be no quintessence. It comes from before the sexual contact between parents. When this sexual behavior is being carried out by parents, when the father has not released his sperm and the mother's blood (an old belief in ancient China that children come from father's sperm and mother's blood) has not enclosed the baby-to-be, when the parents' heart and feeling are totally synchronized, there forms somethingness in the void state that penetrates and connects, that is well integrated and is densely 'misty' and not dispersed. Then the sperm is released and the blood embraces it, and sperm and blood mix and merge, with this point of truthfulness contained. It changes into form with *jing*, *qi* and *shen* residing in the form. Though there are three names (of *jing*, *qi*, *shen*), the nature is actually the same. One is the mergence into oneness (*hunyuan*) and three are the divided miraculous aspects. One is the entity while three are the functions. In the entity of *hunyuan*, that which is pure with nothing else in it is *jing*, that which makes *qi* and blood flow smoothly in the *qi* channels and blood vessels is *qi*, and that which is void, miraculous and with initiative in motion is *shen* (human spirit).

"The three are one and the one is three. Calling them the highest quality medicine of *jing*, *qi* and *shen*) is about the function. To mention the one is enough to accomplish is about the entity. We could not only talk about one, nor should we only talk about three. Practitioners who cultivate *Dao* do not know that three are one and one is three. They either hold the primordial and guard the oneness and incline on the path of sensing void, or refine and practice the three medicines and tend to embrace images. Neither inclining on the path of sensing void nor embracing images is the wonderful truth of returning to the primordial oneness state. To return to the original oneness state means to refine and sense the original *jing*, the original *qi* and the original *shen*. Practitioners who cultivate *Dao* must know that these three things are hidden in

33

hunyuan, the mergence in oneness. What is away from *hunyuan* is not the original entity of prenatal *jing*, *qi* and *shen*. When the original entity is lost, how can a practitioner get the original thing?"

As to what is "the midst of *hunyuan*", there is this poem by Sima Zihui in *Xing Ming Gui Zhi*:

"There is one void acupoint called *Xuan Guan*. It is in the mist of the human body between sky and land. It is 84,000 *li* between the above and the below, with nine, three, six and five distributed along. It accommodates the entire space of the universe without any trace for naked eyes to detect. It is so tiny that, if it is in the midst of dust, naked eyes cannot find it. This is called ancestral *qi* acupoint. It is like one grain of glutinous millet, hanging right in the middle of the human body."

There is also a poem written by Zhang Jinghe:

"The *hunyuan* acupoint is a prenatal point. It is void inside and follows the truth of nature. If one can see what one is like before birth, one will know for sure that one must be a great celestial being."

As for the position of the *hunyuan* acupoint, it is said in *Xing Ming Gui Zhi*:

"It is said in *Dao Jing*, 'Between the extremely highest point in sky and the extremely lowest point on land is 84,000 *li*, and the *hunyuan* acupoint is right in the middle, at the point of 42,000 *li* from the two points above and below between heaven and earth. If the human body is a little universe, the distance between the heart and the navel is also 8.4 *cun*, and the *hunyuan* acupoint is in the midst of the middle, at the point of 4.2 *cun*. This acupoint is right below *qian* (trigram) and above *kun* (trigram). It is west of *zhen* (trigram) and east of *dui* (trigram). The eight extra *qi* channels and the nine acupoints as well as main and collateral meridians are all well connected here. It is an acupoint leisurely in void and hanging like a grain of glutinous millet. It is the very midst of the universe of the human body and the acupoint with the original ancestral *qi* hidden here.'"

Although there are some works particularly on *hunyuan qigong* in the last hundreds of years, most of them lay emphasis on the illustration of practice methods. Comprehensive and systematic statement of the entity and attribute of *hunyuanqi* is still lacking, and the statement of *hunyuanqi* is not as comprehensive and detailed as the statement of *Dao* and *yuan qi*.

The statement of *hunyuanqi* and *hunyuan* acupoint above is helpful for both understanding of theory and practice of *gongfu*. One point that needs to make clear is that the statement quoted so far about *hunyuanqi* is not as concise and clear as the rhymes (rhyming in Chinese) that I learned from my teachers. One of the rhymes is:

"*Hunyuan* contains the entire five elements. It has form and is also formless. *Yin, yang* and the mechanism to generate life are all included. In the opening, closing, gathering and dispersing (of *qi*), all existences are generated."

This rhyme has included all about the entity and the attribute of *hunyuanqi*. The second rhyme is:

"*Hunyuanqi*—it can be penetrated by human spirit. It gathers into form as it is demanded (by human spirit). It dissovles into wind (or *qi*) as commanded (by human spirit)."

This rhyme has given a clear illustration about the relationship between *hunyuanqi* and human spirit. The third rhyme is:

"*Hunyuan* above. *Hunyuan* below. *Hunyuan* outside. *Hunyuan* inside. *Hunyuan* in the middle. *Hunyuan, hunyuan*, the entirely merged oneness."

This rhyme points out the knack in practicing traditional *hunyuan qigong*. These rhymes and theories are what have been used in guiding and creating *Zhineng Qigong* theories.

In addition, a lot of content in the theory of consciousness in the theory of *hunyuan* whole entity comes from *Cittamatra* in Buddhism. The statement of consciousness is very profound and comprehensive in Buddhism. It may well be said that the Buddhist theory of eight consciousnesses has made a thorough analysis about human mental activities and is no less profound or even more profound than modern psychology. Among Buddhist theories related to consciousness, the theory to break away from bias in mind and the theory of turning knowledge into wisdom (The eighth consciousness turns into "Great Mirror" Wisdom, the seventh consciousness turns into Wisdom of Equality, the sixth consciousness turns into Wisdom of Subtle Observation, senses of the eyes, ears, nose, tongue and body turn into the Wisdom of Helping All) have provided materials and a source of reference for the theory of consciousness in the theory of *hunyuan* whole entity. One point to mention is that in Buddhist theories of consciousness, consciousness is considered as the root and foundation of the universe and the world of *QI* (things that have forms) is considered as the manifestation of consciousness. This is similar to Friedrich Hegel's statement of absolute spirit. Karl Marx said that Hegel's dialectic stands on its head, and similar to this, *Cittamatra* in Buddhism also stands on its head.

Section III Reasons for the Establishment of the Theory of *Hunyuan* Whole Entity

Just as it has been pointed out in *A Survey of Zhineng Qigong Science*, *qigong* science is the science that researches, recognizes and reforms the laws of life motion of the human body and also its interrelationship with Nature using the method of *qigong*. It is not only a complicated holistic science that covers the vast realms of the universe (which was called *heaven* by the ancient Chinese), the earth and life, but also a brand new science that concerns the laws and nature of the motion of human consciousness. The special feature of its target of research which is hard for the means of modern science and technology to approach determines that new methods of research must be created and new guiding ideologies and fundamental principles of scientific research must be established in *qigong* science.

Qigong science has established not only its method of using super intelligence as its means of scientific research, but also its methodology of the Inward Seeking Approach which differs from modern science. All this requires its own theoretical foundation—the whole entity theory which is integrated and motional, that concerns the unity of nature and life, of consciousness and body, and of form and function. This theory is a description of the objective world in the state of super intelligence. The holistic theories in modern science such as systems theory, the theory of dissipative structures and synergetics, though they have illustrated many laws that describe the wholeness of things, cannot reach the depth of the realm of life, nor the depth of the realm of consciousness. This is because these theories illustrate the holistic features of things from the level of science that is based on familiar intelligence. To explore the motions of life and consciousness which belong in the domain of super intelligence with these theories is inevitably as incompatible as fitting a square peg into a round hole. This shows that this way of research of using science based on familiar intelligence alone to study life and consciousness is an impasse and another thoroughfare should be sought.

Will traditional *qigong* theories be able to shoulder this weighty responsibility?

We think that traditional *qigong* theories like the *yin yang* theory, the *wu xing* (five elements) theory, the *ba gua* (eight trigrams) theory, the *jing luo* (*qi* channel) theory, the *qi*-and-blood theory and the *jing qi shen* (essence, *qi* and spirit) theory are not only to a certain extent the models of holistic theories, but also the achievements of the ancient Chinese who observed nature and the life of the human body with super intelligence. It has been pointed out in *Xi Ci* in *Yi Jing*,

"In remote antiquity when Bao Xi Shi was ruling the kingdom under heaven, looking upward, he observed the starry configurations; looking downward, he observed the laws on the earth. ...

Attentively perceiving his body near to him and things remote from him, he began to draw the eight trigrams (*ba gua*)." And *yin yang* is the root of *ba gua* which is *yi*. It is said in *Xi Ci* in *Yi Jing*, "Are *qian* (the trigram of Heaven) and *kun* (the trigram of Earth) the gate to *yi*? *Qian*, it is what is *yang. Kun*, it is what is *yin*."

It is further pointed out in *Shuo Gua Zhuan*, "In the past when the sages established *yi*, they were to follow the truth of the nature of life. So the *Dao* for heaven to establish itself is *yin* and *yang*; the *Dao* for the earth to establish itself is gentleness and firmness; the *Dao* for humans to establish themselves is goodness and righteousness. ... *Yin* and *yang* develop from the universe, and gentleness and firmness mutually complement and accomplish."

These theories (*yin yang, ba gua*) seem to have covered the holistic connections between heaven (universe), earth and life.

The theory of *wu xing* (five elements) is an extension of the theory of *yin yang*. It is said in *Su Wen* in *Huang Di Nei Jing*, "There are *yin* and *yang* in heaven and also *yin* and *yang* on the earth. Wood, fire, earth, metal and water are *yin* and *yang* of the earth, while to generate, to grow, to change (transform), to gather and to store are the *yin* and *yang* of heaven."

And the theory of *wu xing* is also an integrated system. From what is included in Table of *Wu Zang Pang Tong*, it can be seen that the content of heaven, earth and life seems to have also been covered in the theory of *wu xing*. Additionally, physiology and pathology of humans in the theories of traditional Chinese medicine are exclusively explained with the aid of the theories of *yin yang* and *wu xing*. The theory of *wu xing* is not a trivial philosophy of metaphysics, as some people have said. It is in fact also a summarization of the ancient Chinese people's application of super intelligence in their observation and practice. As what is pointed out in *Su Wen* in *Huang Di Nei Jing*, "Your servant browses through Volume of the Very Beginning of Heavenly Oneness and knows that the *qi* of *Dan Tian*[1] goes through *Niu Nv Wu Fen*, the *qi* of *Huang Tian*[2] goes through *Xin Wei Yi Fen*, the *qi* of *Cang Tian*[3] goes through *Wei Zhi Liu Gui*, the *qi* of *Su Tian*[4] goes through *Kang Di Ang Bi*, and the *qi* of *Xuan Tian*[5] goes through *Zhang Yi Lou Wei*[6]." That is, the cooperation of the five directions[7] and five colors[8] is based on the practice of

Note:

1. *Dan Tian*: It was in the south in the sky and its color was red.
2. *Huang Tian*: It was in the middle in the sky and its color was yellow.
3. *Cang Tian*: It was in the east in the sky and its color was green.
4. *Su Tian*: It was in the west in the sky and its color was white.
5. *Xuan Tian*: It was in the north in the sky and its color was black.
6. *Niu Nv Wu Fen, Xin Wei Yi Fen, Wei Zhi Liu Gui, Kang Di Ang Bi, Zhang Yi Lou Wei*: They refer to the configuration of the stars.
7. Five directions: The south, middle, east, west and north.
8. Five colors: Red, yellow, green, white and black. These were the colors of the image of *qi* in the sky before the West *Han* Dynasty when Nature was still natural nature. They can no longer be found in this nature that has become humanized.

observing *qi* and its color in the sky.

The theories of *jing luo* (*qi* channel, *qi* meridian) and *qi* transformation are the quintessence of the theories of traditional Chinese medicine. These theories were also established and enriched in ancient Chinese people's practice. The author of this book, when stating the relationship between traditional Chinese medicine and *qigong*, has demonstrated that *qigong* practice is the source and origin of the theories of traditional Chinese medicine in his book *Exploring the Mysteries of Qigong*. Moreover, there is also statement about *hunyuanqi* in traditional *qigong* (though not integrate).

For these theories having been given above, first of all, we think that they were all established by the ancient Chinese people in their *qigong* practice and are the description of the whole entity of humans and the universe in the natural state in ancient times. Yet today when science has become highly developed, the original natural state of both humans and nature has been changed. These theories established in ancient times are no longer fully suitable to describe the situation today. A theory that can reflect the holistic relationship between humanity and the universe today must be established. This is the reason why we sublate traditional *qigong* theories and establish the theory of *hunyuan* whole entity. Let me explain it in detail.

I. To Reflect Accurately the Nature and Laws of the Holistic Connections between Humans and Nature

i. Universality and Narrowness Features of Such Theories as *Yin Yang*, *Wu Xing*, and *Ba Gua* in Traditional *Qigong* Theories

We think that these traditional *qigong* theories like *yin yang*, *wu xing* (five elements) and *ba gua* (eight trigrams) have indeed reflected to a certain degree and in a certain scope the natural conditions of objective existences and can be widely and universally applied. For example, "straightness, pointedness, squareness, thickness and roundness" in the five-form theory of *wu xing* theory have covered all the form elements in all existences of the universe. Roundness is point, straightness is line, pointedness and squareness are different surfaces formed with bended lines, and thickness forms body. Different kinds of point, line, surface and body constitute different shapes and structures in the universe and determine the functions and nature of different substances. In this sense, the theory of *wu xing* is universally applicable. But if we unfold the inductive mode and deductive mode in the content of *wu xing*, they not only lack essential connections between each other but in many aspects do not conform to real life.

The theories of *yin yang* and *ba gua* are by far greater in their universal application than the theory of *wu xing*. For example, the theory of opposition, the theory of interdependence and the theory of transformation between *yin* and *yang* as well as the law of wane and wax of *yin* and *yang* still have their certain philosophical and scientific guidance significance even today.

Examples like this are found in physics in positive and negative electricity, in positive and negative particles and in the north and south magnetic poles and so on. In biology, the viewpoint that *yin* is within and *yang* is without has also been proven in modern science. Take resting potential in the biomembrane of a cell for example: The biomembrane of a cell contains negative electricity inside and positive electricity outside, and the potential difference between the inside and outside is -70mv. As for the theory of *ba gua*, it can be applied not only to the macro realm but also to the micro realm. For instance, it has been found in molecular genetics that the 64 codons shouldering the hereditary task correspond from afar to the 64 trigrams. It is because the theories of *yin yang* and *ba gua* incredibly conform to some scientific laws discovered in the objective world that a vigorous campaign of studying *Yi Jing* is being unfolded across the world.

With regard to this, we think that traditional theories of *yin yang* and *ba gua* have truly grasped some fundamental laws of natural existences. However, only the natural laws of natural existences. They have not entered the depth of society, nor into the realm of human spirit. These theories show undoubtedly and obviously their narrowness compared with the theory of whole entity established in *Zhineng Qigong*. Statements about spirit in traditional *qigong* theories are rather few. As far as spirit is concerned, it is only mentioned in the theory of *yin yang* in such simple words as "*Shen* (human spirit) is when no judgment of *yin* and *yang* can be made (about it)." In *Yi Jing* which expands upon the eight trigrams, spirit is ambiguously stated as "*Shen* has no direction and *yi* has no body." In the theories of traditional Chinese medicine, it is only said that "*Qi* and blood are human spirit" and "*Shen* is when the two essences are acting upon each other." Nothing has ever been mentioned about the nature of *shen*, its laws of motion, its position and function in human life activities as well as its effect upon Nature.

From this we can see that the theories of *yin yang*, *wu xing* and *ba gua* have not clearly illustrated the mystery of human life, particularly the mystery of the activity of consciousness. These theories have but dwelt upon the illustration of the natural attribute of natural substances.

We know that after human beings came into existence, the attribute in Nature began to show subjectiveness of human consciousness—willpower. Human willpower continuously changes the pure naturalness of Nature, and this change will steadily increase with the advancement of humanity. Therefore, the theories of *yin yang*, *wu xing* and *ba gua* which merely focus on illustrating the pure natural attribute of nature begin to show their limitations in meeting the demand of the times.

For example, the viewpoint in the theory of *yin yang* that "the front of the mountain is *yang*, the back of it is *yin*" and "what is on the ground is *yang* and what is under it is *yin*" which illustrates the attribute of Nature is undoubtedly correct in pure natural environment. However, it may not be correct today when great changes have taken place in the appearance of Nature. If we build a big electricity plant at the back of the mountain which is originally *yin* in nature and a brilliantly illuminated city under the ground which is also originally *yin* in nature, we have changed in a great extent their original nature of *yin*. On the contrary, if we build a reservoir in front of the mountain which is originally *yang* in nature and set up an icehouse on the ground which is also originally *yang* in nature, we have changed in a certain degree their nature of *yang*. These changes show that a certain mergence and change (*hunhua*) have taken place in *yin* and *yang* that are opposite to each other.

Another example is: In the theory of *wu xing* (the five elements), the five directions (east, south, middle, west and north) correspond to the five normalities (climate, which refers to wind, heat, humidity, dryness and coldness). To view the five directions and the five *qi*s from the small narrow area centered on the culture of the Yellow River in ancient times, they are undoubtedly correct. But it is apparent that the five directions and the five *qi*s do not have their universality today when traffic and communication have woven the entire earth into one integrated body. If we take a plane and keep flying to the west or to the north, we will eventually fly back to the starting point from an easterly or southerly direction as a result. This shows that the different factors in the theory of *wu xing* can be joined into one entity.

Moreover, no matter it is the theory of *wu xing* or the theory of *ba gua*, they have only provided a set of mathematical models to explain this world. They seem to be quite reasonable and flexible at the rational level when they are applied to explaining the formation and change of all existences in the universe. But as soon as they are used to give guidance in practice or when they are in contact with practice, they usually make people who use them feel puzzled. Even the True Man *Shao Kangjie* who is worshiped in the Daoist School as one who has achieved *Dao* and has become a celestial being, said with a sigh before he left this world, "The truth before I was born I know. The truth after that I still know little of." And the reason lies in this.

As to the fact that qigongists of each generation have taken the theory of *ba gua* as the standard of practicing *nei dan* (one type of Daoist *qigong*), it is really a misunderstanding to do so because the theory of *wu xing* and the theory of *ba gua* in traditional *qigong* are but the tools to express the formula of *qigong* practice as well as the theoretical tools for "returning to the prenatal state". They are not the nature of *qigong* practice as the real content of *qigong* practice is the exercise of *jing* (body), *qi* and *shen* (spirit).

ii. Vagueness and Incompleteness Features of the Theory of *Qi* in Traditional *Qigong* Theories

Although a lot of brilliant expositions about *qi* and *qi* transformation are provided in *Huang Di Nei Jing* and *Yuan Qi Lun*, these theories are found to be lacking in precision, insufficiently detailed, incomplete and not integrate in their statements upon being examined from the altitude of *qigong* science. This is shown in:

i) The definition of *qi* is not explicit. The levels and attributes of *qi* have not been illustrated.

Humans and nature are regarded indiscriminately as one and the same *qi*. The attribute of *qi* in the human body has not been pointed out and the transformation relationships between *qi* and body and between *qi* and spirit have not been clarified.

ii) No description about the laws of formation, motion and change of *qi* in the human body which fits the actual situation has been provided. Although there is a lot of discourse on the subject of *qi* in *Huang Di Nei Jing*, they are not systematic or well-structured. The relationship between *qi* and *jingmai* (major *qi channels*) is particularly ambiguous. There is a total lack of illustration about the exchange between *qi* in the human body and *qi* in the outside world. No wonder some learners and researchers of traditional Chinese medicine would rather ignore the theories of *jing luo* and *qi* transformation.

iii) The complexity in the content of *qi* has not been recognized. *Qi* is regarded only as a kind of single simple substance. Although in *Bao Hun Yuan Xian Shu* human *hunyuanqi* is recognized as the unity of *yuan jing* (primordial essence), *yuan qi* (primordial *qi*) and *yuan shen* (primordial spirit), it is described in light of the inherent nature of *hunyuanqi*. No further and comprehensive illustration of the functions and attributes of *hunyuanqi* have been given except that "If *hunyuanqi* (gathers and) condenses (to a certain extent), it comes into form; if it (expands to some extent and) disperses, it turns into *qi* (or wind)".

It is true that in ancient times when the level of scientific knowledge was low and the knowledge about the objective world was quite shallow, people could merely recognize the general holistic connections between existences and could not reveal more precise and detailed knowledge of the objective world even though super intelligence was applied. When they wanted to illustrate the holistic connections in the world, they could only seek aid from some mathematical models or logical inference. Theories as such were based on the culture established upon the foundation of dispersed and backward agriculture and they were undoubtedly correct and also advanced at that time. But today when large-scale industry has already become rather developed, when science and technology can directly observe the judgment and description of the form and structure of many things in the micro quantum world measured in nanometer, correct results can already be obtained directly without the aid of logical inference. In such a scientific background, the existing models in traditional *qigong* theories become incompetent to illustrate precisely the holistic feature of existences and their laws of motion and change that include all the realms of the micro world, the macro world and the cosmos. This is the objective reason why we establish the theory of *hunyuan* whole entity.

II. To Satisfy the Need of Human Beings' Development—to Become "Free Conscious Humans"

It is unanimously acknowledged in modern science, medical science and philosophy that human beings are the most advanced creatures with consciousness. The fundamental point that tells human beings apart from animals is that human spiritual activities command their life activities. In other words, life activities of human beings are the means to accomplish their will and purposes. Human beings have already been reforming the objective natural world according to their own will in extensive realms to make it serve the human race. However, they have not consciously reformed their own body and spirit in accordance with their own will to make their body and spirit progress to more advanced and more beautiful realms. *Zhineng Qigong* is the very science that emphasizes the reforming of oneself by means of actively applying one's own consciousness. Traditional *qigong* theories, however, cannot provide fundamental theories for this, particularly for the great power of consciousness. The newly-established theory of *hunyuan* whole entity has met this demand and solved this problem.

Let me illustrate this with the example of fracture. It is widely believed in both modern science and medical science that fracture cannot be swiftly or easily healed. The old Chinese saying of "Injury of tendon and hurt of bone need 100 days to heal" is just a description of the difficulty in fracture healing. Why is it like that? You may ask. It has been discovered in modern science that the electricity on the surface of bones in the normal condition differs from the electricity inside them. Once a bone is broken, the two fractured surfaces of the fractured spot which was once the inside of the bone become the outside of it. The electricity of the fractured area becomes positive. The fractured ends are both positive in electricity and electrical charge of the same polarity repels each other, thus making it difficult for the bone to heal. This is the description in modern science of the natural attribute of this natural thing (a fracture in this case) in humans. Viewed purely from the angle of natural phenomenon, this is undoubtedly correct. However, in *Zhineng Qigong*, the power of consciousness is applied to change this natural state. It compels the fracture to swiftly heal. Even traditional *hunyuanqi* theory finds it hard to provide a reliable theoretical foundation for this. This is because "If *hunyuanqi* (gathers and) condenses (to a certain extent), it comes into form; if it (expands to some extent and) disperses, it turns into wind" in traditional *qigong* theories mainly describes the features of *hunyuanqi* in the natural world, while, as far as human *hunyuanqi* is concerned, it is recognized merely as *yuan qi* (primordial *qi*) in humans. Yet the effect of *yuan qi* alone cannot accomplish the target stated above.

The newly established theory of *hunyuan* whole entity has highlighted the consciousness feature of human body *hunyuanqi*. It can even be said that the most fundamental characteristic of human *hunyuanqi* lies in its feature of consciousness, and consciousness is holographic[1] (which means here having infinite, entire and holistic information. See note on next page). Hence it provides a reliable theoretical foundation for the reformation of both the objective world and humans' own life motions through the dynamic function of consciousness. The purpose of practicing *Zhineng Qigong* lies just in enhancing this *hunhua* process of mergence and change between human consciousness and objective existences. In actual fact, many *Zhineng Qigong* teachers can heal fractures in a moment and this provides an example of this theory of *hunyuan* whole entity being successfully applied.

The greatest feature of the creatively-established theory of *hunyuan* whole entity is that it has highlighted all the content of consciousness and has revealed the laws and nature of consciousness activities. It has illustrated the power of consciousness and has pointed out that the appearance of human beings on the earth has changed the original pure natural and objective attribute of the world of nature in the past. It has illustrated the point that human beings in their development process in the days to come will exert to the utmost extent the functions of the familiar intelligence and super intelligence of their consciousness in order to serve the human beings and to accelerate the realization of the human species' nature of being free and conscious and knowing their true self as soon as possible.

Note:

Holography (*quan xi*) in *Zhineng Qigong* Science refers to the *hunyuan* whole-entity information. It contains not only all the features of the reality of existences, but also all the information about the change and development of the past and future of existences. It is the information about the whole entity of time and space. This time-and-space holistic information is not only a kind of attribute of things, but also an objective substantiality. Super intelligence can perceive and interact with it.

Chapter II The Theory of *Hunyuan*

The theory of *hunyuan* is the kernel, quintessence and also foundation of the theory of *hunyuan* whole entity. It illustrates the essential content characteristic of *hunyuan* whole entity, which corresponds to the theory of whole entity that expounds the form characteristic of *hunyuan* whole entity. The theory of *hunyuan* includes two parts in content: the theory of *hunyuanqi* and the theory of *hunhua*[1] (mergence and change).

Section I The Theory of *Hunyuanqi*

Quite a lot of statements about *hunyuanqi* can be found in traditional *qigong* theories, yet they have not formed an integrated theoretical system. Although the term *hunyuanqi* in traditional *qigong* is retained in the theory of *hunyuanqi* introduced here, the term *hunyuanqi* in the theory of *hunyuanqi* and that in traditional *qigong* already differ to a great extent in essence. To make the study of this theory easier, let us begin with the meaning of the term *hunyuan*.

I. The Meaning of *Hunyuan* (Mergence into Oneness)
i. *Hunyuan* in Traditional *Qigong* Theories

The term *hunyuan* hasn't been comprehensively explained in traditional *qigong* theories. There is only fragmentary material about *hunyuan* scattered in different traditional *qigong* classics. A brief introduction about it will be given by the author of this book according to what he knows.

In the literal sense of the word, "*hun*" was considered by the ancient Chinese as "the two and the five are combined (*he ning* in Chinese) and have not manifested themselves yet". That is to say, it is formed with *yin yang* (two) and *wu xing* (five elements) condensed and combined (*ning* here should not be understood as solidify) while the features of *yin yang* and *wu xing* have not shown yet. Now that *hun* includes the condensing and combining of the two levels of *yin yang* and *wu xing*, it cannot be just "mixing together" (*hun he*) but should be understood as

Note:

Hunhua is the process in which two or more than two substances interact with or act upon each to form another kind of new substance. Literally, *hun* means merge, and *hua* means change, which includes quality change and quantity change. *Hunhua* is translated mostly as merge and change in this book. In the context where the mergence state is emphasized in *hunhua*, it is translated as merge and combine. In the context where qualitative change is particularly emphasized in *hunhua*, it is translated as merge and transform. In the context where it means part merges into the whole or a small unit merges into a bigger unit and where transformation also takes place, it is translated as integrate and transform. In Chapter VII, merge and develop are used to refer to *hunhua* according to the context.

mergence and change (*hun hua*). The character "*yuan*" can be understood as one or oneness. In this sense, *hunyuan* means merging into one or oneness.

It is written in Lao Zi's *Dao De Jing*: "You look and yet cannot see it and it is called remote, you listen yet cannot hear it and it is called rare, you touch yet cannot feel it and it is called subtle. The three cannot be further questioned, thus they merge into oneness. It is neither bright above nor dark below. Being endless and beggaring description, it returns into nothingness. This is called the form of no form, the image of nothingness. It is called *huang hu* (literally meaning seemingly obscure state). Facing it, you cannot see its beginning. Following it, you cannot see its end. … This is called the principle of *Dao*."

The "merge into oneness" stated here is *hunyuan* thus called in the later generations. Its entity and attribute have no difference from *Dao*, and the formation of *hunyuan* (i.e., *Dao*) results from the merging into oneness of "remoteness", "rareness" and "subtleness".

The character of "*yuan*" can also be understood as *qi*. It is said in *Gongyang Zhuan*, "*Yuan* is *qi*." From here it can be seen that *hunyuan* and *hunyuanqi* mean the same.

With reference to historic classics, the word *hunyuan* has indeed a lot of different names, such as *hundun* (obscure chaos), *hunyuan* (altogether oneness), *hunlun* (mergence state), *hunlun* (altogether state), *kongdong* (empty hole), *hundong* (merged hole), *hunyuanqi*, *yizi* (oneness according to Chinese character) *hunyuanqi* and *xiantian* (prenatal) *yizi hunyuanqi*. Though given different names, what they refer to all belongs in the domain of prenatal *hunyuanqi*. It is just that sometimes they refer to the *qi* of nature before nature came into being and sometimes the *qi* of the human body before the human body comes into form.

i) *Hunyuan* Refers to the Formless Void *Qi* of the Natural World before Nature Came into Being

It is said in *Yuan Qi Lun* in *Yun Ji Qi Qian*: "*Hunyuan* begins its recording from before *hundun* (obscure chaos) and it is the beginning of *yuan qi* (primordial *qi*)", and "The five *qis* have not formed, the three *cais* (heaven, earth and human) have not divided, the two *yis* (*yin* and *yang*) have not established. This is called *hundun*, *hunyuan* and also *yuan kuai ru luan* (literally meaning original mass of *qi* like an egg) or *wu qi hun yi* (five *qis* merging into oneness)."

In *Zhi Yan Zong*, it is written "The *qi* of *hunyuan* forms somethingness out of nothingness, and the somethingness is called *taiji*." In traditional *qigong* theories, though there are different views as to the origin of the world, such as "the theory of *Dao*" (e.g., *Dao De Jing*), the "theory of *yuan qi*" (e.g., *Tai Ping Jing* and *Zheng Meng*) and "the theory of *yin yang*" (e.g., *Huang Di Nei Jing*), when further studies are made into them, they seem not to have difference in essence. It is stated in *Yun Ji Qi Qian* that "*Dao* is just *yuan qi*", and that "*Yuan qi* is the oneness nature and *yin yang* are the two bodies". In ancient times, there was also the saying of "*Yin yang* is one *taiji.*" From this it can be known that *Dao*, *yuan qi* and *taiji* are in fact different names of prenatal *hunyuan* oneness *qi*.

It is written in *Xing Ming Gui Zhi*, "What is *Dao* exactly? To summarize in one word, it is *qi* actually. Originally being one *qi* which seems to be coiling together, it is vast, seemingly misty, profoundly far-reaching and cannot be fathomed. It is in a motional fine state and contains great initiative and utmost subtlety. And this is *taiyi.* ... The one *qi* is motional and opens and closes in the empty void. ... Somethingness and nothingness attract and accommodate each other, in a mixed, seemingly obscure state. It is harmoniously void and extremely sacred. It embraces oneness with great initiative contained. The initiative can command changes and all existences are established upon this state of chaos... This is called *Dao* giving birth to oneness. It is called the primordial beginning, the *taiji* of heaven and earth..." (Annotation by the author: *Taiyi* is the utmost oneness)

Dao is the root and foundation of all existences in the universe.

ii) *Hunyuan* Refers to *Yuan Qi* (Primordial *Qi*) in the Human Body

Jin Ye Huan Dan Nei Pian in *Dao Shu* states "*Yuan qi* is the *hunyuanqi* in the human body and it is as the root to the tree to the human being. When mental activity rests, *qi* settles. When human spirit (*shen*) moves, *qi* follows." Here it not only points out that *yuan qi* is just *hunyuanqi* in the human body, but also that *hunyuanqi* is subject to the instruction of mental activity.

It is said in *Bao Hun Yuan Xian Shu* in *Xian Shu Mi Ku*:

"What is *hunyuan*? It is *jing*, *qi* and *shen* before one's birth...This is the *yuan* (primordial) *jing*, *yuan* (primordial) *qi* and *yuan* (primordial) *shen*. To call *jing*, *qi* and *shen* before one's birth with *yuan* is because they are the thing-in-itself or original being. Before humans have their bodies, they first have this original thing. After a person gets this original thing, form comes into being from formlessness, and quintessence come into being from where there seems to be no quintessence. It comes from before the sexual contact between parents."

And also:

"...The essence and blood, being merged and containing this point of truthfulness, change into form with *jing*, *qi* and *shen* residing in the form. Though there are three names (of *jing*, *qi*, *shen*), the nature is actually the same. One is the mergence into oneness (*hunyuan*) and three are the divided miraculous aspects. One is the entity while three are the functions. In the entity of *hunyuan*, the pure oneness with no impurity is *jing*, that which helps blood and *qi* channels connect well and flow smooth is *qi*, and that which is void, miraculous and with initiative in motion is *shen* (human spirit)."

From this it can be seen that, in traditional *qigong* theories, human body *hunyuanqi* is born in the process when parents create life. It belongs to the prenatal *qi* of humans and is the most important root and foundation of human life. It is the most essential for the cultivation of *qigong* and *Dao*.

iii) Other Meanings of *Hunyuan*

In Chinese history, Lao Zi was also called *hunyuan*, e.g., *Hunyuan* Lao Zi. Daoist priest Xie Shouhao in the Song Dynasty once wrote a book entitled *Tai Shang Hun Yuan Lao Zi Shi Lue* (*A Brief History of Taishang Hunyuan Lao Zi*) in which he referred to Lao Zi as *Hunyuan*. Emperor Song Zhen Zong once conferred on Lao Zi the title Taishang (Highest) Laojun (Senior Gentleman) Hunyuan (Merged into Oneness) Shangde (Ultimate Virtue) Emperor, and he wrote a preface himself about Taishang Laojun: "Taishang Hunyuan Shangde Emperor (his royal highness in utmost status with ultimate virtue who merged into oneness) is one with the integrate vastness of *qi* and embraces the profound essence. He is in infinity in oneness and has the greatest ability to change. He is characteristic of life breeding and helps all beginning. Therefore, before *taiji* he exists, after *taiji* he does not cease… "(*Hun Yuan Sheng Jing*)

Here, not only was Lao Zi called *Hunyuan* but the connotation of *hunyuan* and *Dao* were mentioned in the same breath. Ever since then, the word *hunyuan* has not only become another name of Lao Zi, but also the symbol of *Dao*. In later generations, some *Dao* cultivation schools in Daoism have been directly called *Hunyuan*.

Different schools in Daoism have been systematically introduced in *Bai Yun Guan Zhi* (*History of White-Cloud Daoist Temple*) and *Xuan Men Bi Tan*, and *Dao Zu Fa Xi* was listed above all other schools while being named as *Dao Zu Hunyuan* School (The *Hunyuan* School of Daoist Ancestor). The inheritance pedigree[1] of the school is "*Hun Yuan Qian Kun Zu. Tian Di Ri Yue Xing. San Jiao Zhu Sheng Shi. Jin Mu Shui Huo Tu. Hong Meng Pan Yin Yang. Tai Ji Bi He Tu. Ren Yi Li Zhi Xin. Sheng Lao Bing Si Ku.*"

Next is the school of *Taishang Hunyuan*. Its inheritance pedigree is "*Yi Yong Tong Xuan Zong. Dao Gao Ben Chang Qing. De Xiang Gong Jing Tai. Yi Yuan Fu Yuan Ming. Hun Yuan San Jiao Zhu. Tian Di Jun Zong Shi. Ri Yue Xing Dou Zhen. Jin Mu Shui Huo Tu.*" Its mnemonic rhyme to pass on the Daoist teaching is: "*Yin Fu San Bai Zi, Dao De Wu Qian Yan, Xi Yi Yuan Wu Yi, Hun Yuan Yi Mai Chuan*" (Three hundred words of *Yin Fu Jing*. Five thousand words of *Dao De Jing*. *Fu Xi* and *Yi Jing* actually have not *yi*. It is but *Hunyuan* being handed down in a continuous stream.)

Note:

Inheritance pedigree: The form of a poem used for giving name to each generation of the Daoist practitioners and successors in the school, with each word being used for and shared by one generation of successors.

There are also quite a few *qigong* methods that are named after *hunyuan*, such as *hunyuan wuxing gong*; *hunyuan* one *qi gong*.

ii. *Hunyuan* in *Zhineng Qigong* Science

Hunyuan in the theory of *hunyuan* in *Zhineng Qigong* Science is a special term which refers to the manifestation of a kind of objective existence perceived in the state of super intelligence and is the holistic existential form of the nature of things. This *hunyuan* whole entity is formed with the mergence and change of at least two kinds of material factors in the state of familiar intelligence. To deduce from this, any independent thing in this universe is not single and simple. Instead, it is the unified *hunyuan* whole entity formed with the *hunhua* of more than two kinds of material factors. For example, the universe is the *hunyuan* whole entity of concrete substance with form and insubstantial (intangible) substance without form. The field substance in modern science is the *hunyuan* whole entity manifested in energy and information. The concrete object or substance with form is the whole entity merged and combined into being with body, *qi* and essence.

From the connotation of the term of *hunyuan* it can be seen that *hunyuan* includes two parts in its meaning: One is the unified whole entity formed with mergence and change, as is called by us *hunyuanqi*; the other is the changing and merging process of each and every material element that contributes to the formation of the unified whole entity, as is called by us *hunhua* (mergence and change). The former is the theory of *hunyuanqi* and the latter is the theory of *hunhua*.

In *Practice Methods of Zhineng Qigong Science*, there are also the terms of outer *hunyuan*, inner *hunyuan* and middle *hunyuan*. These are the terms of *hunyuan* used in the three *qigong* practice phases and belong in the domain of *hunhua*. They refer to the *hunhua* of human consciousness with *hunyuanqi*, and outer, inner and middle refer to the position where the mergence and change takes place. What is expounded in the theory of *hunyuan* is the theory of *hunyuanqi* and the theory of *hunhua* instead of the *hunyuan* in *qigong* practice methods.

II. The Theory of *Hunyuanqi*
i. What Is *Hunyuanqi*

Hunyuanqi is a holistic form of existence that characterizes the nature of existences perceived in the state of super intelligence. If described from the perspective of familiar intelligence, it is the *hunhua* or merged and combined state of *xing* (form), *qi* and *zhi* (quintessence) as stated in the classic holistic concept. Form, *qi* and quintessence here refer to the three fundamental elements of concrete substances. To use forced analogies, *xing* (form) can be compared to mass of matter as described by modern science, *qi* can be compared to energy, and quintessence can be compared to information. Therefore, *hunyuanqi* can be illustrated in modern language as this: *Hunyuanqi* is a special material state that marks the holistic feature of an existence or substance formed with the mass, energy and information of the substance merged and combined.

Hunyuanqi has two existential forms: One is the insubstantial (intangible) existential form that has no form or image. It is usually directly called *hunyuanqi* in the theory of *hunyuan* and is *hunyuanqi*'s fundamental form of existence. It is a special material state with extremely even texture that is integrated and unable to be divided. Here, *xing* (form), this essential material element in the physical world lies in a hidden state. The other is the existential form of substance having form or image (which refers to objects that have evident characteristics of mass, energy and information). It is called substantial (concrete) entity in the theory of *hunyuan* and is formed with formless *hunyuanqi* condensed together. It belongs in the existential form of *hunyuanqi* in the broad sense.

There is *hunyuanqi* both in and around a concrete substance. The concrete substance is the condensed form of expression of its *hunyuanqi* while the *hunyuanqi* around the concrete substance is the expanding form of expression of its *hunyuanqi*, which is called *hunyuanqi* field (and is usually called *hunyuanqi* as its abbreviated name). The greater the density and volume of the concrete substance, the greater the density and the broader the extent of the *hunyuanqi* field outside it. Change in the composition and function of the concrete substance can lead to change in the *hunyuanqi* field suffusing around it. In contrast, when change in the *hunyuanqi* field reaches a certain degree, the *hunyuanqi* field can condense or expand into concrete material entity correspondent to the change in the *hunyuanqi* field and causes a certain degree of variation in the original substance. There exists a close interdependent relationship between the *hunyuanqi* field and the concrete substance. The kind of concrete substance determines the kind of its correspondent *hunyuanqi* field.

A certain amount of time is needed in the transformation taking place between the concrete substance and its *hunyuanqi* field, namely, changes that take place in them are not synchronous. If change in the concrete substance takes place from the inside, the original *hunyuanqi* field outside the concrete substance will remain for a certain period of time until the *hunyuanqi* of the concrete substance that has already changed fills the outside of the substance and changes the original *qi* field. If change begins first from the *hunyuanqi* field, the original features of the concrete substance will also remain for a certain period of time. The phenomenon of the *qi* field not conforming with the concrete entity will occur until the *qi* field completely condenses inward, forming concrete substance in a new state, and a new balance will appear. If change in the *hunyuanqi* field exceeds a certain limit, fundamental qualitative change will happen in the original concrete substance and it will turn into another concrete substance.

Any concrete substance is the unity of its concrete existence and the *hunyuanqi* field outside it, and this unity is called a *hunyuan* entity by us. Different *hunyuan* entities can interact with each other, integrate and transform into *hunyuan* entities at higher levels and can also divide and form lower level *hunyuan* entities.

The holistic features of *hunyuan* entities can be observed and perceived with human beings' super intelligence yet modern scientific and technological means still cannot deal with it. This is because study of a particular feature of things with modern scientific and technological means is usually carried out with the premise of excluding their other features. That is to say, modern scientific and technological means cannot cognize features of things from all aspects. Modern science can only reveal physical and chemical features and so on but cannot reveal the feature of life motion essentially manifested in the form of the *hunyuan* whole entity feature and the reason lies mainly in this.

The two forms of existence of *hunyuanqi* can transform into each other under certain circumstances, namely, substantial entities that have form and image can disperse into formless *hunyuanqi*, and formless *hunyuanqi* on the contrary can also condense into substantial entities having form and image. The universe is just the *hunyuan* whole entity in which the two existential forms of *hunyuanqi* mutually transform and evolve.

What should be pointed out here is that the *hunyuanqi* we talk about in the theory of *hunyuan* seems to have a lot of similarities with chaos theory in modern physics. As a matter of fact, however, they completely differ. Although it has been pointed out in the chaos theory in modern science that substance in the state of chaos is the key reason for the production of substance in a new order, chaos is after all a special state that is formed when the original substance (once in its original order) has evolved into a state of disorder. That is, chaos has lost the features that the original substance once had.

Hunyuanqi is rather different. *Hunyuanqi* is the special holistic manifestation that characterizes the essential nature of all things. Everything has its own *hunyuanqi*. The different features of objects that people with familiar intelligence perceive are but the familiar intelligence characteristics of *hunyuanqi* sensed and perceived in the state of familiar intelligence. The phenomenon of chaos as described in chaos theory in modern times also belongs to the characteristic perceived in the state of familiar intelligence. Although it is still not easy for people to understand, the phenomenon of chaos is not the characteristic of things described in the super intelligence state.

ii. Different Levels of *Hunyuanqi*

All existences in the universe have their own *hunyuanqi* that characterizes their own holistic features, which shows the feature that the *hunyuanqi* of all existences has levels. The *hunyuanqi* of all existences in the universe can be divided into a certain number of complicated different levels. Here we will just briefly introduce five of them.

i)   The Level of *Hunyuanzi* (*Hunyuan* Particle)

"*Hunyuan* particle" refers to the combination of point in three dimensional space and point in one dimensional time that is in progression. It is a state of absolutely no difference, an extremely subtle realm reached by the *hunhua* or mergence and change of all existences in the universe. Therefore, it is also called *hunji* (utmost mergence state). As the *hunyuan* particle is already too fine and subtle to be further divided, the multitude of *hunyuanzi* gathering together cannot be further divided, either. They display a state of total unity in which one is many and many are one, and one and many cannot be differentiated. All the generation and change of both *hunyuanqi* and all existences in the universe are carried out against this background.

Further discussion:

The "point" discussed above differ from the point described in geometry which has merely position but not length, width, or height. The point described here occupies not only a position in space, but also a certain amount in terms of three-dimensional coordinates. This is an amount that is infinitesimal and hard to describe. It is too small to be divided any more, thus exhibiting an image of even state. It is smaller than the Planck length (1.6×10^{-33} cm) and, for the sake of convenience in explanation, we refer to it as "point space".

As for time, in modern science it is not called a line nor is there such a term "point time". It is only for the convenience of expression that we draw one forced analogy between time in a process of passing and a one-dimensional line and we draw another forced analogy between time at a particular moment and the point similar to that stated in mathematics that has limitless length. In fact, the point in time we state here, though rather too small, also occupies an extremely short period of time. It is even shorter than the Planck time (5.4×10^{-44} sec). For the convenience of description, it can be called "point time."

In the sphere of space and time mentioned above, the interaction between point time and point space enter into unanimity, which makes the two of them merge into one, i.e., *hunyuanzi*. To be more specific, the term *hunyuanzi* refers to the idea that the two ends of the length of point time mentioned above bend inward, which endows time with the feature of space, and that the lengths of the three dimensions of point space enclose inward, which gives space the feature of time. So point space and point time become one, i.e., *hunyuanzi*.

Hunyuanzi is in a state of no difference in features of both time and space. They are in a highly even state which seems to be an absolute vibration of time and space. It is a kind of time and space with absolutely no difference in it. There is no difference in mass, energy and information. It can generate primordial *hunyuanqi*. The level of *hunyuanzi* is equivalent to the description of *Dao* in traditional *qigong*.

"You look and yet cannot see it and it is called remote, you listen yet cannot hear it and it is called rare, you touch yet cannot feel it and it is called subtle. The three cannot be further questioned, thus they merge into oneness. ... Being endless and beggaring description, it returns into nothingness. This is called the form of no form, the image of nothingness. It is called *huang hu*."(*Dao De Jing*)

With regard to the name of *Dao*, it was also called "*wuji*" (being in an ultimate nothingness state) or "*hunji*" (being merged to extremity) by the ancient Chinese. To call it *wuji* is to describe it from its entity feature of "nothingness". *Xin Chuan Shu Zheng Lu* states:

"The truthfulness of *wuji* is neither having nor having not. With images of all existences hidden within, it has reason and no forms yet. Being empty and having nothing in it, it can enrich everything though it is void. Being the ancestor of three schools (Buddhism, Daoism and Confucianism), its entity and functions are not shown yet. Having neither within nor without, it connects the deepest, remotest and brightest. It has no thinking or consideration, nor smell or sound. It is still, void and has great initiative. With no difference between outside and inside, the outside and inside are closely connected. Covering both heaven and earth, *Dao*'s motion is never in vain". "The prenatal is obscure and chaotic and *yin* and *yang* are not divided. *Qi* and reason are attached to each other. This is *wuji* or the ultimate nothingness state."

When called *hunji* (being merged to extremity), *Dao* is described from its entity feature of "somethingness". In *Xin Chuan Shu Zheng Lu*, it is written:

"*Hunji* is the image in the vast void. Image[1] and number are contained in it while the clear and the chaotic are not divided, and it has shape and form before *He Tu* and *Luo Shu* appear[2]. With *qi* and mechanism intact, the reasons for *yin yang*, *si xiang*[3] and *ba gua* are already in it in a hidden state. This is *taiji* with its entity not mixed with *yin* and *yang* (as *yin* and *yang* have not developed from *taiji* yet). With those whose form and *qi* already manifest, though there is the difference in *yin yang*, in the four images of *qi* and in the signs of trigrams, there is at the beginning no sound or smell if reasons for why they are so are inquired."

From this it can be seen that, *Dao*, *wuji* and *hunji* so called by the ancient Chinese are actually one and the same thing, namely, they are the root and origin of the universe. In order to further

Note:

1. During the process when formlessness is changing into form, there is an image (or called information structure) in it; when there is image, there is number.
2. Before concrete forms and shapes of tangible substances appear.
3. The four images of *qi*: *tai yin*, *tai yang*, *shao yin* and *shao yang*.

our understanding of the entity and function of substance at this level, I present here an excerpt from an illustration of *wuji* in *Dao Mai Tu Jie*:

"*Wuji* is formless, nameless, infinite and boundless. Being utmost void and most magical, it is motionless heaven of reason in extreme quietness. This reason, though miraculous and chaotically integrated, is actually clearly in order. Utmost nothingness can generate utmost somethingness. Utmost void can defend utmost substantiality. It is the source and origin of all existences in the universe. It is formless in sight yet can produce form. It is soundless in the ear yet can produce sound. It surpasses the highest point of heaven and penetrates down through the earth. Inseparable from *qi* though it is, it does not mix with *qi*. It runs through the midst of *taiji* and surrounds *taiji* as well. Silent and vast, it is independent and steadfast. Deep and remote, it does not decay for innumerable years. It is the beginning ancestor for all existences in generating heaven and earth."

This passage is the most consummate and comprehensive description of *wuji*, also known as *Dao* and *hunji*. It highly agrees with the substance that we describe at the level of *hunyuanzi*.

ii) The Level of Primordial *Hunyuanqi*

This is the primordial level of *hunyuanqi*, so it is called "primordial *hunyuanqi*". It is a kind of special substance even in texture without differences in functions. It fills the entire universe and permeates and penetrates all existences. It is also an integrated whole entity that cannot be divided and can evolve into formless substance at different levels (*hunyuanqi*) and gather together and condense into various existences with forms (substantial entities). It is the root from which our universe evolves.

The level of primordial *hunyuanqi* is equivalent to *yuan qi* (or called *taiji*) in the universe described in traditional *qigong* theories:

"It is one *qi* in motion that opens and closes in the void. The male and the female attract each other and black and white merge and combine. Somethingness and nothingness transform into each other and it is a state seemingly chaotic and obscure. It is sufficiently void and extremely sacred. It embraces oneness with magical function contained. The magical function can command miraculous great changes and all existences are established upon this seemingly 'obscure' state… It is the beginning of heaven and earth."(*Xing Ming Gui Zhi*)

Further discussion:

Primordial *hunyuanqi* "emerges" when *hunyuanzi* 'breaks' away from its inward enclosing feature so point space and point time are separated and *hunyuanzi*'s feature of having no difference in time and space 'decomposes' into a special state containing the content of time and space. Thus 'compound particles of time and space' are formed. Primordial *hunyuanqi* is a special material state in which there is only pure information formed with time and space compound particles while energy and mass are still in a potential state. Different accumulations of time and space compound particles can form different kinds of information sequences. Therefore, we can also say that primordial *hunyuanqi* is the merged and combined (*hunhua*) state of various potential information. Time and space compound particles which are the fundamental element of primordial *hunyuanqi* at this time already differ from *hunyuanzi* as they have acquired the ability to evolve in the opening process. All in the universe actually evolves into being from them. This seems to be very similar to the moment immediately after the Big Bang began. Gravity and electric magnetism at that time were not divided yet and they were still in a *hunhua* state. This was a primitive chaotic state having not entered the physical world.

As to the process and mechanism of how the *hunyuan* particle (*hunyuanzi*) generates primordial *hunyuanqi*, details about it can be found in The Concept of Universe Whole Entity (Section II) in Chapter III The Theory of Whole Entity. Only the gist is stated here: When all the stars in the universe evolved into the black hole, the time-and-space structure in the black hole started to change under the effect of great gravity. When the change reached a certain limit, time and space enclosed inward, formed *hunyuanzi* and made gravity disappear. The black hole broke at that very moment and *hunyuanzi* filled the entire universe. In *hunyuanzi*'s escaping process, time and space separated from each other and formed primordial *hunyuanqi*.

Similar statements can also be found in ancient Chinese writings. In *Jie Yun Bu* of *Hun Yuan Hun Dong* in *Dao Zang*, it is said, "*Yuan qi* is in indiscriminate vastness and outside deep remoteness. It is generated from the empty hole". "In the empty hole, there is no light, no image, no form, no name, no color, no four seasons, no voice, no sound…"

Although the terms used by the ancient Chinese people in their statement differ from the terms we use to express our viewpoints, aren't they rather similar to each other in content when being carefully examined?

The level of primordial *hunyuanqi* is equivalent to *taiji* (*yuanqi*) in traditional *qigong* theories. It is said in *Xin Chuan Shu Zheng Lu*, "*Taiji* is *lingji* (literally meaning utmost miraculous which actually refers to having greatest initiative) gradually becoming remote and void. Profound and thorough, it does not mix with *yin yang*. Clear and bright, it lacks none in its oneness entity. Changes are stored in its state in which there is no settled *ji*[1] yet (literally meaning extremity), while *yin and yang* actually contain their own reasons for reaching settled *ji*s". It is also said that

Note:

There are usually 5 'extremities': *wuji, hunji, yuanji, lingji,* and *taiji*. Hunyuan, wuji, hunji, Dao (equivalent to *hunyuan* particle) and *lingji, yuanji, taiji, yuan qi* (equivalent to primordial *hunyuanqi*) in the quotations of ancient Chinese sages are the statement of their perception and cognition of the third level substance.

"Without the mentioning of *yuanji* (extremity of oneness), *taiji* would be equated with a common substance and cannot be qualified to be the root of all changes. Without mentioning *taiji*, *yuanji* would be reduced to empty stillness and cannot be the source and beginning of all existences. Thus motion and stillness have no end; *yin* and *yang* have no beginning, and this is the very ma-gical use of *taiji*."

It is said in *Dao Mai Tu Jie*, "*Taiji* is the *qi* of *yin yang* combined, the reason for upward and downward *qi* motions, and the reason for the flowing of *qi*".

The entity and attribute of *taiji*, if analyzed from its feature of 'being void and having initiative', is called *lingji*. It is written, "Harboring magical somethingness in void and embracing true nothingness before the appearance of extremities, it divides by itself without the need to differentiate and opens by itself without the need to chisel. It is to generate the starting mechanism of creation and to initiate the beginning acupoint of magical light."(*Xin Chuan Shu Zheng Lu*)

If analyzed from its feature of "having somethingness and being able to change", it is called *yuanji* (extremity of oneness). That is to say, "Rooted in the condensed harmony of the primal beginning, with enveloping mist of the ultimate harmony hidden within, its essence is extremely exquisite and colorless and it contains the oneness *qi* and is formless." "It contains the body of *yin* and *yang*. It has the mechanism of change. It includes three extremities (*wuji, youji, taiji*) and is very significant for the extremity of mergence (*hunji*). It is in such a recurrent process. Isn't it the *qi* mother of *taiji*?"(*Xin Chuan Shu Zheng Lu*)

From this it can be seen that *yuanji, lingji* and *taiji* stated in ancient times illustrate from different aspects the attribute of this special material state only one level below *Dao*. They are similar to the primordial *hunyuanqi* previously talked about.

iii) *Hunyuanqi* at the Level of All Existences

This refers to the *hunyuanqi* of substances in the existential form of concrete entities (atoms→molecules→animals and plants). Concrete substance is the material existential form that can be cognized with familiar intelligence and has a certain structure as well as physical and chemical features. *Hunyuanqi* at this level, as far as this substantial material is concerned, is the holistic material state formed with the mergence and combination (*hunhua*) of its own form, *qi* and essence. Any concrete substance is the expression of its own *hunyuanqi* condensed in form. Within this form is the flowing *hunyuanqi* of the substance, while around it is permeated the relatively sparse *hunyuanqi* of the substance. This is the *hunyuan* entity that has been previously mentioned. Rivers, mountains, lakes and seas; flowers, grasses and trees; animals and human be-ings are all exclusively like this. As far as the formation of concrete substance is concerned, it is that the *hunyuanqi* of simple substance forms the *hunyuanqi* that has the entire holistic feature of complex substance, then the formless and imageless *hunyuanqi* condenses into concrete (com-plex) substance, whereas the *hunyuanqi* of simple substance evolves from primordial *hunyuanqi*.

(For further explanation of this formation process, please refer to "Discussion" afterward).

Hunyuanqi of a concrete substance is the noumenal state of the substance. It is a special state that interacts with super intelligence. Although it differs from the features of the concrete substance perceived in the state of familiar intelligence, it is upon what the concrete substance is based. Various features of the concrete substance cognized with familiar intelligence are but the partial attribute of the substance individually perceived with different sensory organs, whereas in the whole entity of *hunyuanqi* these partial features of the concrete substance alter their partial attribute and subordinate themselves to the holistic feature. Super intelligence only acts upon *hunyuanqi* that has the holistic feature. Therefore, different physical and chemical characteristics that belong in partial characteristics cannot show their special (physical and chemical) features in the whole entity state. Experiments on extraordinary super powers and abilities in the human body show that magnetism of magnets cannot be detected with magnetic needles when they are being teleported. This is the reason why "I see that the mountains are not mountains and that the rivers are not rivers" in Dhyana (Zen) after one "knows one's heart and sees one's nature" (*Ming Xin Jian Xing*, being fully enlightened). Of course, when further progress is made in one's *gongfu*, when one can combine super intelligence with familiar intelligence, one can perceive the normal state and the super state of concrete substances at the same time, and this is the reason for that "I see that the mountains are still mountains and that the rivers are still rivers" in the Zen System.

It should be further pointed out here that the world of substances with forms can be divided into several levels, namely, the inorganic world, the organic world and the world of organisms. The world of organisms can further be divided into the world of plants, the world of animals and the world of human beings. *Hunyuanqi* at these levels is not all the same. For example, for the substance at the level without life (including inorganic substance and organic substance), since the independent existence of a substance at this level does not need to interact with the outside world (If the substance is acted upon, the nature of the substance will be changed), the *hunyuanqi* of the substance maintains the original state and attribute of the *hunyuanqi* that forms the substance. Notice here that in this kind of substance itself there also exists incessant motions. Aside from displacement, there is also its rise and fall and vibration, namely, the motions of opening, closing, going out and coming in of *hunyuanqi*.

The substance in the world of organisms, on the other hand, is different. During the whole process of its existence, the substance needs material exchange (i.e., metabolism) with the outside world. Therefore, *hunyuanqi* of the substance at this level is formed with the combination of the *hunyuanqi* that generates this substance and the features brought about by metabolism (which is also carried out under the effect of the above-mentioned *hunyuanqi*) after this substance comes into existence. That is to say, the *hunyuanqi* of an organism is not only the original *hunyuanqi* that generates this substance.

From this it can be seen that *hunyuanqi* in the world of organisms consists of two parts, namely, the primordial *hunyuanqi* that forms this substance, which was called by the ancient Chinese prenatal *qi*, and the *qi* of substances that this organism obtains in its process of growth, which was called by the ancient Chinese postnatal *qi*. This feature differs between the world of plants and the world of animals. Although there is such change as form, *qi* and essence in the growing process of a plant, the growing and changing process is completely carried out naturally with the outside world and it is mainly the influence of *qi* and the change in form, whereas animals have their initiative and their mental activities (which mainly refer to neural activities). Therefore, instead of being called *xing* (form), *qi*, *zhi* (quintessence or information), essential features in animals are called *xing*, *qi* and *shen* (spirit).

Discussion:

Hunyuanqi of all existences is the holistic expression of both the structural feature and the functional feature of concrete substances. In the state of familiar intelligence, the structural feature of a substance is the positional feature displayed in space while the functional feature is the feature of process displayed in time. It is well known that functional features adhere to structural features. Although there is in this a correlation between partial structures and partial functions, it is essentially the correspondence of holistic structures with holistic functions. It is particularly so in the kingdom of organisms. The holistic feature of *hunyuanqi* is the agreeing expression of the holistic structure with its instantaneous function, or the instantaneous spatial expression of the holistic time characteristic (process) of an object.

Take a small tree for instance. It has a course in time in which its growth and decline are displayed: Being a seed, a sprout, taking root, becoming a seedling, a big tree, blooming, yielding fruit and death. This is also this tree's holistic function. Yet this growing process has different spatial content (structural characteristic) in different periods of time. Familiar intelligence can only receive information regarding structural characteristics of this tree. It cannot perceive information about the time characteristic contained in the structural characteristic. Cognition of the development process (time composition) of things in familiar intelligence is accomplished through thinking. Super intelligence, however, can grasp directly all the content of the time (process) characteristic of things from the material structural characteristic in an instant. This is because in the material content at every moment all its information in the past as well as its buds of change in the future is inherent (Please refer to *hunhua* time or time of mergence and change discussed later in this chapter for detail).

From this it can be seen that *hunyuanqi* at the level of all existences is very similar to *hunyuanzi* (*hunyuan* particle) and the time-and-space compound particle (the *hunyuan* whole entity in which point space and point time are combined). It is also the *hunyuan* whole entity in which space (the space of structural whole entity) and time (the time of functional whole entity) are combined. It is formed with multitudinous time-and-space compound particles which gather together and become combined according to certain frames and structures. The *hunyuanqi* of any specific existence is an integrated unity.

In the long evolutionary chain of all existences in the universe, *hunyuanqi* at this level evolves from primordial *hunyuanqi*. It is the transitional form from formless *qi* to concrete substances. The preliminary base of form, *qi* and quintessence of concrete substances is already in it. It is just as what is written *Lie Zi*, "*Tai Chu* (primal start) is the beginning of *qi*. *Tai Shi* (primal beginning) is the beginning of form. *Tai Su* (primal simpleness) is the beginning of quintessence (information). With *qi*, form and information within and not having separated, it is called *hunlun* (being integrated). "

When our universe was in its preliminary holistic state, it was the primordial *hunyuanqi*. When primordial *hunyuanqi* was being divided into specific things, there also existed a process from being formless to having forms. Once the features of specific things came into being and their bodies had not yet formed, primordial *hunyuanqi* turned into *hunyuanqi* of all existences. When all existences have formed, the *hunyuanqi* of a specific existence and the existence itself form the *hunyuan* entity.

iv) Human *Hunyuanqi*

Although human body *hunyuanqi* also belongs in the domain of *hunyuanqi* of all existences, because human beings already have activities of consciousness and human holistic life activities are carried out under the guidance of consciousness, human body *hunyuanqi* is not only further divided, but fundamental changes have also taken place in its functions. For example, human *hunyuanqi* has the feature of following the intention of consciousness. This will be comprehensively expounded in Chapter IV Human *Hunyuanqi* and will not be described here in detail.

v) *Yi Yuan Ti* (Mind Oneness Entity)

Yi Yuan Ti is a special term used to describe human brain *hunyuanqi* when it has developed into the stage of thinking in terms of concept. It is the important component of human body *hunyuanqi* and is the *hunyuanqi* peculiar to human beings. It is the most exquisite and magical part of *hunyuanqi* in the human body. It cannot only command human body *hunyuanqi* but can also mobilize all levels of *hunyuanqi* in the external world. Many of its characteristics are similar to *hunyuanzi* (*hunyuan* particle) and primordial *hunyuanqi*, yet it already has features of initiative and being able to move at will. It is the highest level of the forms of *hunyuanqi* among all the existent levels of *hunyuanqi* having developed up to now in the universe. It will be thoroughly described in Chapter V The Theory of Consciousness. Hence it is only briefly stated here.

iii. Features of *Hunyuanqi*

The features of *hunyuanqi* stated here are the common features that *hunyuanqi* at all levels share. Special features of human body *hunyuanqi* as well as *Yi Yuan Ti* will be stated in different chapters later.

i)   The Gathering and Expanding Features of *Hunyuanqi*

The features of gathering (including condensing) and expanding (including dispersing) exist in *hunyuanqi* at every level: Gathering in *hunyuanqi* increases *hunyuanqi* in intensity. When a certain extent of expansion is reached, concrete substances can also disperse into formless *hunyuanqi*. This gathering-and-expanding process in the natural world results from the interaction of all existences in the universe and it is going on naturally. In living organisms, this changing process of gathering and expanding in *hunyuanqi* is going on at every moment (For detail, please refer to the theory of *hunhua* as described later in this chapter). In the human body, human consciousness activity can also cause human body *hunyuanqi* to condense and disperse. When mental activity is combined with *q i*, the mental activity that gathers to and concentrates on one point can make *qi* gather and condense; it can even gather and condense *qi* into substance with form. On the contrary, when mental activity penetrates into concrete substance, the instruction of dispersion in mental activity can make the concrete substance disperse into formless *q i*.

ii) The Holistic Feature of *Hunyuanqi*

Every level of *hunyuanqi* is the integrated display of the totality of a particular existence. (Detail about this can be found in Chapter III The Theory of Whole Entity).

iii) The Distribution Feature of *Hunyuanqi*

Primordial *hunyuanqi* is distributed all over the entire universe. It is able to change with the generation and formation of all existences in the universe and becomes the most primordial source material of the structure and function of all existences. *Hunyuanqi* of a specific substance among all existences in the universe condenses into the material entity of this substance and is permeatively distributed around it. Its degree of density decreases as its distance from the center of the substance increases. The distribution of the *hunyuanqi* inside the concrete substance is related to the structural feature of the substance. Generally speaking, there is more *hunyuanqi* in the center and near the surface of the substance.

iv) The Feature of Retaining Information in *Hunyuanqi*

The function of retaining information exists in both primordial *hunyuanqi* and the *hunyuanqi* of concrete substances. For example, if we give a mental instruction of 'gather' or 'expand' or 'push' or 'pull' of *hunyuanqi* to a certain space in the void, a sensitive person can feel the original information about 'gather', 'expand' or 'push' and 'pull' in our mental activity in this space though our mental activity has already stopped.

v)   The Compatible Feature of *Hunyuanqi*

Hunyuanqi between different levels can display compatibility. Lower level *hunyuanqi* can move smoothly through higher level *hunyuanqi* or concrete substances. Primordial *hunyuanqi* is the most primordial and original *hunyuanqi* in the universe. It can smoothly run in and through substances at all levels in the universe and serves as the source material that constitutes *hunyuanqi* at all levels. Organisms, after having absorbed this *qi*, prosper in their life force.

It is true that primordial *hunyuanqi* can penetrate from "low" level *hunyuanqi* to "high" level *hunyuanqi* one after another, whereas human consciousness activity (the activity of *Yi Yuan Ti* which is the holistic synthesis of human brain *hunyuanqi*) can penetrate from "high" level to "low" level one after another. It is because of this that practicing *qigong* can combine with different levels of *hunyuanqi* through practitioners' mental activity and takes the *hunyuanqi* at all levels back into the practitioners' bodies for their own use.

iv. Re-Understanding of Some Concepts in Modern Science in Light of the Theory of *Hunyuanqi*

i)   Re-Understanding of Such Concepts as Point, Line, Surface and Zero

In mathematics, only zero represents that there is nothing at all; a point has only position but no length, width, or height; a line is a one-dimensional existence that has position and length but no width or height; a surface is a two-dimensional existence with length, width, but no height. The definitions of zero, point, line and plane mentioned above are the abstract descriptions in mathematics and geometry, which actually cannot possibly be true in reality as there is no zero state of time and space in the universe. This, therefore, leads to the incomprehensibility of the definitions of point and line. Please just use your imagination: Is it possible to have a position in the universe that does not have any measurement (length, width and height)? If such kind of points existed, then it would be zero space, and the absolute nothingness of the zero space can by no means become somethingness in reality. In physics, a moving object is at a certain point and at the same time not at this point at a particular moment. If the length of point in space is really that "zero" state in which nothing exists, how can the sequence that represents the zero points become something real? Hence is born the saying that motion itself is a contradiction.

In the theory of *hunyuanqi*, zero refers to the absolute even state of *hunyuanzi* (*hunyuan* particles). The position that a point takes is *hunyuanzi* which also has a certain measurement. It is simply that the measurement is too tiny (It is shorter than the Planck time in time and the Planck length in space). Therefore, the line formed with points in motion (or called the sequence of points) has not only length, but also measurements of length and width of the line itself. Likewise, a surface has not only length and width, but height as well.

ii) Re-Understanding of the Three Essentials of Mass, Energy and Information

According to modern science, mass, energy and information are three separated yet interrelated fundamental essentials that constitute concrete substances. To scrutinize it further, mass is such source materials as electrons, protons, neutrons or atoms and molecules that constitute a substance; energy is the special functional display of the substance when it interacts with the outside world; information is the state and mode of motion (and existence) of the substance and is sometimes defined as the orderly manifestation of things.

It has already been proven in modern science that mass is the condensed state of energy, and there is the mass-energy relation of $E=mc^2$. According to the theory of *hunyuan*, mass, energy and information are an inseparable whole entity. Mass is the content of the spatial composition of *hunyuanqi*. Energy, or function, is the content of the time composition of *hunyuanqi* (the mode and function of which will not be shown until it interacts with the outside world). Information is the holistic feature of the time-and-space composition of *hunyuanqi*.

The degree of how apparent or hidden the three essentials are varies with different levels of substances. For example, with concrete substance, it is the feature of mass that is in the apparent state while energy and information are affiliated to it. Since the feature of mass which is in the apparent state can be cognized with familiar intelligence, energy and information that are attached to mass can also be cognized with familiar intelligence. In the micro-cosmic world, at the level of substance in the form of energy field, energy is in the apparent state while mass and information are affiliated to it in a hidden state. Although mass of substance that exists in the form of energy field is in a hidden state, its feature of "energy" can not only be detected with instruments, but the mass can also be calculated with the help of the theory of transmutation between mass and energy. Thus, the information that is attached to energy can also be distinguished and understood by familiar intelligence.

At the level of primordial *hunyuanqi*, however, information is in the apparent state whereas mass and information are both in a hidden or potential state and are affiliated to information. This state is a special existence that is beyond the cognition of familiar intelligence. It should be pointed out that what we call hidden and potential is termed in light of the cognitive function of familiar intelligence. With the development of science, the scope of cognition will continuously increase and the border of hidden and potential state will also continuously change. If viewed from super intelligence, *hunyuanqi* at any level is a kind of *"hunyuan* substantial entity" formed with the *hunhua* or mergence and combination of mass, energy and information.

iii) New Understanding of Information

According to the theory of *hunyuan*, information, among the three fundamental essentials of mass, energy and information, is the root and foundation of all existences in the universe which determines their generation and change. Information is divided into "simple pure information" and "compound information" as well as "natural information" and "spiritual information" (or "man-made information").

At the level of *hunyuanzi* (*hunyuan* particle), information is just simple, and this is the feature of *hunyuanzi*. Such information may as well be called "information point" because, in the theory of *hunyuanqi*, no matter it is time or space, when they are tiny to a certain degree, they cannot be further divided and will close inward into an even state that displays a special feature of wholeness. This is a state of absolute evenness without any difference. In such a state, there seems to be no mass, energy or information. However, if analyzed in the rational sense, it still has a certain mass and energy; it is simply that they are too fine and slight to show any function. So we say that mass and energy are in a hidden or potential state. This shows that in this feature of *hunyuanzi* there is still its information content. The only thing is that it is the information of the simplest kind.

At the level of primordial *hunyuanqi*, though differences in the features of time (function) and space (form and quintessence or information) occur, the existential form is still an independent existence, in which there is only the "line" nature that extends along time and space while combinations between different units have not yet happened. Therefore, the whole entity still shows a state with no difference in it. So it still belongs in simple pure information and can be called an "information line".

At the level of all existences, things are quite different, however. Different combinations of information have appeared in simple pure information and form different features of time and space. They can be called "information planes" or "information bodies". Information like this is generally called compound information.

It is true that, in the formation process of all existences, content accompanies form. Content and form are inseparable. However, in the microcosmic world or the super-microcosmic world, the kind of time-and-space structure in the constitution of basic material elements determines the particular substance to be formed. It seems to emphasize that information determines the entity and nature of the entire substance. Information can gather and condense formless primordial *hunyuanqi* that has no difference within into various kinds of apparent energy. It can further condense energy into apparent mass as well. This changing process goes on naturally in the natural world and such information is called natural information.

Human spiritual activity, once it possesses the holistic characteristic of time-and-space structure in the realm of human consciousness, can be displayed in human language and actions and can be copied and stored with voice, symbols (including written language) as well as material objects. This we call spiritual information or man-made information. Human spiritual information in motion can gather energy and condense mass. Yet these two processes of gathering energy and condensing mass in the familiar intelligence state differ widely from those in the state of super intelligence.

In the state of familiar intelligence, the gathering and condensing processes are carried out from particular aspects of things. For example, in the process of making a table that is used to uphold weight or on which objects are placed, there should firstly be a blueprint of the table and a sequential plan of how the table is to be made (which is actually the time-and-space structure). Then the raw materials needed are processed according to the plan, and this is the process in which energy is added (i.e., gathered) to the table that is being made according to the time-and-space structure of the table and also the process in which the end product (mass) of the table is gathered into form.

In actual fact, gathering energy into materials to form combined energy with the aid of information is the critical point in making any object. During this process, energy follows the requirement of information to combine with concrete materials and is transformed into part of the newborn whole entity. Here lies the truth of how labor creates wealth. Since the process of making a table is a compound process that uses various raw and processed materials, people neglect the nature and function of gathering energy and mass according to its time-and-space structure (i.e., information).

Many things accomplished in the state of super intelligence are accomplished through the direct effect upon the holistic time-and-space structure (information) of things. Teleportation is just a display of this function.

Section II The Theory of *Hunhua* (Mergence and Change)

The theory of *hunhua* (mergence and change) is a theory about the process of how the *hunyuan* (mergence-into-oneness) whole entity is formed—it is formed with the mergence and combination of at least two kinds of material essentials. This is a theory about the general law of the motion and change of *hunyuan* substances and includes the theory of material *hunhua* and time-and-space *hunhua*.

I. The Theory of Material *Hunhua*

As it has been stated in the theory of *hunyuanqi*, any independent existence is not simple and single, but a *hunyuan* whole entity formed with the mergence and change (including combination, transform) of at least two kinds of substances. Therefore, every existence has its own *hunhua* formation process and, after its formation, continues to merge and change with the external world. All existences evolve from primordial *hunyuanqi* and continue with their new evolution against the grand background of primordial *hunyuanqi,* thus making the time-and-space structures in the universe show fantastic changes with each passing day and month. How do they change? That is to say, what are their modes of motion? The fundamental motions of *hunyuan* whole entities are as follows:

i. Open and Close, Gather and Expand, Go out and Come in, Change and Transform

The four modes of motion are ubiquitous whether it is in the motion of a *hunyuan* whole entity itself or in its merging and changing motion with other substances. They will be respectively stated as follows:

Open and close (*kai he*): This is the most fundamental mode of motion that *hunyuan* whole entities at all levels share. All the other forms of motion are based upon it. "Open" refers to the expanding outward of the boundary of *hunyuanqi,* while "close" refers to the closing inward of the boundary of *hunyuanqi.*

Gather and expand (*ju san*): Gather refers to the gathering inward of every component element of *hunyuanqi.* It is at first the condensing of the density between each component element. When a certain degree is reached, transformation will happen in the phase state. The changing process of "forming somethingness out of nothingness" in the stage of "condensing into form" is just the condensing and gathering inward in the change of *hunyuan* phase state. Expand refers to the expanding in the density between each component element of *hunyuanqi.* It can be the general expansion between each component element of *hunyuanqi* and can also be dispersion in the change of phase state. The transformation process of "somethingness turning into nothingness" in the stage of "dispersing into wind (*qi*)" of concrete substance is just the dispersion that takes place in the change of *hunyuan* phase state.

Go out and come in (*chu ru*): "Going out" refers to the motion that the *hunyuanqi* of an existence overflows and goes out, and this usually functions on the basis of opening and dispersing. "Coming in" refers to that the *hunyuanqi* of someone or something or the *hunyuanqi* outside someone or something gathers and turns inward. It usually functions on the basis of closing and condensing.

Change and Transform (*hua*): This refers to changing into form or bringing into life. It is said in *Huang Di Nei Ning* that "*Hua* is when things are born". From this it can be known that *hua* is the change in *hunyuanqi* that is caused by *kai he* (opening and closing), *ju san* (gathering and expanding; condensing and dispersing) and *chu ru* (going out and coming in). It concerns the changing process in the nature of substance. The nature of *hua* is the reorganization or reconstruction process of the time-and-space structure of substance. The motion of *hua* covers a relatively broad scope and usually causes fundamental change in *hunyuanqi. Hua* is the abbreviation of the term *hunhua*.

Among the four modes of motion mentioned above, *kai he* (opening, closing) is fundamental while *hua* (change, transform) is critical. The reason why *kai he* is fundamental lies in that *ju san* (gathering, expanding; condensing, dispersing) and *chu ru* (going out, coming in) are established upon the foundation of *kai he*. For example, expanding and opening are closely related and condensing and closing are closely related. There is neither condensing without closing nor expanding without opening. It is the same with *chu ru*, and the case in *chu ru* is even more complex. For *hunyuanqi* to exit, there must be opening and expanding to accompany it; for *hunyuanqi* to enter, there must be inward condensing and closing to accompany it.

However, the three kinds of motion (*kai he*, *ju san* and *chu ru*) are also different. Opening and closing mainly refer to the opening and closing of the boundary of the *hunyuan* whole entity. Gathering (including condensing) and expanding (including dispersing) refer to the gathering and expanding of the *hunyuanqi* of the *hunyuan* whole entity itself. Going out and coming in refer to the going out and coming in of the *hunyuanqi* of the *hunyuan* whole entity, in which exchange with the external world is involved.

The reason for saying *hua* (change and transform) is critical is because, in all the processes of *kai he*, *chu ru* and *ju san*, a certain degree of change is generated. It is simply that, in the *hunyuanqi* of different things and at different levels, the degrees of change caused by *kai he*, *ju san* and *chu ru* differ, and the content or type of change also differs widely.

For example, at the level of *hunyuanzi*, there is only the *kai he* mode of motion of opening outward and closing inward of *hunyuan* particles. Change thus caused is the forming of primordial *hunyuanqi* when *hunyuanzi* opens, and the closing of *hunyuanzi* occurs only when all existences return to *hunyuanzi*. The "birth and death" of the universe is in fact the opening and closing process of *hunyuanzi*.

At the level of primordial *hunyuanqi*, since there are two motion modes of *kai he* (opening and closing) and *ju san* (gathering and expanding) in it, the content of *hua* (change and transform) becomes accordingly diversified. For example, the combination of different time-and-space features among time-and-space compound particles (*fu he zi*) that have the features of time and space can form different *lines* of information (This is also the process in which time and space separate and become independent). From the lines of information, *hunyuanqi* that is energy-state substance condenses into being, which then forms the *hunyuanqi* of all existences.

At the level of all existences, all the motions of opening and closing, gathering (condensing) and expanding (dispersing) and going out and coming in begin. Since *hunyuanqi* is already fixed by bodies in different shapes and correspondent *hunyuan* entities are already formed, the *hunhua* motion in inorganic substance is not evident (For example, electrons in atoms can exit and enter atoms in equal numbers without causing any change and *hua* is shown in their chemical combination with other substance—this can be used as a forced analogy to illustrate this point). In the substance that has life, however, it is rather different. The *hunyuanqi* of the substance is carrying out the motion of mergence and change (*hunhua*) all the time with *kai he*, *chu ru* and *ju san*. It is not only carrying out the *hunhua* motion in itself, but also the mergence and change (*hunhua*) with the *hunyuanqi* outside itself.

The question then is: How does a *hunyuan* whole entity merge and change with the outside world?

ii. *Hunhua* at the Level of All Existences

Here we mainly talk about the mergence and change (*hunhua*) of concrete substances. All changes in concrete substances result from mergence and change of their *hunyuanqi*. The mergence and change of concrete substances include two modes—simple substances merge and transform into complicated substances and complicated substances divide (which is also *hunhua*) into simple substances. Although this *hunhua* process is the change that takes place between concrete substances, it is in nature a complicated *hunhua* process in which somethingness and nothingness mutually generate. Let me try to explain this with the process of change in inorganic substance.

For example, sodium metal (Na) becomes sodium hydroxide (NaOH), which is highly corrosive, when it meets water (H_2O). The overall features of NaOH, such as being white, solid, soluble and corrosive, are not formed with a simple sum of Na^+ and OH^-, but result from the mergence and change of their *hunyuanqi* as the existence of Na+ and OH^- alone does not necessarily lead to the formation of NaOH. In aqueous solution of NaCl or Na_2SO_4, for example, there are both Na^+ and OH^-. It is because *hunhua* does not happen that they cannot form NaOH. However, electrolyzing the solution of NaCl gives out hydrogen from the negative electrode with NaOH formed beside it and Cl_2 from the positive electrode. This is because in the electrolysis process, Na^+ has experienced *hunhua* with OH^- and H^+ has experienced *hunhua* with the electrons of Cl^-.

The *hunhua* process of substances is a process in which the original features of the substances involved in the mergence and change are changed and, meanwhile, the new unified whole entity (or entities) is formed. During this process, some substances vanish and some substances are newly formed. This, according to the theory of *hunhua*, is by no means only simple chemical reactions, but the result of the mergence and change of the *hunyuanqi* of substances. There exists a complicated process of mutual change between somethingness and nothingness in the material change in this process, which is even more evidently shown in the metabolic process of organisms.

The assimilation process in metabolism, for example, is a process in which the independently existing substance absorbed into the body of the organism becomes part of the whole entity by means of changing its independent nature. To unfold this process, it is like this: Independent concrete substance changes into the *hunyuanqi* of this concrete substance, which then turns into *hunyuanqi* that has lost its independent nature and condenses into part of the whole entity (and becomes a non-independent existence).

Likewise, the alienation process in metabolism is that the organism releases from its whole entity part of the content in the whole entity, making it lose the holistic feature of the organism and changing it into an independent substance that has its own material features. If this changing process is unfolded, it is like this: The component in the whole entity of the organism changes into the *hunyuanqi* of the component having the holistic feature of the organism, which then turns into *hunyuanqi* of an independent substance having lost the holistic feature of the organism and condenses into an independent concrete substance.

To sum up, the process of simple substances merging and changing (*hunhua*) into a complex substance is the breaking of the original whole entities and *qi* fields and the combining of two simple substances into one holistic *qi* field sharing one center. On the contrary, the process of a complex substance dividing into simple substances is the breaking of the original whole entity and *qi* field and splitting it up into multiple units, each of which reorganizes its own central whole entity and *qi* field and forms a new independent individual entity.

Then, what are the mechanisms and modes of mergence and change like?

iii. Mechanism and Mode of *Hunhua* of All Existences

Although in the mergence and change of all existences there exists a difference of simple objects (*hunyuan* entities) merging and transforming into complex *hunyuan* entities and complex *hunyuan* entities divided into simple *hunyuan* entities, the crucial point for both of them lies in *hunhua* or the mergence and change of the holistic time-and-space structures as far as their mechanisms of change are concerned. The *hunyuanqi* of all kinds of things is actually interconnected and mutually infiltrated. So long as the correlation in their time-and-space structures (information) reaches a certain degree, their original time-and-space structures will be broken and become new holistic time-and-space structures to gather energy and form mass. This is the process and mechanism of forming new *hunyuan* whole entities through *hunhua*. When a complex substance under the effect of the *hunyuanqi* from the outside world has its original holistic time-and-space structure broken, which makes each of the component elements in it have relative independence, the component elements will *hunhua* with each other and form new time-

and-space structures that gather energy and mass and thus form new independent *hunyuan* whole entities.

From a comprehensive view of the accomplishment of the *hunhua* process of all existences in the universe, it can be concluded that the modes of mergence and change are divided into basically three forms.

i) Energy Is Used as the Dynamic Power to Make Time-and-Space Structure Merge and Change

This is mainly shown in the mergence and change in time-and-space structures caused by various random motions in inorganic substances in the natural world. The various physical and chemical changes caused by scientific and technical means, according to the theory of *hunhua*, also belong in the mergence-and-change mode of "energy changing time-and-space structure". The feature of this *hunhua* mode is to change the balanced state of energy, to deprive the time-and-space structure of the stable state that it keeps with its original energy and make new mergence and change take place. Since the time-and-space structure is closely combined with the energy and the mass that are controlled by it, to break this balance from outside is rather difficult. In order to break the fixed time-and-space structure, a great amount of energy must be applied, though the forms of energy may differ. This feature can be sufficiently proven by the difficulty in chemical synthesis using modern science and technology. Isn't it so? A simple chemical synthesis not only needs specific pressure and temperature, but should also go through complicated processes. This is undoubtedly applicable in the inorganic world. It cannot be done, however, in the complicated compound of the organic world.

How is the complicated *hunhua* process carried out in the organic world, then?

ii) Time-and-Space Structure Is Used to Induce Mergence and Change in Time-and-Space Structure

As it is well-known, a myriad of changes of chemical resolution and synthesis are going on continuously at every moment in living organisms, particularly in advanced living organisms. All these belong in the process of mergence and change and are accomplished in conditions of normal temperature and pressure. That is to say, rather little energy is consumed in the accomplishment of this mergence-and-change process. This is evidently more economical than the mergence and change mentioned above carried out by means of the power of energy because this *hunhua* process is accomplished not by means of energy but by "inducement of time-and-space structure". This is the basic mode to carry out the *hunhua* process in living organisms and it can be divided into two categories:

a. Complementary Inducement in Time-and-Space Structure

Complementary inducement refers to the time-and-space structure of an existing object which is structurally complementary to the time-and-space structure of the object newly generated. It is to induce from the time-and-space structure of the original object a new time-and-space structure complementary to it. This is the essential form of the production and function of all kinds of enzymes in the energy metabolism process in organisms. It has been revealed in biochemistry that the chemical changes taking place in the metabolic process are all accomplished under the effect of enzymes, while the start of enzymic function or the producing of an enzyme needs the catalyst (enzyme) or the coenzyme. From this it can be seen that every chemical change in the organism involves a series of generation and change of substances. Such changes are accomplished in an extremely short time and are far beyond what synthetic technology in modern science can achieve. This, according to the theory of *hunyuan* whole entity, results from the complementary inducement shown in the time-and-space structure of the substance, part of which is experiencing change under the holistic effect of the organism. Its reaction process is also basically holistic.

b. Same Structure Inducement in Time-and-Space Structure

Same structure inducement refers to that the time-and-space structure of the original object is the same as the time-and-space structure of the object newly generated. The time-and-space structure of the new object is formed under the effect of the inducement of the time-and-space structure of the original substance or object. The processes of the re-generation of DNA and the duplication of RNA in bio-genetics are typical examples of same structure inducement.

iii) Consciousness Is Used as the Impetus to Make Time-and-Space Structure Merge and Change

All the *hunhua* modes stated above are the mergence-and-change processes naturally going on among all existences in the natural world. In addition to this, human spirit can also bring about complicated mergence-and-change processes. This is a mode of mergence and change that has not been realized and recognized by modern science.

All human consciousness activities are exclusively *hunhua* activities. This we will further discuss in Chapter V The Theory of Consciousness in this book. Here we just give a brief introduction to consciousness when it is used as the power to change time-and-space structures of substances.

Using consciousness as the power to carry out *hunhua* is, in fact, not to use the force of consciousness directly, but to use the time-and-space structure constructed with consciousness to induce the real structure of time and space. To say that consciousness is used as power is because the driving force of consciousness is needed when consciousness is applied to construct a form or structure and there must be the holistic start of consciousness when mergence and change (*hunhua*) is taking place. This is the very reason why familiar intelligence people cannot use this method.

The difference between time-and-space structure inducement and this process of using consciousness as the impetus lies in that one is the carrying out of inducement *hunhua* with the time-and-space structure of a genuine object or substance, while the other is the carrying out of gathering-and-merging *hunhua* with the time-and-space structure constructed with consciousness. The latter is more direct than the former, so its speed to accomplish *hunhua* is swifter. This functional ability exists both in people with familiar intelligence and people with super intelligence, and there is only a difference in degree. For example, external *qi* dispels a tumor and heals a fracture in an instant—these are the typical examples of using consciousness to change the time-and-space structure.

Up to this point, readers probably can't help asking, in the mergence-and-change process brought about with time-and-space structure inducement, though there exist such differences as complementary inducement and same structure inducement, there is at least an objective foundation of time-and-space structures of real substances. In the inducement brought about with consciousness constructing form, however, there is neither the foundation of objective and tangible time-and-space structure nor even the obvious process of consciousness constructing form. How, then, does this mode accomplish the mergence-and-change process?

It is pointed out in the theory of *hunyuan* whole entity that the accomplishment of the *hunhua* process with consciousness as the impetus is determined by the entire information feature of consciousness activity and the holistic feature of life activity. All the information (both inside the human body and outside it) in the process of life activity can not only be deposited in the life entity, but can also be stored in consciousness. Whether the information can be withdrawn or not depends upon the sensitivity and capability of *Yi Yuan Ti* (Mind Oneness Entity). *Qigong* practice is intended to strengthen the ability of withdrawing various stored information.

The constructing-form process in consciousness sometimes cannot be felt when consciousness is being applied in carrying out mergence and change. This is because consciousness motion is an extremely swift motional process. The occurrence of any simple idea or image in the mind needs an extremely complicated process to accomplish and all this is hard to be perceived. For detail, please refer to The Theory of Consciousness (Chapter V) in this book.

Some readers may wonder: Now that time-and-space inducement and consciousness-constructing-form inducement in organisms are both based on realistic time and space, how was *hunhua* realized in the entire evolutionary history of organisms?

iv. *Hunhua* Is a Unity of Heredity and Variation

Some modern biologists emphasize that biological heredity is the fundamental reason for the multiplication of living things. They base their opinion upon the localization and function of the time-and-space structure in bio-genetics (DNA). Some emphasize that environment is the fundamental reason for the multiplication of living things and they base their opinion on the idea that environment can change the hereditary features in living things. In fact, both of these ideas are two interrelated aspects in the process of *hunhua*. It is only when the two of them merge and combine into one that the multiplication of living things can be realized. That is to say, neither of them can accomplish the mission of multiplication alone.

For example, molecular biologists think that the foundation of the life of a cell is determined by the genetic information in DNA. As a matter of fact, it is not that simple because the DNA in the nucleus of a cell is only one aspect of the entire time-and-space structure that carries out the life activity of the cell. There must also be correspondent content of the time-and-space structure in the environment. It is only through mutual *hunhua* of them both that a new *hunyuan* whole entity can be formed.

If the time-and-space structure in the environment agrees with the organism's DNA and the two of them combine into one with the original features retained, this will display the mergence and change (*hunhua*) with heredity taking the leading place. If the time-and-space structure in the environment disagrees with the organism's DNA, the opening and closing and going out and coming in of *hunyuanqi* in the mergence and change will undoubtedly change the content and form of the original *hunyuanqi* to a certain extent. Those that are changed slightly will usually survive but shows a certain degree of variation; those that are changed too drastically might lose their life force and decease, and this is the reason for the "survival of the fittest".

The universe gradually develops step by step like this through the mode of "self organization", from primordial *hunyuanqi* to all existences today. It is true that, with every step forward in the development of all existences, change in the *hunyuanqi* of existences will condense inward into substance that has form and is deposited and stored, which then becomes the time-and-space structure that supports the change in the existences themselves and carries out new mergence and change with *hunyuanqi* in the outside world.

The influence of mergence and change upon heredity and variation of living things seems to be not so easy to immediately understand from the *hunhua* process of monoplast alone. If we make an analysis of change in the cells of advanced organisms, the effects of *hunhua* will be evident. Take clawed frogs for an example. It has been reported that when the cell nucleus of a clawed frog's cell in the small intestine is transplanted into its ovum with its cell nucleus removed, a tadpole can grow from it as a result. Some of the tadpoles even grow into adult frogs. The cell nucleus is the same. In the cytoplasm of the small intestine, it grows into a cell of the small intestine, whereas in the cytoplasm of the ovum, it grows into a tadpole. Change in environment determines its direction of development.

This sufficiently shows that such time-and-space structure of DNA in cell nuclei that can control life activities must draw support from the correspondent time-and-space structure in the outside world (e.g., the cytoplasm) in order to truly display the ability it has to control life activities. It is only after *hunhua* of them both that real life activities can take place. In reality, the living process of any individual life form is actually the living in interaction with the environment on the basis of inheritance and heredity. An individual organism, therefore, continuously changes its own time-and-space structure to a certain extent and passes it on to the next generation. Here, the process of mergence and change make heredity and variation unified.

In brief, the existence and change of anything, esp. organisms, are based on the holistic feature of its innate time-and-space structure. Concentrated display of this holistic function in the body of organisms is what was called by the ancient Chinese "*shen ji*" (spiritual mechanism), or in other words, the fundamental information about life. It is the foundation for the generation and change of organisms and the reason for their inherence. But this alone is not sufficient for the generation and change to take place. It is only through such motions as opening and closing, going-out and coming-in, condensing and dispersing as well as transforming which make them merge and change with the *hunyuanqi* outside that organisms can become *hunyuan* entities in independent existence. This external environment is also the reason for the generation and change of things, which was called "*qi li*" (*qi* establishment) by the ancient Chinese. It is the external reason for the generation and change of existences. From this we can know that *shen ji* and *qi li* represent respectively half of all the necessary conditions for the living whole entities and life activities. It is only when the two of them (*shen ji* and *qi li*) merge and combine into one that it is an integrated *hunyuan* whole entity.

Here heredity and variation are unified in the mergence-and-change process of things and they make evolution in species continue. This process was naturally going on before humans came into being. After humans came into existence, change in the natural world began to be impressed with the mark of humanity. This will be comprehensively illustrated in Chapter III The Theory of Whole Entity, hence it will not be stated here in detail.

From the exposition of the theory of *hunhua*, it can be known that mergence-and-change motion is continuously going on in the world of nature and the universe, which makes the universe progress from simplicity to complexity. This viewpoint just contradicts the theory of heat death (a maximum-entropy state) in modern physics. The reason for the contradiction lies in that the theory of heat death does not recognize the function of the time-and-space structure in the external environment during the *hunhua* process. In fact, the theory of self-organization in modern science also challenges the theory of heat death.

v. The Motion of *Hunhua* Makes All Existences in the Universe Display a Unity of Opposites

Each of all the existences in the universe has its antithesis that exists in unity with it. In the world of physics, particles and virtual particles, positive charge and negative charge and the north and south poles of magnets are typical examples of the unity of opposites. Male and female in the world of organisms, synthesis and decomposition in organisms, receptor and gamete in biochemical process and so on are examples of how opposites complement each other. The two opposite factors are interdependent and it is only when the two of them combine that a unified whole entity can be formed.

This has puzzled Chinese philosophers as well as philosophers in other countries. There is such a wealth of creation in this world. How can it be so coincidental that, when there is one substance or creature, there must be another substance or creature opposite and complementary to it? The theory of *hunyuan* has found the theoretical solution to this question. In sum, this is the result of the mergence and change (*hunhua*) of *hunyuanqi*.

It is pointed out in the theory of *hunyuan* that anything independent is formed with the mergence and change of at least two factors. This means that in the generation and formation of an existence there must be at least two systems that participate in the mergence-and-change process. Let me give you an example to illustrate this: System A interacts with system B. As a result, a, being part of system A, is pulled out and merges with b that is a part pulled out from system B and together they form the new *hunyuan* (merged-into-oneness) whole entity ab. Meanwhile, a trail of a is left in system A and a trail of b is left in system B. The *hunyuan* whole entity ab is not only dependent upon these two trails to a certain extent, but can also exert influence upon the two trails through *hunyuanqi* field, which makes these two trails of a and b acquire relative independence respectively in Systems A and System B. If the *hunyuanqi* of trail a and the *hunyuanqi* of trail b condense inward and form a concrete substance that exists independently, two existences that are opposite and complementary to each other are formed. This is one case. Another case is that the split of a *hunyuan* whole entity turns one *hunyuan* whole entity into two opposite and complementary parts. The third case (which mainly refers to the microscopic particle world) is that there are actually two forms of expression of time-and-space structure in the *hunhua* process during which an integrate individuality is formed. One is the time-and-space structure or form dominated by space, which is the substantiality; the other is the time-and-space structure or form dominated by time, which is the virtuality. This is the special phenomenon in the process of which somethingness forms out of nothingness.

II. The Theory of Time-and-Space *Hunhua*

Time and space are both considered as the special form in which substance exists in the philosophy of dialectical materialism which belongs in the realm of familiar intelligence science. However, nothing has ever been mentioned as to what kind of form it actually is. In the theory of *hunyuan*, the view of time and space has its special content.

According to the theory of *hunyuan* whole entity, time and space are two inseparable marks of *hunyuan* whole entities. If scientific language is used to give a brief account of them, time is the successive effect of the function of existences, i.e., the process of change, while space is the extension process of the position of existences. However, the display of the successive effect of function, i.e., the process of change, needs the aid of the extension process of the existence in order to become reality. Likewise, the accomplishment of the extension process of position must draw support from the successive effect of function to become reality. Time and space are a *hunyuan* whole entity in which they permeate into each other and are interwoven with each other. Its nature (i.e., the nature of *hunyuan* time and space or merged-into-one time and space) is still hard to express with the language of modern science. This is because it is the holistic expression form of *hunyuanqi* and in it is "inlaid" the content of *hunyuanqi* at different levels.

In the time and space of the universe today, there exist at least the time and space at the level of *hunyuanzi* which is the absolute time and space, the time and space at the level of primordial *hunyuanqi* which is the holistic time and space, the time and space at the level of all existences which is the relative time and space and the time and space at the level of *Yi Yuan Ti* (Mind Oneness Entity) which can be thoroughly connected with the previous two kinds of time and space of primordial *hunyuanqi* and all existences.

In daily life, it is customary to say that space is thoroughly empty and time is insubstantially void. In fact, space is not empty and time is not void. They both are special substantial state. As it has been previously mentioned in the new recognition of zero in *hunyuanqi*, in the realm of time or space in the universe, there is no zero in which nothing exists at all, namely, there is no genuine emptiness in space or insubstantial void in time. This is because time and space are both the expression form and content of *hunyuanqi*. Considering that both time and space have the material content of *hunyuanqi*, they inevitably have features similar to material mergence and change.

i. The Mergence and Change of Time

As it has been stated above, time is the successive display of the function and change of existences. The past, present and future of everything are the aggregation of the display of its innumerable instants, which is very similar to the idea that a line in mathematics is the aggregation of points. This is perhaps the reason why people say "time is one dimensional". According to the theory of *hunyuan*, a momentary display of anything is not isolated, nor is it absolute, but the result of the mergence and change of the state of its past, present and future. Let me try to explain it in detail.

i) Trails of changes in the past of things are deposited in their state at present. This can be illustrated with the forced analogy of a kind of infrared photography that can restore images. For example, the state of somebody in a particular place at a particular time can be photographed and shown in pictures taken with this type of method within a certain period of time after this person has left this place. What is photographed is the trail of change which is already past about this person and is left in space. Likewise, the trail of change in the past can also be deposited in the body of a concrete existence.

ii) Buds of future change are contained in the present change of existences. This can be illustrated with the forced analogy of a kind of biofield photography. For example, before the leaves of trees come out of their branches, their rudimentary forms can be photographed and shown with the biofield photography. This shows that the scene of future development is already contained in the branches of trees in which undeveloped leaves are already growing.

On account of the two points mentioned above, it is believed in the theory of *hunyuan* that the momentary image displayed in everything is the *hunyuan* state of its past, present and future. Therefore, when super intelligence is applied, changes of the past and future of specific things can be perceived or known.

ii. The Mergence and Change of Space

Taking into account that space is the existential form of *hunyuanqi* and considering it in light of *hunyuanqi*'s mutual accommodating feature, space is bound to be the *hunhua* state filled with all kinds of *hunyuanqi*. Since *hunyuanqi* has the function of storing and retaining information, in the primordial *hunyuanqi* in space there must be the *hunyuanqi* information about all existences preserved in it. It is because of this that people with super intelligence can not only perceive attributes of other people and things contained in specific concrete existences (e.g., they can perceive from those who come to visit what their family members are like), but can also perceive the attributes of specific existences from the void.

Chapter III The Theory of Whole Entity

The theory of whole entity is a theory that expounds the characteristics of the existential form of all existences in the universe. The theory of whole entity and the theory of *hunyuan* are two aspects that are inseparable and interrelated. In the concept of substance, the concept of change and the concept of time and space in the theory of *hunyuan*, the feature of their mergence into oneness is stated, and the "oneness" reached through mergence and change is the whole entity. This is also the content of the concept of whole entity. Of course, the theory of whole entity and the theory of *hunyuan* should not be confused. The theory of whole entity lays emphasis on the illustration of the holistic feature of things as well as the interrelationship between their constituent parts. This theory includes three parts: The concept of universe whole entity, the concept of human-and-universe whole entity and the concept of human body whole entity. For the convenience of study, the general concept of whole entity will be first introduced.

Section I Synopsis of the Theory of Whole Entity

The theory of whole entity as part of the greater theory of *hunyuan* whole entity seems, at first sight, to have a lot in common with the concept of wholeness in modern systems science. Fundamental differences lie between them, however. For the convenience of study, let us begin with the meaning of *hunyuan* whole entity and the formation of the whole entity.

I. A Brief Introduction to the Whole Entity
i. What Is the Whole Entity

In the literal sense of the term, a whole entity is an integrated individuality. The term integrated as a concept in mathematics means undivided. It is pointed out in the theory of *hunyuan* that any individuality is formed with the mergence and change (here referring to combination), or *hunhua*, of at least two material essentials. This *hunhua* substance can show an even merged-into-oneness *qi* state (*hunyuanqi* state) of absolute sameness and also a *hunyuan* substantiality state in the form of a concrete existence. The former is admittedly a whole entity. For the latter, though its constituent parts can be of any form and its functions of different parts may differ in numerous ways, the *hunyuanqi* of this concrete substance is all around the whole substance, permeating both the inside and outside of it, which makes its components form a harmonious whole entity in which the constituent parts are organically connected.

The whole entity talked about here refers to the time-and-space structure that forms *hunyuanqi*. It is the root and foundation that maintains *hunyuanqi*'s independent and special features. The whole entity can also be understood as the structural mode of *hunyuanqi*. For example, let us suppose that a mode is a special triangle. Since the nature of this specific form and structure is not affected by how big or small a position it takes, once the mode of time-and-space structure is formed, not only every part of the entire substance or existence is in the same structure, but this holistic feature can permeate into each part of this substance and make the components in each part of this whole entity embody the whole entity. This is the fundamental difference between the components in a whole entity and such components as independent substances. Take electron transport coefficients for example, it is 10^{-2}-10^{2}Msec in the inorganic system yet 10^{7}-10^{8}Msec in the mitochodria in the human body. There is such an evident difference between them.

From this it can be understood that a whole entity is by no means a simple sum of all its constituent parts, but the new special nature generated through *hunhua* or mergence and change. This new special nature is determined by the *hunyuanqi* that is generated through *hunhua*. From this it can be seen that the holistic feature is also the fundamental attribute of *hunyuanqi*. Therefore, the whole entity is sometimes directly called *hunyuan* whole entity.

ii. How Is the Whole Entity Formed

Although whole entities differ in types and are not all the same in their formation processes, the crucial point in the formation of the whole entity is, in general, the integration function of information (i.e., time-and-space structure). Whole entities can be divided into two categories— the whole entity naturally formed and the whole entity processed and made by human beings. They will be respectively expounded as follows:

i) Formation of the Whole Entity in the Natural World

There are innumerable existences in the natural world and it is difficult and unnecessary to give enumeration one by one. Illustration of the formation of whole entity is given according to different levels of *hunyuanqi*.

a. Formation of the Whole Entity at the Level of *Hunyuanzi*

As *hunyuanzi* (*hunyuan* particles) is the direct combination of the structures of time and space and the most fundamental and exquisite independent *hunyuan* unit with no difference in it, there is actually no difference between the individuality and the entirety of *hunyuanzi*. At this level, one is many and many are one. To say that it is one *hunyuan* particle is because it does take up a certain phase of time and space—it is a time-and-space structure too tiny and subtle to be differentiated. To say that it is an entirety is because it does fill the entire time and space—it is the holistic time-and-space structure that is impossible to be differentiated. Here, the formation of the entirety or whole entity and the formation of *hunyuanzi* are identical. The moment *hunyuanzi* is formed, the whole entity also comes into being. Although *hunyuanzi* is too tiny and subtle to be differentiated and its mass and energy seem to have both dissipate into time-and-space structure, all this is still the result of the special integration of time-and-space structure. Without the inward enclosing motion of the time-and-space structure, formation of *hunyuanzi* is impossible.

b. Formation of the Whole Entity at the Level of Primordial *Hunyuanqi*

When the time-and-space structure of *hunyuanzi* dissociates into the "the two poles" of time and space, not only the chaotic evenness state of *hunyuanzi* is destroyed so it displays the features of time and space, but it also acquires differences in "force" and "polarity features" in this process. In order to deepen our understanding of the nature of this process, knowledge from mathematics will be adopted as a forced analogy in the explanation as follows:

Hunyuanzi in which time and space are combined decompose into time-and-space compound particles in which time and space are separated (The holistic state of compound particles is called primordial *hunyuanqi*). Time can be expressed with the imaginary number of $\sqrt{-1}=i$ in mathematics. Space can be expressed with the real number in mathematics. Suppose time is 2i and space is 3y, the compound particle can be expressed with the coordinate in Fig. 3.

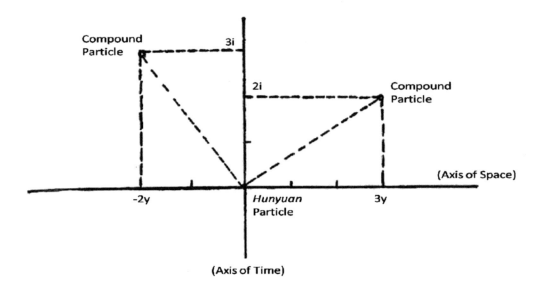

Fig. 3 Analogical Diagram of Polarities in Time-and-Space Compound Particle

According to the expression of the complex number (the imaginary number plus the real number) in mathematics, a compound particle can either be 3y×2i, the "substantial existential" display of the compound particle essentially in the form of space, or 3i×(-2)y, the "insubstantial existential" display of the compound particle essentially in time (Notice that the expressions above are not the real algebraic symbols of a compound particle, but a forced analogical illustration). This is the "polarity feature of time and space" of the compound particle as it is so called in the theory of whole entity, which leads to the dual features of "secular somethingness" and "genuine voidness" in the object of the physical world.

From the perspective of arithmetic operations, the two kinds of expressions above are actually different. Yet from the perspective of the theory of *hunyuan* time-and-space whole entity, the two of them are in fact one *hunyuan* whole entity. In the state of super intelligence, they are one merged-into-one or mergence-into-oneness (*hunyuan*) substantiality.

In the ancient *yin yang* theory, tangible space is *yin* and formless time is *yang*. But people are accustomed to calling tangible space structure *yang* and formless time structure *yin*. This is rather similar to the terms of positive and negative electrical charges given to electricity by people that are contrary to the positive and negative nature of electricity. This is the falsehood caused by not knowing the truth of genuine facts.

Since the number of compound particles is innumerable and also random, the whole entity of primordial *hunyuanqi* formed with the mergence and combination of the multitudinous compound particles display a genuinely primordial chaotic state. As the mass and energy of compound particles are too subtle and weak to function independently, it is hard for them to show the feature of turbulent flow in the physical state even though they are in a state of chaos.

It is apparent that, in the whole entity of primordial *hunyuanqi*, compound particles may arrange into different time-and-space structures in accordance with their "polarity features". In other words, compound particles may permute all the various time-and-space structures in the universe. However, the reality of the various time-and-space structures has not been permutated yet. Only simple merging between time-and-space particles is going on in it.

c. Formation of the Whole Entity at the Level of *Hunyuanqi* of All Existences

In the whole entity at this level, it is firstly the formation of simple whole entities from primordial *hunyuanqi*, and then the mergence and change (transformation) of simple whole entities into complicated whole entities. Since the latter has already been talked about in the theory of *hunhua*, here the emphasis is laid on the explanation of the former in light of modern scientific knowledge.

In the random motional process of compound particles of primordial *hunyuanqi*, once the permutation of the time-and-space structures of gravity force and radiometric force appears, specific gravity force and radiometric force begin to display in the primordial energy of primordial *hunyuanqi*. Then various particles are formed until the formation of atoms, thus primordial *hunyuanqi* enters the realm of cosmology and astrophysics in the physical world. In modern science, nuclear physics, quantum dynamics and the theory of relativity are used to trace back and to deduce the formation of the universe. Therefore, the emphasis is on the discussion of gravity, radiation as well as their evolution. We explore the evolution of the universe from the perspective of the concept of *hunyuan* whole entity. So we illustrate emphatically from the combination of time-and-space structures as well as the integration of the primordial energy of primordial *hunyuanqi*.

Although primordial *hunyuanqi* is in the original holistic chaotic state, this chaotic state is relatively even and is characteristic of isotropy, which is determined by the feature of its compound particles. If viewed from the level of the world of quantum, it is simply a highly even state that beggars description. Therefore, once the random motions of compound particles form the specific time-and-space structure (e.g., the primordial frame and structure of the universe), the coordinated effect instantly occurs and the corresponding phase change takes place. The entire frame and structure of the universe is thus founded. All the evolutionary processes afterward are carried out on this basis until we human beings and the human spirit come into existence—all these are determined by the fundamental time-and-space frame and structure at the beginning. Therefore, the universe recognized by human beings today is an interrelated and unanimous whole entity. It is true that the random motions of the compound particles of primordial *hunyuanqi* must have been able to permutate the time-and-space structures different from the time-and-space frame and structure of our universe. Yet the universes formed thereof differ fundamentally from our universe. They not only have nothing to do with us human beings but are also beyond our perception.

In the process in which simple substances merge and transform into complicated substances at the level of all existences, the function of the time-and-space structure in the world of organisms is even more evident. For example, in the seed of a tree is contained the inherent content of all the time-and-space structure of the tree. So long as certain conditions are provided in the outside world that combine with it, all the content of the tree will merge and change into being according to its time-and-space structure (further discussion about it is omitted here).

ii) Man-Made Whole Entity
a. The Whole Entity Processed and Produced with Familiar Intelligence

All products made or manufactured with labor are integrated individualities. Although they differ from things naturally formed, they are also based on the specific time-and-space structures prescribed by humans (i.e., man-made information) that integrate various basic materials and form the whole entities through processing and "implanting" energy into them.

Take the making of a table as mentioned in Chapter I for example. This involves carrying out the working procedures according to the drawings created by designers and integrating various raw materials according to requirements. This process is not a simple process of piling up the raw materials, but a process of making and constructing according to the specifications of each part and of adding to it the energy according to the prescription of the time-and-space structure so that every component part can closely and firmly combine together. Therefore, the process of processing is also a process in which holistic "information combination energy" is formed that turns external energy into internal combining energy. When the time-and-space structure and the energy added to it are combined, different components in the whole entity form a *hunyuan* whole entity having specific features and functions, thus making the components submit to the whole entity.

b. The Whole Entity Processed and Produced with Super Intelligence

Information sent with super intelligence can not only be used in teleportation, in hastening flowers to bloom, in changing the infra-red spectrum of water, in changing the laser Raman spectrum and r decay counting of radioisotopes, but can also carry out chemical synthesis in the normal state (in normal temperature and under normal pressure, etc.). When practitioners' *gongfu* improves to a higher level, they can "make something out of nothing" (and this was called in ancient times "Mind gathering *qi* and *qi* shaping form"). All this belongs in the scope of the whole entity processed and made with super intelligence. The process is also absorbing and gathering *hunyuanqi* in the world of nature (both *hunyuanqi* of tangible substantial entity and formless *hunyuanqi*) with specified time-and-space structures to form *hunyuan* whole entities.

From the formation of whole entities at each level mentioned above, it can be seen that the crucial point lies in the time-and-space structure acquiring energy or force. This process can be naturally accomplished and can also be endowed with human power.

II. Features of the Theory of Whole Entity

The theory of *hunyuan* whole entity is a holistic concept of super intelligence. All features expounded in it are established upon the foundation of this point.

i. The Substantiality Feature of the Whole Entity

The whole entity stated in the theory of *hunyuan* whole entity is not only the nature of existences, but also a kind of special real existence. It is the genuine existence that can interact with super intelligence. It can permeate and be embodied in each part of concrete sub-stances.

ii. The Level Feature of the Whole Entity

According to the theory of *hunyuan* whole entity, any *hunyuan* whole entity is formed with the *hunhua* of at least two kinds of material features, which makes the *hunyuan* whole entity display a feature of levels. This can be shown both in the levels of the evolutionary process of *hunyuan* whole entity and in the levels of the same existence or substance. Each level has the common features of the whole entity. This has something in common with "fractal structure" in modern chaos theory. Yet it has developed into an upward spiral compared with the viewpoint in traditional *qigong* theory which adheres to the existence of a *taiji* in everything—"Everything is born with a body and in every body there is naturally a heaven and earth (*qian kun* or *tian di* in the Chinese language)".

For example, the whole universe includes the level of primordial *hunyuanqi* and the level of concrete substance (*hunyuan* entities). The latter can further be divided into the level of inorganic concrete substance and the level of organic concrete substance (including plants and animals). The human body, for instance, can be divided into such levels as physical motion, chemical motion, the motion of life and the motion of consciousness, and each level is a whole entity itself. Meanwhile, a *hunyuan* whole entity is formed with the mergence and change of each of its constituent parts (or each of its stages).

For this, we must note that:

i) The target being perceived should be considered as a whole entity in order to study its laws of change. It should also be considered as part of a whole entity at a more advanced level and should be investigated against the background of the more advanced whole entity. The whole entity of the target being perceived is just the whole entity of the part to which it belongs. It is formed with the different partial features and the features born in the interaction of these partial features together with the influence from the surroundings.

ii) Just as Sakata Shyoichi, the famous Japanese theoretical physicist, has pointed out, "Modern science has developed a concept, that is, the world of nature is constituted of innumerable levels which differ from each other in nature and each of which follows its own laws." Therefore, for things at different levels, familiar intelligence uses different methods—methods corresponding to different levels of things to know them. Thus were born such different methods as physical method, chemical method and biological method in science. In super intelligence, though it is not necessary to be as varied as the methods in modern science since super intelligence can directly grasp the nature of the whole entity, it should be known that the laws of things at different levels are not all the same, either. This requires not only different levels of *gongli* (the power of *gongfu* obtained in *qigong* practice), but also the grasp of relevant laws.

iii) Each level and part in any whole entity submits to the whole entity. Therefore, when dealing with problems at the level as a whole, the effect of bringing out the essentials will be shown so long as the foundation of the holistic feature is grasped, and there is no need to get entangled in each specific part.

iv) Different levels can mutually transform. Lower levels can leap into higher levels, as is shown in the evolution between different levels previously mentioned. Higher levels may also lose the high level features and degenerate into lower levels. For example, a human being may degenerate into the level of common animals (e.g., the wolf child and the tiger child) if this person cannot acquire consciousness peculiar to humanity shortly after birth. Animals can return to the level of inorganic matter when they die; concrete substances return and are changed into the form of energy. Every level has its special prescriptive feature in essence and is not a simple sum of lower levels.

v) The *hunyuan* whole entity of an organism is gradually perfected in its process of development. The *hunyuan* whole entity of an organism refers to the holistic state when all the characteristics of the organism are displayed. As far as a higher plant is concerned, it is when the root, stem, leaves and flowers of the plant are all grown and, in the process when seeds are formed, all the time-and-space features of the plant have been successfully displayed. It is also at this moment that all the features of the time-and-space structure are internalized into the seed. Although the *hunyuanqi* with all the time-and-space features of the original plant is contained in the seed, this *hunyuanqi*, when it is condensed inward into form and becomes deposited into the tangible substance (the seed), becomes part of, but not all of, the *hunyuanqi* that will grow into a new individuality (a new plant). Therefore, the *hunyuanqi* of the seed has only all the information about a seed that may grow into a new individuality, but not all of the information about the new individuality to be formed. This is because all the time-and-space structure of the new individuality is the entire information (including the internal and external *hunyuanqi* of the body of the plant) about the integrated time-and-space structure formed through putting out sprouts, taking root and blooming of the seed. It results from the mergence and change (*hunhua*) of the *hunyuanqi* of the seed with the *hunyuanqi* of the external world (e.g., the *hunyuanqi* in air, water vapor, water, nutrients and sunlight). In other stages, though the plant itself is also a whole entity, its *hunyuanqi* is not able to accommodate all the time-and-space features of this plant but only the feature of a particular stage, with the holistic feature in a potential state. This is also the special relationship between *hunyuan* whole entity and part of *hunyuan* whole entity.

iii. The Holistic Feature of Time and Space (Also Called the Concept of Holistic Time and Space)

According to the theory of *hunyuan* (mergence-into-oneness) whole entity, both time and space are the *hunhua* (merged and combined) whole entity of *hunyuanqi*. That is to say, both time and space are the *hunhua* manifestation of existences. There exists no absolutely empty space without anything (*hunyuanqi*) in it, nor is there time separated from existences. We call the time and space at the level of *hunyuanzi* absolute time and space merely because it is a *hunyuan* or merged-into-oneness state that is absolutely even and unchanging (and it is not that there is nothing in it). Once change takes place in it, it enters into the time and space at the level of primordial *hunyuanqi*.

i) The Concept of Space Whole Entity

As it has been mentioned above, any space is a realistic display of the structural characteristic of the *hunyuan* whole entity as well as its extension. For example, the original space of our universe is the realistic display of the structure of primordial *hunyuanqi*. When it entered into the level of all existences, space becomes the realistic display of the *hunyuanqi* of all existences. The space of our universe now is a comprehensive totality of all the spaces that have once been existent. Although all existences now take up a certain space, the certain space each of them takes is also part of the overall space; so it displays features of partial space.

a. The Indefinite Feature of Partial Space

The concrete substantial aspect of any specific *hunyuan* whole entity has the boundary of a certain sphere. Yet the boundary of its *qi* field is not definite and can merge into the space around. Take the smell of some substances for instance. People may not be sensitive to the smell of these substances. Yet flies can smell the stale smell from quite afar and fly to the spot where the smell originates. It is so with the smell of concrete substances; one can well imagine the indefiniteness feature of formless *qi* field.

b. The Holistic Feature of Partial Space

On the one hand, any partial space is part of the space of *hunyuanzi* and primordial *hunyuanqi*, so it contains the features of these whole entities. On the other hand, the *hunyuanqi* of all existences permeates into each other, affects and associates with each other. Although we cannot say that any partial or limited space has all the content of all existences, the space of a certain sphere and the *hunyuanqi* of all existences do mutually accommodate to a certain extent.

ii) The Concept of Time Whole Entity

The display in space of the actual structure of anything has a successive changing or motional process, and this successive motional process is time. We know that the change of anything results from the interaction of this something with a multitude of factors in the environment. In order to fully restore things that have changed to what they once were, there must be the total restoration of the interaction between what is inside and what is outside, which is impossible. This is the reason why time is irreversible. It is because of this that things cognized with familiar intelligence are all their momentary display in reality. It is hard for familiar intelligence to cognize holistic time. In fact, in the reality of anything at any moment are contained its past and future. This we have illustrated in the theory of *hunhua* of time and space. Here further explanation will be given from the perspective of the holistic feature.

For example, the momentary display of a tree we see in fact contains all its time structure in the past and its possible time structure in the future. Let us analyze this with the aid of familiar intelligence common sense knowledge: If we cut the tree we see horizontally and also vertically from the midpoint, we can see different annual rings on the cross section and vertical section. Each annual ring is the condensed display of the *hunyuanqi* of this tree in a certain year. People who know about trees can not only know the age of the tree, the size of the trunk each year and so on from the annual rings, they can also know the effect of such factors as weather conditions (e.g., drought or excessive rain) on the tree each year. People whose super intelligence has reached a certain level can perceive directly such content of the *hunyuanqi* of the tree without the need of cutting the tree open to see its annual rings. We seem to be able to picture this with the aid of familiar intelligence imagination. That is, in the tangible body of the realistic tree, there is an intangible illusion-like time structure which seems to be overlapping and encasing stereo pictures combined together.

It seems difficult to comprehend and imagine that future changes are contained in things in reality. This is a special psychological state caused by the habit of our familiar intelligence which is only able to have contact with things in reality. As a matter of fact, time structures of things also have their indefinite feature in time boundaries. In the instant when we see the reality of a concrete substance, the formless *hunyuanqi* and the *hunyuanqi* field of the substance have already gone through this time boundary and entered its future time scope. As all existences are a whole entity with this forwardness feature, changes in reality at the moment are but the apparent concrete display of interactions between the forward *hunyuanqi* of all existences.

In the world of nature, this changing process is going on naturally. However, this indefinite feature in time, like the indefinite feature in space, does not limitlessly extend. Therefore, we always think that we cannot say the moment of the reality of things at present has contained all the content of their entire time structures. This is because *hunyuanqi* in the outside world can have an effect on the time-and-space structures of things and make them deviate, thereby changing their development process.

iii) Time and Space Are a *Hunyuan* Whole Entity

From the elaboration of the concepts of time whole entity and space whole entity above, it can be seen that time and space are a *hunhua* whole entity that is interdependent and inseparable. Time structure of an existence is the succession of change (motion) in its space structure. The formation of a space structure of an existence results from change (motion) in its time structure shown in its taking a position in space. The noumenon of any *hunyuanqi* is the unity of its time structure and its space structure. Space structure is related to its exposition in taking position in space, and time structure is related to its exposition in functions. Time separated from space and space separated from time are both impossible.

iv. The Holistic Feature in Subject and Object of Study and Perception

Subject here refers to the researcher's own consciousness activities (to be more exactly, super intelligence). Object refers to the target being studied, which includes the *qigong* practitioner's own life activities and the objective substantialities in the external natural world under study. According to the concept of *hunyuan* whole entity, there is a great conformity between subject that researches and cognizes and object being studied and cognized, which therefore manifests the holistic feature between them. This is shown in:

i) When *qigong* practitioners observe and perceive their own internal life motions, the subject that can observe and perceive (with one's whole being), i.e., super intelligence, is part of the *qigong* practitioners' holistic life activities, while the practitioners' own life activities being observed and perceived (with one's whole being), the target being studied, is also part of the *qigong* practitioners' holistic life activities. Great agreement exists between the two of them. In this research process, the subject to study and the object being studied are both unified in the holistic life motion process of the practitioners' own *qigong* practice.

ii) When super intelligence is applied to research into an external objective target, the holistic function of super intelligence is exerted upon the holistic feature of the object. It is so in cognizing an object. It is as well so in changing things or existences. It is only when the holistic feature of the subject (*Yi Yuan Ti*) of cognition is combined with the holistic feature of the objective existences that the entire time-and-space features of the object can be reflected and cognized, that functions can be exerted upon the time-and-space structures of the object so they can be changed. This process of cognizing and changing object is a process in which the subject and the object are combined.

iii) When *Yi Yuan Ti* observes and cognizes *Yi Yuan Ti* as the target of cognition, it is the direct combination of subject to study and object being studied. If the instruction to observe and the process of observation can be combined, if it can further combine with the process of thinking, the practitioner will understand in an instant that what gives instruction and what receives instruction, what can observe and what is being observed, what can think and what is being thought of are in fact a whole entity in which there is not any difference. And this is sudden enlightenment (prajna) as it is often so called.

v. The Concept of *Hunyuan* Whole Entity Is a Materialistic Monistic Concept in which Spirit and Substance Are Unified

According to the theory of *hunyuan* whole entity, spirit (consciousness) is the motion of *Yi Yuan Ti* which is a special form of *hunyuanqi*. That is to say, consciousness is a mode of motion of a special substance. It does not only agree with human body *hunyuanqi* but also with the *hunyuanqi* in the natural world. It can not only have effect on the *hunyuanqi* in the human body and cause change in it, but can also have effect on the *hunyuanqi* outside the human body and make it change. This not only lays the theoretical foundation for all the content in *Zhineng Qigong* Science, but also provides a theoretical foundation for "refining essence into *qi* and refining *qi* into *shen* (spirit)" in ancient *qigong*—if *shen* and *qi* are two totally different things, how can *qi* be refined into *shen*?

vi. Human Beings' *Yi Yuan Ti* Has Developed into an Upward Spiral Compared with Primordial *Hunyuanqi* of the Universe according to the Concept of *Hunyuan* Whole Entity

Primordial *hunyuanqi* has no form or image. Nor does it have mass or energy. It is an indescribable even-state substance without any difference in it. Yet it can evolve into all existences that have mass and energy, including human beings. Human consciousness activities are activities in *Yi Yuan Ti* and *Yi Yuan Ti* is formed when the *hunyuanqi* of the brain cells has become condensed and integrated to a certain extent. It is closely dependent upon brain tissues (concrete substance), yet it has also a certain independence. *Yi Yuan Ti* has no form or image, nor energy or mass in its texture. When consciousness activities calm down, *Yi Yuan Ti* shows no difference in itself. When motions occur in *Yi Yuan Ti*, it can gather energy and even cause change in mass. This is very similar to primordial *hunyuanqi*.

Moreover, *Yi Yuan Ti* can not only reflect external existences, but can also reflect its internal activities. It can bring about all kinds of material change. It already has initiative and can make the world of nature change according to human subjective wishes. This, compared with the complete natural change of primordial *hunyuanqi*, has undoubtedly evolved to a more advanced level. Primordial *hunyuanqi* is the ancestral beginning and foundation of all levels of substances in the universe and can penetrate each level from below to above. In addition to this, the laws of change in different celestial bodies in the universe also display differences due to the appearance of human beings. The fortune of the planet on which human beings live begins to vary with the degree of freedom that human beings attain.

Section II The Concept of Universe Whole Entity

There is a big difference between the concept of the universe in modern science and that in philosophy. In dialectical materialism, space is the extension of the three dimensions of length, width and height and is infinite. Time is the succession through the ages and is also infinite. In modern scientific thought, both space and time begin with the Big Bang. As to whether there is an end to them or not, scientists' opinions are widely varied.

The concept of the universe in the theory of *hunyuan* whole entity is a further development based on inheriting the concept of the holistic universe in traditional *qigong* theory. It is believed in the concept of the universe in the theory of *hunyuan* whole entity that the universe (*yu zhou*) is a unified whole entity of space (*yu*) and time (*zhou*). It has neither beginning nor end in time and is boundless and limitless in space. In this sense, the universe is identical to *hunyuan* particles (*hunyuanzi*). As the universe is occupied by primordial *hunyuanqi* as well as its changes, when we talk about the concept of universe whole entity, we are in fact talking about the holistic concept of *hunyuanqi* in the universe as well as its changes. We think that changes of all existences in the universe are changes of the entirety of universe *hunyuan* entity in different stages with different content. All the celestial bodies as well as all existences on the earth now came from the evolution of primordial *hunyuanqi*. They will become the form of primordial *hunyuanqi* again after long period of evolution.

A deductive illustration is to be given in light of related modern scientific knowledge as follows:

I. The Holistic Feature in the Time Structure of the Universe

According to the theory of *hunyuan* whole entity, the universe is a *hunyuan* substantial entity in a state of continuous change. All existences in the universe today came from the evolution of primordial *hunyuanqi* and they are continuously experiencing new evolution and new change.

Cosmology and astrophysics in modern astronomy have become specialized branches of learning. These subjects have, according to the achievements in physics and chemistry and in light of the relevant data from astronomers' observations, put forward models and theories of the formation of the universe. According to the most appreciated Big Bang Theory, the universe came into existence from the explosion of a particle (a singularity). Although it gives no illustration as to how the particle exploded and what it was like the moment the explosion took place, it has given relatively convincing exposition with regard to a series of questions, such as the formation of galaxies and stars. It has expounded that all the celestial bodies (including the earth and the solar system) have a process of birth and death. Our flourishing earth and the ever-changing solar system are also in their process of "dying". When the hydrogen in the sun becomes almost completely exhausted by burning, it will collapse with the increase of gravity in the center of the sun. Although it will explode again and become a red giant with a gigantic volume, it will eventually contract and form a white dwarf which is extremely high in density with a huge mass yet a relatively small volume (Atoms are ionized, and electrons become degenerate electron clouds that mix with nuclei) and then become a neutron star (formed when electrons and protons disappear and neutrons are annexed).

When that time arrives, differences between each celestial body will vanish and preliminary unity in them will occur. The form (*xing*), *qi* and quintessence (*zhi*) of all existences will be highly concentrated into a special state. Meanwhile, the density and the degree of expansion of the *hunyuanqi* around the celestial bodies will increase accordingly. When celestial bodies cannot resist the pull of gravity and further collapse inward, black holes will be formed as are described in astronomy.

Gravity in black holes is so strong that even radiant rays cannot escape from them. Materials absorbed into them by gravity seem to have entered a place of nowhere. There are two results for the black holes: One is that small black holes may evaporate on account of the change of their entropy and then explode. The other is that bigger black holes cannot evaporate, yet on account of the effect of the great pull of gravity, when they have absorbed all the substances near them (including dust and small celestial bodies), they may get themselves attached to nearby bigger celestial bodies and thus obtain greater power and explode and become part of the bigger celestial bodies. Then the big celestial bodies continue to evolve until they become black holes. Changes like this continue until the whole universe becomes a grand collapsar.

Density in the collapsar is so high that it is almost beyond imagination. Outside the collapsar, all kinds of matter specified in modern science has disappeared (not even radiation field) and there is only *hunyuanqi* with no difference that belongs to the collapsar. The further the collapsar condenses inward, the greater its density becomes and the broader the scope of the *hunyuanqi* around it. When the distance (or called density) between what is contained in the collapsar has become so extremely short (e.g., shorter than the radius of gravity), the *hunyuanqi* of the collapsar will expand to the scope of the entire universe.

When substance in the collapsar absorbs inward to an extremely subtle tiny scale (e.g., shorter than the length of Planck) under the effect of gravity, time-and-space compound particles will be formed. Then time and space enclose inward, which makes time and space merge into one and form *hunyuanzi* (*hunyuan* particles). The moment *hunyuanzi* are formed, gravity becomes immediately vanished, the collapsar disintegrates, and *hunyuanzi* flee and fill the entire universe. This is the Big Bang so termed in cosmology.

Although details about the Big Bang still cannot be clearly explained, several points can be deduced: a) When substance in the collapsar is strongly affected by the great pull of gravity, there will also be an opposite force of repulsion in the substance. Although this force of repulsion cannot make substance in the collapsar spill out, it becomes the impetus for the *hunyuanqi* in the collapsar to expand outward. b) Since substance in the collapsar has become highly unified and differences between substances have vanished, the *hunyuanqi* outside the collapsar has not only formed time-and-space compound particles, but also displays a highly orderly distribution pattern of concentric spheres, the realm of which almost fills the scope of the original universe. c) It has been pointed out in the theory of *hunyuanqi* that changes of the *hunyuan* body and changes of the *qi* field outside it do not happen synchronously. So the moment *hunyuanzi* are formed out of the substance in the collapsar, the time-and-space compound particles outside the collapsar are still in a highly orderly state. d) When substance in the collapsar changes into *hunyuanzi*, which is similar to the second order phase change of the equilibrium phase change in synergetics, namely, when it suddenly enters the phase state of *hunyuanzi* that has not any difference within, gravity disappears, so repulsion takes effect and the collapsar disintegrates. As time-and-space compound particles outside the collapsar are in a highly orderly state, *hunyuanzi* can spread all over the universe in an instant. When *hunyuan* particles are spilling outward, on the one hand, they distort the original order of time-and-space compound particles; on the other hand, their own enclosed time and space disintegrate on account of repulsion and form time-and-space compound particles and primordial *hunyuanqi*.

With the interactive collision in the explosion, time-and-space compound particles connect into different time-and-space structures and gradually form gravity, electromagnetic radiation as well as various particles and enter the stage of the formation of nebulae and galaxies. Then the solar system, the planets and all existences on the earth are formed until the appearance of human beings, as makes the change in the natural world start to step into the self-recognition stage of the universe.

From the changing process stated above, it can be seen that change of the universe, evolving from simple substance into the form of multiple substances, has an integrate time structure. If viewed from the perspective of super intelligence, this time structure is but an existence of super-intelligence-state substantial entity which actually does not have that much difference in it. This, according to traditional *qigong*, is an absolute substantiality beyond description. The explanation we give in light of modern science is but to draw an analogy, the purpose of which is to deepen our understanding that our universe is a whole entity changing in accordance with *hunyuanqi*.

People might ask how it can be that the universe has a time structure and may feel it hard to comprehend. For this, we might as well draw support from rational thinking. For example, the electromagnetic waves observed in astronomy in the celestial body 15 billion light years away were in fact sent out 15 billion light years ago; the electromagnetic waves observed in the position of a celestial body 10 billion light years away were actually emitted 10 billion light years ago. It is the same with the electromagnetic waves of the celestial bodies 50 light years away, 10 light years away and 1 light year away and so on, until the solar system in which we live today. Isn't this series equivalent to the time holistic structure of the universe?

II. The Holistic Feature in the Space Structure of the Universe

If we say that the holistic feature of the time structure of the universe illustrates the holistic structure of change in the "ordinate" of the universe, the holistic feature of the space structure of the universe to be introduced here will talk about the holistic feature of a transverse section in this "ordinate". According to the theory of *hunyuan* whole entity, everything in the universe now comes from the evolution of primordial *hunyuanqi*. They change in accordance with the same time-and-space structure and are supposed to share rigid identity. If it is observed from the point of view of modern science, it is indeed like this. Although the universe is vast and boundless, the constituent parts of the universe have their certain common points according to data shown in various astronomical subjects.

i. From the perspective of structure, over one billion galaxies have been discovered so far that constitute the universe, and galaxies generally make up super galaxy clusters (or called group of galaxies). Being constituted of galaxies different in number, super galaxy clusters are the biggest constituent unit of the universe. Galaxies, on the other hand, are constituted of multitudes of suns. Our Milky Way consists of over 100 billion suns. Suns are the essential concentrated form of expression of current substances in the universe. The sun will go through the main sequence stage (the stage in which hydrogen burns to emit white light), the red giant stage (the stage in which helium burns) and then develop into a white dwarf, a neutron star before turning into a collapsar. The sun in our solar system is right in the middle of its entire life span of the main sequence (which is about 10 billion years). Planets are cool condensed celestial bodies revolving around the sun. There is atmosphere in different forms around most of the planets. From the composition characteristics of the celestial bodies in the universe found out by astronomers, it can be seen that there exist not only great similarities among different constituent parts in the holistic space structure of the universe, but they are all subject to the laws of gravity and electromagnetism.

ii. From the element composition of celestial bodies: With the advent of spectroscopy and radio astronomy, it is known that there are generally the same substance elements (eg., heavy elements and light elements) in the stellar space as on the earth. Ammonia (NH_3), water vapor (H_2O) and even more complex molecules such as formaldehyde, methonal, acetaldehyde, ethanol, methylamine and methylacetylene are also found. Particularly the discovery of hydrogen cyanide, the substance very important for the synthesis of amino acid in the laboratory, offers further evidence that different celestial bodies in the universe follow the same laws of mergence and change (*hunhua*).

We have given a simple illustration above as to the time structure and the space structure of the universe with the aid of knowledge in modern science. Content as this sounds seemingly quite reasonable and able to give people a sense of whole entity. However, further inquiries find that every link in the content stated above is an *x*. For example, it seems that there is not much disagreement as to the formation of galaxies and the evolutionary process of suns. This is in fact that we still lack genuine understanding about them, namely, there is still not much content from practice to check the truth or falsehood of them. If you disagree, please take a look at the theory of how the solar system (including the earth) is formed and you will find that this theory cannot justify itself. Every current theory has its insoluble contradictions.

The reason for this, we think, lies in the fact that the universe is a huge complicated whole entity, or rather, a super huge system. If only the laws discovered so far in physics are adopted to deduce and trace back to the origin of the universe, we will undoubtedly simplify the universe. If the universe is indeed that simple, how can there be such highly complicated and intelligent human beings? Therefore, the super huge system of the universe and the multiple-level huge system of the human beings must be connected when we consider the whole entity of the universe.

The anthropic principle was born in this background. It discusses from the reality of the existence of human beings that the various kinds of coincidence (including the agreement between large dimensionless numbers) in every stage and at every level of our universe ever since its beginning are destined to be so, for without these coincidences, none of the laws (particularly the laws in physics) cognized so far by human beings could exist and intelligent human beings would not have come into existence. This theory traces the evolution of the universe and finds that the evolution of the universe has to be so from the reality of the existence of human beings. Though it hasn't solved any practical problem, it provides us a plot of the theory of whole entity, that is, from the "creation" of our universe to the evolution of the intelligent human beings today, it is an inevitable result of the evolution of the time-and-space structure of the universe. We intelligent human beings are the product of the whole entity of our universe in its most advanced stage, and this is the result of the change of the whole universe.

We might as well trace back like this: The earth on which the human beings live is a special environment under the comprehensive effect of the sun, the moon as well as the planets, and the solar system keeps revolving around the nucleus of the Milky Way. That is to say, the solar system is under the effect of the Milky Way, while the Milky Way is one of all the galaxies in the universe that influence each other. All these influences come from the chaotic state when our universe was being created. We can imagine as this: In the chaotic state when the universe was being created, order parameters of infinite universes were formed. The order of our universe is but one of them. This initial condition determines the holistic feature of the primordial universe and all the changes afterwards take place under the control of this major order.

Modern chaotic theory has revealed the special feature of "fractal structure", i.e., any part of the chaotic whole entity seems to repeat the feature of the whole entity. Yet it is only similar instead of real repetition, with each repetition having its new content. This has, to a certain extent, revealed the law of evolution of the whole entity.

After we know the feature of the chaotic fractal structure in chaotic theory, it is not hard for us to understand the various magical coincidences in our universe. Now it seems apparent that, from the beginning of our universe to the emergence of us intelligent human beings on the earth—this is the principal sequence of our universe (i.e., the creation of the universe→the Milky Way→the solar system→the earth→the human beings). This principal sequence is not isolated but connected with the entire universe. Due to the limitation in modern science, the law of this whole entity has not been revealed yet, which shows that intelligent human beings need further improve and perfect their own intelligence. The process in which human beings improve and perfect their intelligence is also a process in which they consciously establish a harmonious and perfect whole entity with nature and all existences.

III. The Holistic Feature in All Existences in the Universe

According to the theory of *hunyuan* whole entity, all existences in the universe are an interrelated whole entity. It should be relatively easy to understand this point after the study of the concept of the universe as a whole entity. It is simply that people are accustomed to habitual thinking and deem it somewhat "cruel" to regard all existences (including human beings) different from one another in numerous ways as one whole entity. In order to further illustrate this truth of the whole entity of all existences, let us make a specific analysis by tracing the clues of the universal whole entity. The whole entity feature in all existences will be expounded from three aspects as follows:

i. The Whole Entity Feature Shown in the Evolutionary Process from Primordial *Hunyuanqi* to Human Beings

Generation and evolution from primordial *hunyuanqi* to all existences, including human beings, were not accomplished all at once. All existences and human beings gradually came into being after a long period of evolution and change. Let us illustrate this long evolutionary chain in different stages.

i) The evolutionary chain from primordial *hunyuanqi* to the appearance of organisms: primordial *hunyuanqi* in intangible existence→particles→concrete existence (atoms, molecules)→inorganic matter→organic matter→viruses→cells. There are two crucial points in this chain of evolution. One is particles in the transition from intangible existence to concrete existence. The other is viruses between inanimate matter and animate matter (strictly speaking, a virus should belong to life substances because it already has genetic structure and function, except that it cannot independently accomplish this process by itself).

ii) After cells came into existence, life substances experienced a very long process before they divided into two branches: One branch is plants. The other branch is animals, then higher animals until human beings, as forms an evolutionary chain of species. Here we merely talk about the evolutionary chain of animals: monoplasts→multicells→coelenterates→arthropods→proto-chordates→chordates→primates→ human beings with consciousness.

To take a comprehensive look at the long evolutionary chain from primordial *hunyuanqi* to human beings, we can find that this is a complete process of the negation of the negation from simplicity to complexity and from complexity to simplicity again. Primordial *hunyuanqi* in which there exists no difference but infinite information is negated by the *hunyuanqi* of complex all existences. Then, human *Yi Yuan Ti* (Mind Oneness Entity which also has no difference in it but has infinite information) negates the *hunyuanqi* at the level of all existences. They seem to have gone through a cycle of evolution. If we make a careful analysis of this cycle (or long chain), we will find that it can be divided into many links, each of which experiences a development process from simplicity to complexity and from complexity to simplicity again.

From primordial *hunyuanqi* to the generation of gravity, electromagnetic field and then numerous particles (Those having the longest life span are electrons, protons and neutrons)—this is from simplicity to complexity (One is the complexity in variety; the other is that the structure of each category is more complex than primordial *hunyuanqi*). All this takes place at the level of insubstantial or intangible existence. Once the level of substantial or concrete existence is reached, the element of hydrogen which is the simplest in structure at the level of substantial matter is formed (It has only one electron and one proton). Then again the structure becomes complex step by step. There are not only many kinds of elements (including isotopes of each element), but also compounds which are very rich in content until the macromolecules of organic substance (e.g., the structure of chain or ring of hydrocarbon, alkane, organic acid, alcohol, aldehyde, sugar, benzene, anthracene). This is also a process of development from simplicity to complexity. However, once the level of life is reached, the new and simple form of individual life, cells having the metabolic function, begin.

If we say that the form of a hydrogen atom is a negation of formless *hunyuanqi*, then the form of a cell which has life is a negation of the level of inorganic substance which has no metabolic activities. Organisms begin their long process of evolution from single cells, from the lower level organism to the higher level organism, from the simple organism to the complex organism, and form the two great complementary systems of plants and animals, thus displaying a world as brilliant as brocade flourishing in its great varieties.

When the stage of higher animals is reached, there are not only complicated concrete structures, but also tissues and organs explicit in division of labor, varied in function and united in the whole entity. Animals have the nervous system which renders a certain degree of initiative to their life activities. However, their nerve activities are combined with their entire life activities and are subject to their complicated life activities. That is to say, the *hunyuanqi* of animals, including the *hunyuanqi* of their nervous system, is combined with concrete substance (their bodies). This has evidently developed the level of the *hunyuanqi* of all existences to a summit. Nevertheless, once the level of human beings is reached, fundamental change takes place, and *Yi Yuan Ti* with such even texture as primordial *hunyuanqi* is formed. Up to this point, it is not only the negation of animals—average organisms, but the negation of the *hunyuanqi* of the entire level of all existences.

It is true that, after the formation of *Yi Yuan Ti*, there is also a process of double negation, and this is the very task being shouldered by *qigong* science.

ii. The Whole Entity Feature Viewed in the Evolutionary Replay in an Individual Human Being

If we say that the evolutionary chain from primordial *hunyuanqi* to all existences mentioned above shows that every specific existence is part of the whole entity of all existences in the universe, the evolutionary replay in an individual human to be talked about here will illustrate that each individual human contains the integrate content of this whole entity of all existences. In every negation-of-negation link and process in the long chain of evolution, there exists an evolutionary replay process at this level about the levels having been negated. Here let us have a look at the evolutionary replay in the individual of human beings who are the most intelligent of all existences to see how humans evolve from a monoplast (which belongs in the category of low-level organisms) so as to better understand their holistic relationship with all existences. Let me illustrate the development process from a zygote into a human fetus with an evolutionary chain as follows:

The zygote (equivalent to a monoplast)→the morula (equivalent to a multicell)→the diploblastic stage (the endoderm and ectoderm are equivalent to a coelenterate; when the prestreak appears, it is equivalent to an arthropod)→the notochord formed in the triploblastic stage (equivalent to a protochordate)→the nerve canal formed in the somite stage (equivalent to a chordate)→the accomplished period of embryonic development (with characteristics of human fetus displayed).

From the development chain of an individual human being we can see that human beings, as the most intelligent creature among all existences, are repeating the evolutionary process of the biological system in their embryos' development. This is just as what has been pointed out by Ernst Haeckel, the well-known biologist:

"The History of Evolution of Organisms consists of two kindred and closely connected parts: Ontogeny, which is the history of the evolution of individual organisms, and Phylogeny, which is the history of the evolution of organic tribes (that comes from the same origin as individual organisms). Ontogency is a brief and rapid recapitulation of Phylogeny, …"(*General Morphology*, 1866)

It is because humans evolved from lower organisms that in the holistic time-and-space structure of humans all the information in the development process from lower organisms to human beings are deposited. Likewise, lower organisms evolved from inorganic matter and inorganic matter evolved from particles and primordial *hunyuanqi*. Therefore, in the substantial entity of the human body is also deposited all the information in the long evolutionary chain from primordial *hunyuanqi* to human beings. This is also the reason why human *Yi Yuan Ti* (Mind Oneness Entity) can contain infinite information—so much so that it reaches the state of Chaos.

iii. The Whole Entity Feature in the Mutual Effects between All Existences

In the two points above, the whole entity feature of all existences is respectively expounded from the two aspects of historical evolution and realistic accommodation. Here the whole entity feature among all existences is to be expounded from their realistic interactions.

It has been pointed out in the theory of *hunhua* in the greater theory of *hunyuan* whole entity that all changes taking place in all existences result from the *hunhua* or mergence and change of their own internal and external *hunyuanqi* and this process is itself a process in which all existences interact. Changes between all existences, if viewed from the standpoint of the universe, are but the different combination processes of matter, energy and information in the universe; if viewed from the standpoint of all existences, they are the transfer of matter, energy and information in different degrees. Transfer as this reinforces connections between all existences (individualities) and makes them embody the whole entity feature.

It is also realized in modern science that nothing in the universe exists in isolation. For example, the celestial bodies far away and independent from each other are all huge objects with their own specific content. Yet, between these celestial bodies, there exists not only the mutual influence of gravity, but also the mutual effect of radiation. The light we see from the celestial bodies is in fact the difference between the radiation they receive and the radiation they give out. If the quantity of radiation received and emitted by a celestial body is the same, it cannot be seen by us.

Take the earth for another instance. The climatic circumstances of the earth are not only affected by the sun and the moon (*yang* and *yin*), but also by the five big planets (Venus, Jupiter, Mercury, Mars and Saturn). The theory of five *yun*s (referring to the five elements of wood, fire, earth, metal and water) and six *qi*s (referring to the six climatic factors of wind, fire, hotness, humidity, dryness and coldness) in traditional Chinese medicine based its establishment upon these traditional viewpoints. In this theory, the influence of the five *yun*s and six *qi*s on the normal changes of birth, growth, decadence and death in plants and animals (including humans) on the earth has been comprehensively illustrated and it has been greatly valued by famous TCM doctors of past dynasties in history. In addition to this, all existences on the earth are even more dependent upon each other and transform into each other, thus forming a picture of natural cycle: natural inorganic substances (all kinds of elements and moisture content)→absorbed by plants and synthesized into various nutrients→animals exhaust them and release waste→microorganisms decompose→return to the natural world. Even intelligent and wise human beings have not escaped from this picture of circulation of substance. Beside, plants and animals can also reciprocally satisfy their respective needs in oxygen and carbon dioxide.

We might as well envisage this in the manner of science fiction: If we could make a tracing mark on the material elements in every object (including the human body), it would not take very long for us to discover that each and every individuality exists in an interconnected way, in a manner that you have some of me in you and I have some of you in me, and the world displays a grand picture that all existences mutually accommodate. Isn't our reality just like this? It is simply that we cannot show this with the tracing mark technology.

Finally, it needs to be pointed out that although modern science and traditional Chinese medicine have already understood the general holistic connections among all existences, they have not recognized the comprehensive and thorough connection of *hunyuanqi* among them (What is realized in modern science is but part of the holistic content of *hunyuanqi* displayed in the world of physics). For example, *hunyuanqi* can not only affect the birth, growth, decadence and death of animals and plants, but can also affect the entity, nature and characteristics of inorganic substances. This is already a concrete fact that has been repeatedly proven by the experiments in *Zhineng Qigong* Science. Now that this function of *hunyuanqi* can be brought into play in experiments, it proves that such functions of *hunyuanqi* do actually exist objectively. It is simply that these functions have not been recognized, grasped and applied by human beings. Once the holistic connections of *hunyuanqi* between all existences are grasped by human beings, we will inevitably establish a new super intelligence culture.

Section III The Concept of Human-and-Heaven Whole Entity

Heaven here refers to the world of nature and society, and human refers to human beings themselves. In the concept of human-and-heaven whole entity, the holistic feature in the connections between humans and nature and between humans and society is expounded. Since humans are super animals with consciousness and have formed the dynamic subjective world, they set the subject of human beings themselves (including the subjective world of consciousness) against the objective world of nature (including society), thinking that humans are the most intelligent creature independent of the natural world. This is a historical mistake which will be gradually corrected with the development of science. The holistic correlated laws between humans and nature will be gradually revealed and grasped. The concept of human-and-heaven whole entity in *Zhineng Qigong* Science will guide people out of the labyrinth that confines their subjective world and help them achieve human-and-heaven harmony in which the subjective world and the objective world are unified.

As humans are both humans of nature and of society, our concept of human-and-heaven whole entity will also be illustrated from these two aspects.

I. Humanity and the World of Nature Are a *Hunyuan* Whole Entity
i. The World of Nature Created Humanity

We say that the world of nature created humanity. This is because:

i) Both humans and all other existences originated from primordial *hunyuanqi* and are an inseparable whole entity. As has previously been mentioned, human beings were created on the basis of all existences in the natural world.

ii) The world of nature provides human beings with material insurance which guarantees the successful continuance of the human body's normal activities. The entire life process of humans is a process in which human beings continuously absorb the *hunyuanqi* from the outside world, make it merge with their own *hunyuanqi* and let it become the *hunyuanqi* of their own bodies. This process is carried on at three levels as referred to in familiar intelligence, that is:

a. Material exchange: Humans absorb substances from the outside world, assimilate them, making them become substances in the human body, and excrete the waste from the human body.

b. Energy exchange: We know that the essential source of energy on the earth comes from the sun. Human beings directly or indirectly (e.g., through chemical energy) absorb solar energy, then consume the energy and send it out of the human body.

c. Information exchange: The meaning of information that we talk about here refers to the time-and-space composition feature of existences (which is constituted of the three elements of information matrix, number and form). Everything has its own special information. Human beings living in the natural world exchange information with the world of nature all the time. This information exchange is more subtle than the exchange of matter and energy mentioned above and is not easy to be noticed by people. Yet it has effects on the human body. For example, some people naturally feel comfortable, relaxed and physically and spiritually joyful when they come to a certain place. This is not necessarily the effect of anions all the time, as many places do not have the condition for the generation of anions, while some places may make them feel agitated, uncomfortable or even depressed. Of course, the reception of information is closely related to the internal condition of each person.

In sum, the world of nature provides human beings with the life materials and living conditions of clothing, food and shelter, which maintain human lives and make people able to multiply. However, the world of nature does not simply carry out material exchange at different levels as mentioned above with humans to form the human body that has shape. Instead, it carries out all-dimensional mergence and change of *hunyuanqi* with the human body, and the influences of different environments upon human tissues and organs as well as their functions are not all the same.

For example, in the sphere of the whole globe, we can say that the evolution from anthropoids to human beings is basically the same. However, the color of skin, the characteristics of physique as well as some of the physiological functions in races of people distributed in different areas differ widely. Still more widely different are their characteristics of culture, their customs, habits and moral standards and so on. How did all these variations come about? According to the concept of *hunyuan* whole entity, human beings live in the natural environment, and the mergence-and-change effect of *hunyuanqi* of all existences upon the human body *hunyuanqi* is going on all the time. Any great change in any factor in the external environment has an impact on human body *hunyuanqi* and thus causes the *hunyuan* entity of the human body to change. In the course of time, various kinds of variation take place. As for this, Karl Marx pointed out a long time ago:

"The first and foremost premise of any human history is undoubtedly the existence of individual human being who has life. Therefore, the first specific fact that needs to be made sure is these individual humans' flesh and tissues and their relationship with the natural world which is controlled by their flesh and tissues... and the geological, geographical and weather conditions as well as other conditions that people meet with". This is because "these conditions not only control human beings' flesh and tissues naturally born at the very beginning, particularly the racial differences between them..." (*The Complete Works of Marx and Engels*, Vol. 3, p. 23)

Friedrich Engels clearly pointed out:

"We together with our flesh, our blood and our brains all belong to and exist in the world of nature."(*Dialectics of Nature*)

This is an exact description of the fact that humanity and the natural world are one whole entity.

iii) The world of nature provides humans with the object of spiritual activities and enriches vari-

Note:

Some of the quotations by Karl Marx and Friedrich Engels and all of the quotations by Vladimir Lenin in this book are translated by the translator of this book from Chinese into English. Marxism is used in this book in the pure philosophical sense.

ous human functions. It provides human beings with not only the material living condition, the object of labor and the working condition, but also the object of their spiritual activities. All human consciousness directly or indirectly reflects the objective world. Without this objective world, there would be no human spirit. When human beings just separated themselves from the world of animals, the functions of their different sensory organs and the thinking function of their brains were all in a chaotic state. There was almost no thought in their brains and they were still in an uncivilized chaotic state. Different tissues and organs in the human body, particularly the functions and shapes of hands and feet, also needed further development.

Our ancestors, drawing on the power of collectivity and relying on the living environment bestowed on them from nature, created their own material life and spiritual life. Human beings' culture and arts were all established in the process of cognizing nature. This process also gradually enhances the functions of different human sensory organs. For example, the visual distinguishing ability of the eyes, the audial distinguishing ability of the ears, the taste distinguishing ability of the tongue, the touch distinguishing ability of the skin, the labor skills of the hands, the thinking ability of the brain as well as the ability to appreciate beauty and so on all continuously evolve.

In brief, human beings' various functions now are already greatly different from their functions in the period of anthropoids. These differences in functions also result from the *hunhua* of human *hunyuanqi* with the *hunyuanqi* of the natural world in the process that humans cognize and reform nature. It is precisely because of this that we say nature created humanity.

ii. Humans Are Also Creating the Natural World

As it is well known, before human beings came into being, the world of nature existed in a natural state. After the emergence of human beings, the subject human beings with subjective consciousness antagonistic to the object of nature was divided from the natural world, and this adds new content to the diversity of nature.

We know that, before the emergence of human beings, the world of nature changed according to the law of nature. The emergence of human beings deprived the world of nature of its complete natural state in the past and impressed the mark of human will on it everywhere. When human beings come in contact with the natural world, they always change the original state of the natural world with the means of labor in accordance with their subjective will and make the natural world change in the direction beneficial to humans.

Humans' process of labor, whether it is to reform the environment or to absorb materials necessary for their own lives, is a process in which human consciousness activities are realized and healthily developed. It is a process in which consciousness is externalized from the subjective world into the objective world and becomes materialized and deviated. Human information and energy are implanted in objective existences and humans and nature are closely integrated into one whole entity. In a certain sense, the natural world has become a humanized natural world. Particularly with the development of modern science and technology, human beings have acquired more power to "gallop" in the natural world. From the high mountains to the vast seas, from the land around to the sky above, in the scope of the earth and even in the solar system, human information is everywhere, which closely integrates humanity and nature.

iii. Humans Are Part of the Whole Entity of the Natural World and the Human Body Can Embody the Whole Entity of Nature

Although humans are in the most advanced stage in the evolutionary chain of the cosmos, they are still part of the whole entity of nature in the universe. It is because of this that the human body can reflect the holistic feature of nature and the universe (essentially the earth). For example, the human body contains different material levels in the universe as well as their motion modes.

i) From the perspective of the level of substance, the human body contains all the levels of substances in the universe, e.g., the level of organisms, the level of compounds (which includes organic substance and inorganic substance), the level of elements and the level of *hunyuanqi*, etc.. Let me make a tentative analysis about this. As far as the level of organisms is concerned, systems, organs, tissues and cells that constitute the human body represent the characteristics of the entire world of animals, and the development process of human embryo can display and replay the evolution of animal species. As far as the level of compounds is concerned, the human body has not only sugar, fat and protein which belong in the category of organic matter and the various derivatives in the metabolic process, but also only 20 kinds of amino acids in the natural world. There are also many kinds of compounds such as sodium, potassium and calcium which belong in the category of inorganic matter within the human body. The proportion of water in the human body is the same as the proportion of water on the earth. As far as the level of elements is concerned, not only are there such macroelements as hydrogen, oxygen, carbon, nitrogen, phosphorus, calcium, iron, sulfur, sodium and potassium in the human body, but the proportion of most trace element content is also similar to that of the earth. As far as the level of *hunyuanqi* is concerned, human consciousness (the content and process of the motion of *Yi Yuan Ti*) is similar to primordial *hunyuanqi* in entity and attribute.

ii) From the perspective of motion modes, in the human body there exist all kinds of motion of matter in the universe. According to modern science, they can be categorized in the forms of physical motion, chemical motion, biological motion and consciousness motion. According to *Zhineng Qigong* Science, the human body contains all the *hunhua* motions stated previously in Chapter II.

It is because the human body can reflect the entirety of Nature that the research into the human body made by *qigong* science is a key to revealing all the mysteries in Nature.

iv. Humans and Nature Are a Unity of Opposites

It is generally believed that humans as subject with thought are antagonistic to natural things as object and that they are two absolutely different kinds of existences. Particularly in modern science, humans are often regarded as the royal king of the world of nature, the dominator of the natural world. There is quite a tendency for humans to be considered as the prized son of God. The development of such thinking as this, as a result, destroyed the harmony in the natural world and endangered humankind. Thus it makes some people go to the other extreme, thinking that humans are just natural humans who have no essential difference from natural things and that humans should return to nature.

We think that both of the viewpoints are partial, hence not desirable. The relation between humans and nature should be both opposite and unified. To say that humans and nature are opposite is because, on the one hand, human beings as subject already have subjective spiritual activities that can not only reflect the existence of nature, bestow on nature its significance of objective existence and has changed nature's attribute in the past of being isolated and meaningless existence, but can also reflect the objective existence of the human body itself. This differentiates the human spirit as a special kind of existence different from common natural things. On the other hand, the human spirit as subject can also command its own life activities to interact with nature, to change the innate look and attribute of nature and to duplicate human subjective intention in the natural world so as to prove that humans are the creature with consciousness and that spirit also objectively exists. That is to say, the world of nature proves its objectivity with human spirit and human spirit proves its objectivity through the natural world.

Humanity as subject and nature as object seem to be the two opposite poles, yet the two poles are also unified because the world of nature itself actually contains humans in it. This has already been extensively discussed so far in this chapter. Here we will merely talk about their unity from the perspective of epistemology.

Object in epistemology refers to the objective existence independent of people's consciousness. However, this objective existence is connected with human consciousness because any object can be nothing else but the object recognized by human beings. This object is the foundation for the creation of the subject of humans. Therefore, humans' knowledge about the world of nature is in essence the natural world having reached the stage of knowing itself. It is true that, before the emergence of human beings, the natural world was also in existence, but it was only an independent existence for itself. It is only after human beings have derived from the world of nature that the difference between subject and object appears. Even all human thinking activities are, after all, the attribute peculiar to the world of nature when it has developed into a certain stage. The activities of humanity as subject are but a special case among all existences in Nature.

Therefore, what is emphasized in the theory of *hunyuan* whole entity is the unity between humanity and the natural world. Meanwhile, it is also pointed out in this theory that human *Yi Yuan Ti* is the display of the most advanced level of *hunyuanqi* of this world. The purpose is to reveal profoundly that humanity and nature are an inseparable whole entity and that humans should not unscrupulously abuse nature. However, human beings should not be equated with nature, nor should they return to what they once were in the world of nature. Instead, we should follow the guidance of the theory of *hunyuan* whole entity based on super intelligence and establish a harmonious human-and-heaven relationship, the center of which is human beings.

II. Humanity and Society Are a Whole Entity

Human life, as far as each human being is concerned and according to the concept of heaven-and-human whole entity, is embodied in individual human being's relationship with other people and in the restrictions to which the individual has to conform in society. A human being completely and truly isolated from human society is not a human being in the real sense. All human life activities are carried out in the whole background of society. This background of society includes not only interpersonal relationships, but also the most fundamental activity of human beings—production activity as well as production environment and so on. All these are the external environment interacting with human beings. Therefore, they are the important conditions that interact with the *hunyuan* entity of the human body and cause change in it. It is on the premise of continuous improvement of production environment and production condition that human beings develop from ignorance to civilization.

i. Humans Created Society and Society Also Created Humans

It has been previously stated that humans are humans of nature. However, the more important attribute of human beings is their social attribute. It has been pointed out by Marx, "Human nature is not something abstract that is innate in an individual human being. In fact, it is the sum of all social relationships."(*The Complete Works of Marx and Engels*, Vol. 3, p. 5)

Why is the species-being (or nature) of human beings their social attribute? First let me make an analysis of the similarities and differences between human beings and the natural world (particularly other species beings or existences in the biological world). The feature that the biological world has in common is metabolism, that is, the species beings exchange material, energy and information with the world of nature in order to maintain their own existence. However, the three types of species beings of plants, animals and humans have their respective characteristics. Plants assimilate inorganic substance and sunlight (energy) and combine them into organic substance to embody their characteristic as a species existence; animals essentially absorb organic substance and air and transform them into energy to show their characteristic as a species existence. Although in the two species beings mentioned above there is such a difference in their content of metabolism as energy changing into substance and substance changing into energy, they are both natural beings.

In human beings, though there is also the natural existence of the human body's metabolism, this natural existence in the human body is the natural existence under the control of human consciousness. Human life activity is a process to realize human consciousness activity. Therefore, the characteristic of the human race as a species being is human consciousness activity as well as the language that expresses consciousness. All this can only be displayed in the connections between people, i.e., society. Without society, there would be no human nature.

Some readers might ask: In that case, which came into being first, human nature or society? The answer is that the nature of the human species and society were formed simultaneously with the emergence of language and consciousness in ancient anthropoids' process of labor and collective life. Human nature and society are inseparably related to each other. Once society came into existence, it became human beings' living environment. The production condition, the level of science, technology and culture, the customs and code of morality in life, and particularly the language that expresses such content have all become the living condition for the new generation of humanity. This not only determines the formation of the new generation, but also affects the individual development of the new generation. It is true that the new generation continuously changes the old environment in their life. The development of society and the improvement in the level of human nature always keep apace with each other.

Readers may also ask: The development of individual humans results from satisfying the individuals' physiological needs (which are mainly all kinds of demand for materials). Why do you say that society can also influence an individual human's development? This question can be answered from two aspects. One is the formation process of consciousness as well as its impact on individual human development. Engels once pointed out that "the normal state of humans corresponds with their consciousness and is in fact created by human beings them-selves."(*Dialectics of Nature*) The effect of consciousness upon human beings' life activities as well as its influence upon an individual human being's development will be talked in detail in Chapter V The Theory of Consciousness. Here we only remind you of the fact that a new-born baby who lived in a pack of wolves developed into a wolf child due to being unable to receive information about humans in society, and from this we can comprehend the influence of society on human nature.

On the other hand, the physiological needs of human beings and those of animals seem to be the same in mechanism on the surface. In fact, they widely differ. Not only are human physiological needs the needs in consciousness established on the natural physiological needs of animals, but humans clearly know the necessity of their own needs and they consciously create their own needs. Human beings' means and processes to satisfy their own needs are carried out consciously and purposefully according to plans. Because of this, human beings must contact the world of nature and this process of contact is accomplished through the medium of society. Even with a person who works alone, the tools that this person uses and the knowledge that this person has acquired are also society's.

Moreover, emotions, morality, science, arts and so on that are peculiar to human beings are both the common social consciousness of humans in society and the individual consciousness of everybody, none of which is not the "product" of society. Let me try explaining this with the example of emotions. As it is well known, a newborn baby usually has no emotions. Emotions in a baby are formed gradually in the contact with people after the baby's birth. They are established upon the foundation of the baby's simple physiological needs. For example, when the joyous feeling of getting rid of hunger is combined with the person who feeds it, happiness is shown in the baby whenever it sees the feeder. This is both the beginning of emotions and the start of morality (because this is already the change in the inner world of the baby who begins to deal with the relationship with people). This seems to be the same as conditional reflex in animals. Yet the development after this shows the unique feature peculiar to humans, i.e., all this can be combined with language and the accomplishment of this process needs the information about people in the surroundings (namely, society) to merge (interact) with the baby and cause change (*hunhua*) in it. So the fundamental attribute of humans is their social attribute.

It should be pointed out that the natural attribute of humans should not be equated with the natural attribute of animals because the natural attribute of humans as well as their natural physiological needs are also realized in society. That is to say, all human natural attribute and their natural physiological needs have been endowed with social attribute. When a human being comes into contact with another human being (namely, when they enter the realm of society), it is still the interaction between a natural being and another natural being. However, there is only one subtle difference as follows between the communication among people and the interaction among things: The former is the interaction carried out with initiative between the natural beings that have the function of consciousness (i.e., initiative), while the latter is the interaction naturally carried out between the natural beings that have only natural or instinctive responses. Yet this subtle difference is the very fundamental difference between the nature of each species.

The essential point that tells humans apart from animals of other species is their feature of consciousness and their social attribute. All human material life is satisfied through the medium of society. In other words, the connection and unity of humanity and nature are realized through the medium of society. This is the critical point that differs from the overall connection between animals and nature.

Unfortunately, human beings have not yet truly realized the real attribute of human nature. In many ways they still confuse humans with animals. Human beings begin to recognize and are in the process of recognizing that the world of nature is the root and foundation to satisfy human physiological needs and that humanity and nature are one integrated entity. They have not truly realized, however, that society is also the root and foundation to satisfy the needs of the nature of the human species. As a matter of fact, human society is a natural phenomenon when the universe and nature has developed into the stage of human beings. What is the difference between the connection between human beings and that between humans and natural things? It is simply that the former is the connection between the natural beings having consciousness while the latter is the connection between the natural beings having consciousness with the natural beings having no consciousness. From this it can be seen that both the natural world and society are the environment for human beings' existence, and the latter is the more important environment that guarantees the realization of the nature of the human species.

Society and human beings are developing and advancing. Human beings today are already different from human beings in the past and societies today also differ from what they once were. In sum, society (environment) is creating humanity, and humanity is also creating society (environment).

ii. With Human Beings' Development, More and More Natural Things Have Entered Society

With the continuous development of science and technology and with their efficient entrance into the field of industrial production, natural things in the world of nature have not only highly efficiently turned into goods to satisfy people's demand in life (including physiological needs), but have also become machines that function as the substitute for human organs and have entered the realm of society.

As it has been pointed out by Marx, "The world of nature has not made any machines. It has not made locomotive, railway, telegram, mule spinning machine and so on. These things are the products of human labor. They have become the organs with which human will controls nature or the natural substances of the organs with which human beings carry out their activities in the natural world. They are the organ of the human brain created with human hands, the materialized power of knowledge. The development of fixed assets has shown that general social knowledge has to what great extent become direct productive forces, and to what great extent the condition of social life process itself is controlled by general intelligence and reformed according to this intelligence."(*The Complete Works of Marx and Engels*, Vol. 46 (II), pp. 219-220)

Particularly with the fantastic spurt of information technology in recent years, computers are entering all kinds of fields and artificial intelligence is realized in many fields of production and society. All this accelerates the progressive course of human material civilization and spiritual civilization. Isn't it like this? Transportation and communication are becoming increasingly developed; various kinds of buildings are becoming increasingly impressive; daily necessities are becoming increasingly abundant; and culture and arts are becoming increasingly colorful. Natural substances are humanized and have entered every corner of society, which makes the simple and pure mode of communication between people in the past diversified and complicated and humanity, society and nature are more closely unified. The look of nature is changing in accordance with human will. The look of society is changing with the progress of human beings and with the degree of integration between humans and nature. All this facilitates human knowledge to advance toward the whole entity feature and has created the material condition for forming the holistic concept, with the three of humans, society and nature in a unified entity.

iii. Transmission of Consciousness Is Fundamental for the Holistic Feature of Society

Consciousness is transmitted by two means: One is consciousness transmission through communication with spoken and written language. This is the display of human beings' familiar intelligence and is the intelligence consciously studied and used at will by every human being. The other is consciousness transmission through the mode that consciousness receives and sends holistic consciousness information instead of using spoken and written language. This is the display of human beings' super intelligence (For detailed information, please refer to The Theory of Consciousness in Chapter V in this book). It is the intelligence that people still have not consciously studied, mastered or used at will. Super intelligence is also innate human intelligence. For example, correct recognition about things born of inspirations is just a display of having received holistic information.

Human beings' familiar intelligence and super intelligence play different roles in the spread of consciousness in human beings' different historical stages. For example, in ancient times when transportation was backward and communication was limited, human beings in different areas on the earth successively invented hardware, brassware and iron tools and so on almost in the same stage in history, and this still makes people wonder now how the information was transmitted by the ancient people. It was in fact that the holistic information sent directly by consciousness was received by other people. Since the consciousness activities of the ancient people were relatively simple, super intelligence tended to be easily shown in some of them. Once an epoch-making scientific invention is ascertained as being real, it will become the target of people's consciousness and knowledge. When such knowledge and information become sufficiently intense in their birthplace, the holistic consciousness information can be unconsciously received by people with super intelligence. Therefore, in places other than the original cradle of the invention, the same invention can also be independently accomplished. This is the reason why some significant inventions and creations which were very difficult at birth, once they were accomplished, could be relatively easily duplicated by other people even though the original achievement was kept in extreme secrecy. It was once like this in ancient times for the ancient people. It is still like this for the people today.

People already have a relatively deep understanding of the significance of expressing thoughts, transmitting consciousness and ensuring the integrative connection of human society with language. However, they still know nothing about the function and significance of super intelligence. The development of *qigong* science will, however, gradually unfold super intelligence. With super intelligence becoming widely grasped and extensively applied by people, the holistic connection between people will be forwarded to a new stage.

Section IV The Concept of Human Whole Entity

Humans are part of Nature. Each human being himself or herself is also an independent individual entity. In each individuality, there are many constituent parts, each of which can embody the holistic feature, while the holistic human feature is maintained through the interaction with the *hunyuanqi* in the outside world. All these belong to the concept of human whole entity.

I. A Human Being is a Whole Entity of the Trinity of *Jing*, *Qi* and *Shen*

According to the concept of *hunyuan* whole entity, the human body is a whole entity with *jing* (form), *qi* and *shen* (consciousness) in a trinity. In this whole entity, *jing* is the concrete material state condensed into shape with human body *hunyuanqi*. It is where the life activity of the human body is attached. *Qi* is the intangible material state of human body *hunyuanqi* and is the manifestation of holistic life activity. It functions as what enriches and nourishes the human body and human spirit. *Shen* is the motion in the special manifestation form of human body *hunyuanqi*—*Yi Yuan Ti*. It is the commander of the human body's life activities. Widely different in their forms of expression though the three of them are, they are but the different existential forms of human body *hunyuanqi* as well as the functions displayed. They depend upon each other, transform into each other and help form an organic whole entity.

We can regard an integrate human being as an integrate system, in which there are the order parameters of three interrelated sub-systems. They interact with each other and form through cooperation and competition a unified order parameter. In different conditions, the unified order parameter formed varies. Does this mean that the unity of the human body can have different unified forms? The answer is yes. In the *hunyuan* whole entity of the human body, there are not only such differences as *shen* (spirit) and *qi* unified in *xing* (form), *shen* and *xing* unified in *qi* and *xing* and *qi* unified in *shen* , but also the differences in familiar intelligence level, *qigong* practice level and super intelligence level. They will be illustrated respectively as follows:

i. At the Level of People with Familiar Intelligence
i) *Shen* and *Qi* Unified in *Xing* (Spirit and *Qi* Unified in Form)

In the life activity process at the level of normal-state intelligence people, spirit and *qi* usually passively follow *xing* and this is a low-level unity. No matter what life activity a human being is carrying out, attention (spirit and *qi*) of this person should be concentrated on the related motional process. The formation of motion or action is the exertion of the commanding function of spirit and mentality on the basis of instinct. The concentrated manifestation of the entire process, however, is that one's spirit and *qi* are adherent to one's body. And this concentration process is not to consciously concentrate one's mind, but a process being naturally carried out.

Now let me analyze it with the example of picking up a cup with one's hand to drink water and readers will understand. The motivational instruction of picking up a cup is because of the need to drink water. The process of picking up the cup is to command the action of the hand with the mental activity of "take". Therefore, there is also *qi* assembled to the position in the human body where the action takes place. Yet this process of concentrating spirit and *qi* is carried out automatically. It is not that one consciously guides one's spirit and *qi* to concentrate on one's hand. All the activities of familiar intelligence people are carried out basically like this.

ii) *Shen* and *Xing* Unified in *Qi* (Spirit and Form Unified in *Qi*)

In familiar intelligence people's familiar intelligence state, it is hard to unify *shen* and *xing* into *qi*. Since *shen* and *qi* can influence each other (i.e., "*qi* unified mobilizes emotion" as the ancient Chinese said), *qi*, when dramatic change has taken place in it, can make spirit and form passively mobilized. In traditional Chinese medicine, "blood and *qi* that adversely go upward make one become easily angered"(*Si Shi Ci Ni Cong Lun* in *Su Wen*) and "anger harms the liver"; when essence *qi* is "concentrated in the lungs, one easily gets sad"(*Xuan Ming Wu Qi Pian* in *Su Wen*) and "sadness harms the lungs"—this kind of state is in fact that changes in *qi* influence emotions, concentrate on the human body (actually the internal organs) and cause abnormal change.

Why can't a familiar intelligence person in the familiar intelligence state achieve body and spirit unifying in *qi* but can only show this in an abnormal state? The crux lies in that people still do not understand the reasons for this and they have not learned to use this function. However, this function of change does exist in the human body. When abnormal stimulation has exceeded a certain limit, this kind of change unconsciously takes place, thus displays the abnormal phenomenon.

iii) *Xing* and *Qi* Unified in *Shen* (Form and *Qi* Unified in Spirit)

In familiar intelligence people's familiar intelligence state, it is even harder to unify form and *qi* into spirit. But when the human spirit is in a special state—being highly concentrated in the mind in a pathological state, e.g., in the state of mental disorder, the phenomenon of form and *qi* following *shen* is shown. The super abilities of running and jumping and moving things with extraordinary strength and so on in a mental patient belong in the unification of form and *qi* in *shen* in a morbid state.

ii. At the Level of *Qigong* Practice

The unification of form, *qi* and *shen* can be achieved through *qigong* practice. Although far more miraculous abilities can be displayed at this level than familiar intelligence people's functional abilities, the innate characteristics in *shen*, *qi* and form at this level have not surpassed the frame of familiar intelligence people.

i) *Shen* and *Qi* Unified in Form
Shen (spirit) and *qi* are concentrated on *xing* (form) in practice and various applications of form (human body) are emphasized so that various functions of the human body are enhanced. The popular exercise of body building is a practice method of unifying *shen* and *qi* into form. It is simply that the practitioners have not realized the mechanism in it. If they can practice with the mechanism of *qigong*, they will achieve more progress with the same or less effort. The elementary level and intermediate level in martial arts exercises and in hard-form *qigong* practice are mostly unifying *shen* and *qi* in form so as to make the functions of the human body surpass those of the familiar intelligence people. For example, holding up two tons of weight, allowing an automobile to run over one's body, bumping and breaking the stone tablet with the head, piecing the throat with a spear, etc., all result from the unification of *shen* and *qi* into form.

ii) *Shen* and Form Unified in *Qi*
Spontaneous action *qigong* that appears in the process of *qigong* practice belongs in unifying spirit and form into *qi*. When the *qi* in the human body is generated and mobilized through a certain *qigong* practice method, an internal *qi* current is formed, flowing and rushing along the locomotorium (Notice that it is not the *qi* in *jing mai* or the *qi* channel, but the *hunyuanqi* in the membranes). At this moment, *shen* is combined and moves with *qi*, which causes the motion and action in the human body. This kind of motion in the human body is formed mainly because of the motion in *qi*. It is the result of *shen* and form moving with (instead of disobeying) *qi*.

When practicing the *qigong* that can bring about spontaneous actions, practitioners must keep their spirit in a very bright state during the whole practicing process. They should both keep the innate commanding function of their spirit and follow the motional mechanism of *qi*, to be in the right state of neither leading nor lagging the *qi*. Otherwise, if the spirit becomes the slave of *qi*, or if the practitioners have illusions about *qi* and unduly guide *qi*'s flow, they will "lose fire and enter the evil state" (lead *qi* into the wrong direction and result in mental illness) and will be in incessant big actions beyond the practitioners' control. This is somewhat similar to the behavior of an average mental patient whose form and *qi* are unified in *shen*. But the nature is not the same in the two cases.

iii) Form and *Qi* Unified in Spirit

The unification of form (human body) and *qi* into *shen* (spirit) can be reached only when the practitioners have reached a rather high level in *qigong* practice. In this high level state, form and *qi* are completely under the command of *shen* and such a state is reached—when *shen* starts to move, form and *qi* move with it and when *shen* calms down, form and *qi* accordingly calm down with it. This is hard to achieve without arduous practice. Stopping one's breath and heartbeat generated by the super tranquil state achieved in one's spirit in quietude *qigong* practice (i.e., the realm of the stop of mental activity, breath and obvious pulsebeat mentioned by the ancient Chinese) and jumping high up in the sky in super martial arts, for example, all belong to this kind of state.

iii. At the Level of Super Intelligence

This is achieved when an advanced level is reached in *qigong* practice, when the practitioners have become thoroughly different than the state of familiar intelligence poeple in form, *qi* and spirit. When such an advanced level is reached, form, *qi* and *shen* cannot be strictly differentiated and the true unification of form, *qi* and *shen* is realized. If a differentiation has to be made according to the categories of form, *qi* and *shen*, we can only give a simple description as follows: *Shen* and *qi* unify in form and the form shows itself as an immortal body; *shen* and form unify in *qi* and the *qi* freely flows without trail; form and *qi* unify in *shen* and one is free to show one's image or not at will. Of course, in the unification at the super intelligence level, there is also a changing process from low level to advanced level.

To sum up, the concept of human whole entity in the theory of *hunyuan* whole entity includes both the whole entity of familiar intelligence and the whole entity having entered a super intelligence state, and the emphasis is on first realizing through *qigong* practice the whole entity at the level of *qigong* practice in order to enhance normal functional abilities and then gradually enter the whole entity realm at the super intelligence level.

The concept of whole entity concerns the entire *qigong* practice process in *Zhineng Qigong*. If the life activity and thinking activity at all levels in the human body can be understood more deeply, it will be easier to exert the subjective dynamic function of consciousness during *qigong* practice and the relationship between different levels of life activity and thinking activity with *Yi Yuan Ti* will be enhanced. In order to make it easier for beginners to understand the human whole entity, "Diagram of Human Holistic Function and Structure" (Fig. 4) (See page 113) is provided as fol-lows which is illustrated with the aid of modern medical scientific knowledge.

There are altogether two parts and seven levels in this diagram. The two parts are the Mind Oneness Entity (*Yi Yuan Ti*) and the life entity, the two of which overlap in distribution. Human holistic functional activities are carried out at seven levels: The level of *Yi Yuan Ti* which is where human nature exists. *Yi Yuan Ti* includes the domain of thinking which is engaged in thinking (omitted in this diagram) and the domain of life which is engaged in the commanding of life (essentially shown in this diagram). *Yi Yuan Ti* cooperates with the brain and sends either excitation or inhibition instruction (Level two) and carries out the instruction through the motor nerves and sensory nerves (Level Three), while in nerves there are the difference between autonomic (involuntary) nerves and somatic (voluntary) nerves (Level Four). Nerves carry out their activities at the level of tissues that connect muscles or the terminal organs. These activities inevitably cause changes in matter and energy in tissues (Level Five), and the change in matter and the change in energy always accompany each other, which is shown in the synthesis and decomposition of matter and the emission and storage of energy in cells (Level Six). The nature of all the energy change and matter change in cells is still not fully understood in medical science. Yet medical science has recognized part of the content of synthesis and decomposition in matter as well as part of the phenomenon of transmission, transition and storage of energy. They are basically carried out at the level of biomacromolecules (Level Seven). These seven levels should be all included in the domain of *Yi Yuan Ti*. As activities at the level of biomacromolecules can go on naturally (e.g., isolated cells can continue their metabolic activity if put in appropriate nutritious liquid), in the state of familiar intelligence people, *Yi Yuan Ti* still has not truly gained control over this level. Therefore, in this diagram only half of the activity at this level is enclosed in *Yi Yuan Ti*. The life entity includes all the content below the level of the brain and it overlaps with the life realm of *Yi Yuan Ti*.

In the holistic activities of a familiar intelligence person, only voluntary motion follows this person's active command. Aside from this, all the other motions are carried out naturally. Therefore, human holistic life activities are still at a low level. *Qigong* practice aims at improving the degree of controllability of holistic human life activities and implementing voluntary instructions into each and every part and level of the human body. This is an enormous systematic project which cannot be realized just in one step. If we can plan rationally, we can attain more effect with the same or less effort. Now let's standardize the procedures of realizing the human holistic project as follows:

i) To begin with Level Four and practice the controllability of voluntary motions, namely, to

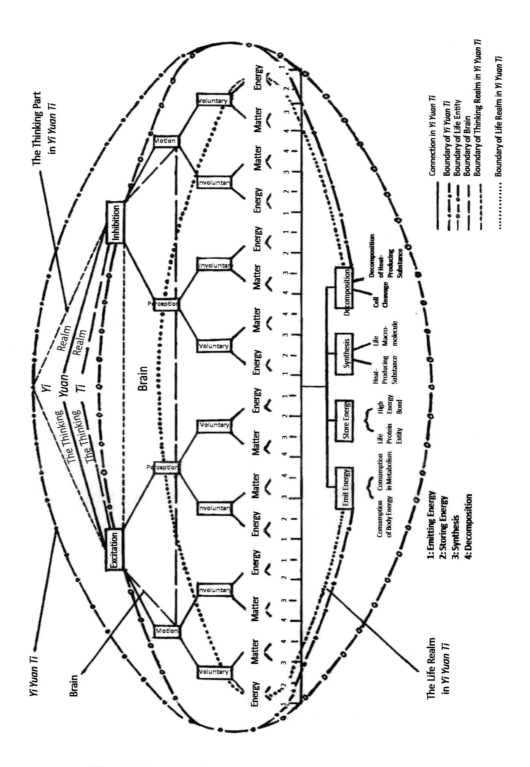

Fig. 4 Diagram of Human Holistic Function and Structure

accurately realize and exhibit the motion instructions of excitation and inhibition (*Xing Shen Zhuang* or Body Mind Form is used to practice this).

ii) To practice the controllability of perception. By concentrating one's attention, one chooses the target of perception, rids oneself of distraction and improves perceptive sensibility (*Peng Qi Guan Ding Fa* or Holding *Qi* up and Pouring *Qi* down Method is used to practice this. The full name of *Peng Qi Guan Ding Fa* is The Method of Holding *Qi* up and Pouring *Qi* down from Top of the Head).

iii) To apply voluntary instructions to the involuntary realm, such as the internal organs and the blood vessels, and this is carried out on the basis mentioned above (*Wu Yuan Zhuang* or Five in Oneness Form helps to practice this).

iv) To apply voluntary instructions to the change in energy and matter at the fifth and sixth levels. Of course, none of the activities mentioned above are carried out in isolation.

II. Relationship between Part of the Human Body and the Whole Entity

In the whole entity of humanity, there exist many constituent parts and levels. From the perspective of histology in Western medicine, there are the levels of cells, tissues, organs and systems. From the perspective of traditional Chinese medicine, there are the internal organs inside the human body with the four limbs, hundreds of bones as well as the seven apertures (eyes, ears, nostrils and mouth) and *jingluo* (*qi* channels) outside them. From the perspective of traditional *qigong*, the human whole entity is essentially *jing* (body, form), *qi* and *shen* (spirit). Although Western medicine, traditional Chinese medicine and traditional *qigong* emphasize different points, they all believe that this whole entity is a unity under the command of human spirit.

According to the theory of *hunyuan* whole entity, the whole entity of the human being is a unified entity constituted of each part of the human body. Human body *hunyuanqi* is the specific expression of this whole entity. The concrete substance of the human body is the condensed exposition of human body *hunyuanqi*, with the formless *hunyuanqi* filling the inside and outside of the human body, which makes every part of the human body contain human holistic characteristics.

Is this holistic feature in contradiction to the partial features of each part of the human body? The answer is no. Firstly, human holistic feature contains the features of each part of the human body. Without the features of each part of the human body, there would be no human holistic feature. Secondly, none of each part of the human body is an absolutely limited part independent of the whole entity, but rather a limited part that is premised on the whole entity feature. This situation can be explained from the level of cells.

We know that genes contained in the DNA of every cell are the same. It is simply that the gene segments switched on in the cells in different tissues are different. That is to say, every cell contains all the information about the life of the human body. It is simply that some information segments have experienced a *qi* mergence-and-change effect in the correspondent *hunyuanqi* in the cytoplasm and acquire a particular vital force, thus displaying different features peculiar to these cells. In every cell, there are relevant gene segments that have not been switched on. The same kinds of cells combine and form tissues, and tissues combine according to certain laws to form organs.

From this we can know that every part (including tissues and organs) inevitably contains the information about the whole entity. This is one of the features of *hunyuan* whole entity, and the human body is no exception. It is indeed like this in reality. Every part of the human body displays the human body in miniature. We have given some examples from traditional Chinese medicine in the classic concept of whole entity in Chapter I. Now let's give some other examples to deepen our impression.

i. The Nose

There are corresponding points representing each part of the whole entity of the human body both on and near the nose. These corresponding points are distributed on the five lines that are on and near the nose (with two lines on each side of the nose and one line on the nose). The line in the middle on the nose (which begins from the middle of the forehead and ends at the acupoint of *renzhong* under the nose) is related to the *zang* internal organs stated in traditional Chinese medicine (there are altogether nine points). The first line close to it (which is closely near the bridge of the nose and extends to the lower part of the side of the nosewing) is about the *fu* internal organs stated in traditional Chinese medicine and there are altogether five points. The second line near this line (which begins from the inner side of the eyebrow and reaches near the side of the nosewing along the outside of the first line) has nine points and they are the corresponding points of the trunk of the body. This is shown in Fig. 5. (See page 116)

ii. The Ears

The ear is equivalent to a human fetus sitting on its head. The crus of helix extending to the cochlea is equivalent to the diaphragm, with its end equivalent to the stomach. Those that are below the "diaphragm" are the heart, lungs, esophagus and trachea; those above it are the small intestine, large intestine and bladder. Along the edge of the cochlea are the spleen, liver, pancreas, gallbladder and kidneys. On the outer side of the helix are the upper limbs; on the inner side of it are the lower limbs. The earlobe is the head. For details, please refer to Fig. 6. (See page 117)

iii. The Feet

The soles of the feet reflect the entire trunk (including the head and brain) as well as the in-

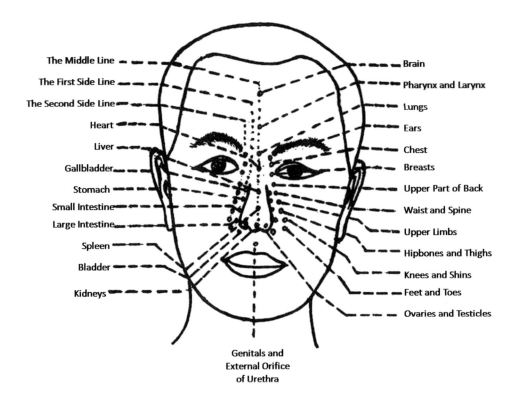

The Middle Line — — — — — Brain
The First Side Line — — — Pharynx and Larynx
The Second Side Line — — — Lungs
Heart — — — Ears
Liver — — — Chest
Gallbladder — — Breasts
Stomach — — — Upper Part of Back
Small Intestine — — Waist and Spine
Large Intestine — — Upper Limbs
Spleen — — — Hipbones and Thighs
Bladder — — — Knees and Shins
Kidneys — — — Feet and Toes
— — — Ovaries and Testicles

Genitals and
External Orifice
of Urethra

Fig. 5 Corresponding Points of Each Part of the Human Body
Whole Entity Reflected on and near the Nose

ternal organs in the body cavity. The left foot and the right foot respectively represent the left half and the right half of the human body. Please refer to Fig. 7. (See page 118)

iv. The Hands

The fingers on the side of the palm reflect the five *zang* internal organs and the six *fu* internal hollow organs stated in traditional Chinese medicine while the back of the hands reflect the limbs, waist and back. (Fig. 8) (See page 119)

On the back of the fingers of each of the two hands, the side that faces the radius on the thumb side reflects the upper limbs, with the bottom joints reflecting the shoulders, the middle joints reflecting the elbows, and the top joints reflecting the wrists. The side that faces the ulna on the little finger side reflects the lower limbs, with the bottom joints reflecting the hipbones, the middle joints reflecting the knees, and the top joints reflecting the ankles. The foot meridians sharing the same names of the hand meridians are in the correspondent positions in the upper and lower limbs where they flow past. The back of the hands represents the waist and back.

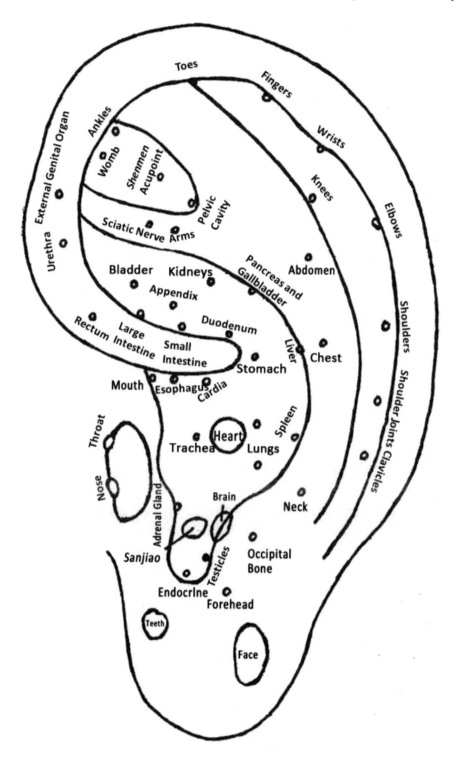

Fig. 6 Corresponding Points of Each Part of the Human Body
Whole Entity Reflected on the Ear

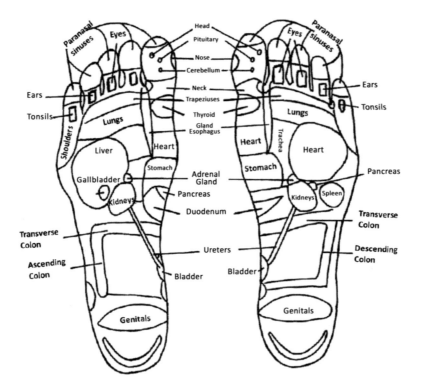

Fig. 7 Corresponding Points of Each Part of the Human Body
Whole Entity Reflected on Soles of Feet

v. The Holistic Feature in the Microcosmic Field of the Human Body

Chromosomes in each cell contain all the information about human holistic life activity. Although cells in different parts of the human body widely differ, the genetic substance in the cell nucleus is basically the same. The reason for the difference in form and function in cells lies in the difference in the cytoplasm in every cell. This has been proven in experiments on lower animals. The example mentioned previously that the cell nucleus from the clawed frog's small intestine is transplanted into its ovum with its nucleus taken out and a tadpole grows from it is a sound proof.

Upon taking a comprehensive look at the illustrations above about every part of the human body reflecting the human whole entity, we can know that the whole entity of the human body is rather exquisite and exploring this whole entity feature from the perspective of *qigong* is rather significant. This is because the purpose of *qigong* practice lies in adjusting and improving the holistic functions of the human body, and this can begin with either the whole human body or part

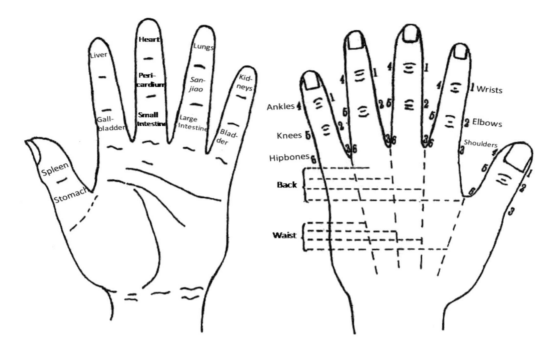

Fig. 8 Corresponding Points of Each Part of the Human Body
Whole Entity Reflected on Fingers

of the human body. Healing diseases with *qigong* is also like this since in every part of the human body is included the content of the human body's whole entity. If we can more accurately grasp more of these laws of the human body whole entity, our freedom in using *qigong* will be undoubtedly increased.

III. Formation of the Human Whole Entity and Maintenance of the Balance of the Human Whole Entity
i. Formation of the Complicated and Complete Whole Entity of Humanity

From the perspective of the evolutionary history of the human race, the whole entity of humanity evolved from single celled organisms. We know that in unicellular organisms there exists not only metabolism and genetic substance but also cell stress response and myosinogen in the cell to accomplish necessary motional process. Simple as the time-and-space structure of the single-celled organism is, it already possesses the characteristics of the basic elements of complicated organisms.

In the *hunhua* process with external *hunyuanqi*, the *hunyuanqi* of the monoplast changes its time-and-space structure once it is affected by the *hunyuanqi* outside it, which makes the *hunyuanqi* of the cell further integrate and transform and become gradually more complicated. When this changing process becomes internalized and deposited in the genetic substance, it

119

gradually makes the monoplast evolve; the various functions of the cell itself are also gradually differentiated accordingly and become fixed by corresponding form and structure.

Since any *hunhua* process results from the mergence and change of the two parts of internal and external *hunyuanqi*, the result of *hunhua* will inevitably affect the two systems involved in the mergence and change, making them become two systems both opposing and complementary to a certain extent. This is the reason why substances at every level are both antagonistic and unified. It is true that this evolutionary process is rather long and many details and links in it are still unknown. Yet from the similarities between existing species, we can analyze and know that this evolutionary process does exist.

As it has been previously talked about in the biological evolutionary chain, since the complicated whole entity of the human body gradually differentiated from one single cell, the human body retains the original common characteristics of the cell in each part of it on the one hand; on the other hand, it also stores the various varied features in the genetic substance that have been acquired in the evolutionary process. And the genetic substance in every cell of the human being is marked with the process and content of the evolution of this time-and-space structure which will be continuously passed on to the next generation through genetic inheritance.

From the perspective of the evolutionary history of human individuality, the entire life of a human being originates from a zygote. Although a zygote is also a monoplast, it already differs in essence from the monoplast in the evolutionary history of the human race. This is because the genetic substance in a zygote already contains the time-and-space structure of the entire features of a human being and the development of the entire zygote is carried out under the directed guidance of this integrate time-and-space structure. Therefore, a zygote can accomplish the growth from a monoplast to a complicated human body in a short period of a few months, as forms a striking contrast to the lengthy evolutionary process from a unicellular organism to a human being.

As to how the zygote develops into the whole entity of a human being, deep understanding about its mechanism is still lacking in modern science. According to the concept of *hunyuan* whole entity, the *hunyuan* entity of the zygote's cell nucleus keeps merging and changing in the surrounding of the *hunyuanqi* of the original ovum's cytoclasm. Each step of mergence and change will make the zygote's time-and-space structure itself display its due functions according to the natural sequence.

When the zygote divides into blastomeres, when it divides from two cells, four cells to eight cells, the environment of each blastomere begins to differ. What is around each cell is not all the *hunyuanqi* of the original ovum's cytoplasm, but the contact between the cells of the embryo, namely, the *hunyuanqi* of the cells of the embryo becomes the mutually enhancing environment for each other. This change is the significant condition for different cells of the embryo to develop in different directions. Those sharing the same condition will develop into the same cell group. During this process, genes are certainly the critical factor, but, if there were no difference in the environment, it would be difficult for the zygote to display its multiplicity in the direction of development.

At the beginning, e.g., when it is the tetrad, the condition for each cell is almost the same, therefore, each cell retains almost all of their original information. Yet, difference occurs with further division, as has been proven in the experiment on the development of animals' embryos (A lower animal can still grow from a cell in the diploid or tetrad with artificial separation, but when it is the octaploid, to grow a lower animal from it is out of the question). With the development of the embryo, the division of cells gradually becomes directed and fixed. Then different tissues and organs are formed until the entire human being is grown.

Although human structures and functions of different tissues and organs are all different, they all divide and develop from one zygote, so every cell retains the most original information in common. This is the foundation that constitutes the *hunyuan* entity of the human body, and the genetic substance of every cell of a human being contains the entire human holistic feature. The different *hunyuan* motions going on in different cells are the very display of the complexity of the human body's whole entity, whereas the special feature of part of the human body is just decided by the holistic feature. If the function of any organ or tissue is lacking, the holistic feature of a human being will be affected. People view part of the human body in an isolated manner, and this is in fact because not enough is understood about the human holistic nature.

ii. How Humans Maintain the Balance of the Human Whole Entity

When humans have separated themselves from the world of animals, and particularly when they have evolved into civilized humans, the whole entity feature of humans reaches a relatively stable stage as far as overall human beings are concerned, namely, the functions and shapes of different tissues and organs in humans are in the stage of relative stability.

However, for an individual human being, human holistic life activities are carrying on the motion of *hunhua* all the time and are experiencing the changing process of birth, growth, decadence and death. How does a human being maintain his or her own holistic balance in the limited process of life and avoid suffering from an early death then? As it has been previously talked about, human *hunyuanqi*, the fundamental substance that marks human holistic feature, manifests in such different parts as essence, *qi* and spirit in the overall life activities of a human being. Therefore, maintaining human holistic balance includes accordingly such different content as adjusting the balance of essence, *qi* and *shen*. People in their normal life are often abundant or insufficient in *xing* (essence), *qi* and *shen* (spirit). In order to keep their holistic balance, they must make correspondent adjustments.

The adjustment of their own holistic balance in familiar intelligence people is still in the natural stage, most of which consists of maintaining the existing balance instead of improving the level of human holistic balance to a higher level. They are even far from exerting their commanding power of consciousness over their life activities, to say nothing of making themselves enter the realm of free humans who know their true self. Let me try to analyze this.

i)   The Adjustment of Spirit

Familiar intelligence people usually entrust their spirit to things in the external world in order to attain peace in their hearts. This is, in fact, a covert act of worshiping gods. However, it is something inevitable when human beings' material life has not reached a certain level yet. Living in such an environment, people are easy to be affected by various stimuli and become imbalanced and troubled in spirit.

Such emotional responses in humans are actually human instincts reminding themselves that they should pay attention to the protection of themselves and adjust their own balance, that they should nourish the brightness of their own spirit and reduce their contact with the outside world in order to decrease stimulation so that they can improve their own holistic qualities.

However, average people cannot do this. When they feel troubled, they try their best to find ways to give vent to their grievances. This seems to give their spirit a kind of comfort, but it actually reduces their power of self-control in spirit and meanwhile makes them suffer a certain loss in inner *qi*. A general balance seems to be achieved, but this is a low-level balance.

When something they feel happy with has happened, their internal *qi* in their body has actually been generated to a certain extent and the degree of excitement in their spirit has been improved. At this very moment, average people do not know that they should "hide" (preserve and store) their *shen* and *qi*. Instead, they consume and drain their *shen* and *qi* through expressing their feelings or showing off. This brings their internal *qi* back to the low-level holistic balance in the familiar intelligence state.

From this it can be seen that emotional release therapy currently advocated is not desirable at the high level of *qigong*.

ii)   The Adjustment of *Qi* and *Jing* (Essence)

Normal-state intelligence people only know that they should take in more nutrients when they become short of *qi*. They do not know that they should practice *qi* and nourish their spirit to replenish themselves. When they become sufficient in their internal *qi*, they do not know that they should adjust their *qi* to nurture their spirit and apply more energy to their career. Instead, they tend to show off to others what they have done or achieved, become arrogant and domineering and waste their *shen* and *qi* in vain.

Abundance in *qi* will usually generate *jing* (essence for life) and make *jing* sufficient in the human body. Sufficiency of *jing* can easily lead to sexual impulse and increase all kinds of excitement degree in the human body. In a condition like this, the consciousness of average people is usually affected by sexual excitement and loses its ability of self-mastery. Instead of transforming and elevating this energy and making it serve their own health, they tend to give vent to the *jing* and *qi* and seek sexual pleasure. They unfortunately do not know that this momentary pleasure has deprived them of the *jing* and *qi* having been accumulated in their body and reduces them to a low-level balance. This is just like exhausting the wealth accumulated and returning to poverty.

The pity is particularly in the loss of *jing* (essence) of reproduction which differs from common cells and proteins as the genetic information in the reproductive cells (sperm and ova) is all-dimensional and the life function endowed on them by the human body is also all-dimensional. So its exhaustion of the human holistic function is high-powered and frequent loss of reproductive essence affects human health.

If the dominating power of consciousness is reinforced when *jing* and *qi* become plenty and the *hunyuanqi* used to generate the productive cells is conveyed to every part of the human body instead, it will enhance the life function of tissues in every part of the human body and will strengthen the genetic information in every cell, and the *hunyuanqi* in the human body will be under greater command of consciousness. The science of *qigong* is orientated to the enhancement of *jing*, *qi* and *shen* and strives to improve human life to high-level balance. This contrasts average people who give vent to or exhaust what they obtain to keep a low level balance.

Chapter IV Human *Hunyuanqi*

Human beings can be considered as the most intelligent creatures among all existences in the universe. The fundamental difference between humans and all other existences, including other forms of life, is that humans are endowed with high level motion of consciousness.

Human consciousness is formed, on the one hand, on account of human beings' natural attribute: Humans have relatively systematic and integrate physiological organs and functions, e.g., the different functions of the sensory organs of the eyes, ears, nose, mouth, tongue and body, the psychological function, the thinking function in the cerebral cortex, and the function of hormones that adjusts both the physiological activity and metabolic function. All of these constitute the most complicated and intricate functions and tissue systems of the human body. On the other hand, human consciousness is formed because of human beings' social attribute. That is to say, human life motions cannot be separated from their communicative activities with society, such as the communication with social communities, with state or government institutions, and with people of all walks of life in different realms of production and life. Hence human *hunyuanqi* has to accommodate the features of these two big aspects.

The ancient Chinese people often mentioned human true *qi* or *yuan qi* (primordial *qi*). The so-called true *qi* is as described in *Ci Jie Zhen Xie* in *Ling Shu*, "True *qi* is bestowed from heaven. It combines with the *qi* of grains to replenish the human body." It means that true *qi* is the combination and mergence of the prenatal *qi* with the postnatal *qi* of water and grains and fills the whole human body. The so-called *yuan qi* is what a person is bestowed with before birth. It develops with the nutrition received after birth. Coming from the prenatal quintessence, being generated from *mingmen*, it is hidden in *dantian* and runs all through to reach the entire human body by the channels of *sanjiao*. It is the power that makes the entire human body grow and change. This also means that *yuan qi* is formed with the mix and mergence, the combination and change of the prenatal *qi* and the postnatal nourishing *qi* of water and crops. From this it can be seen that the ancient Chinese people regarded both true *qi* and *yuan qi* as the mergence and combination of the *qi* before birth and the nourishing *qi* from water and crops after birth. In neither of the statements about true *qi* and *yuan qi* recorded in history have they mentioned humans' social attribute, namely, the features formed in humans' communication with society. Whereas it is believed in the theory of *hunyuanqi* that human beings' *hunyuanqi* that marks the holistic characteristic of humans should also include the social attribute of humans, namely, the features formed in social contact in addition to the previously mentioned *qi* before birth and the nourishing *qi* obtained after birth from water and grains. This is very important in the recognition of the nature of human *hunyuanqi*.

Moreover, according to traditional Chinese medicine, there is also *jingluo qi* (the *qi* in *qi* channels), *zangfu qi* (the *qi* in the internal organs), *ying qi* (nutritious *qi*) and *wei qi* (protective *qi*), etc., in the human body. *Qi* such as these all refer to the *qi*s in the human body that have specific functions. For example, *ying qi* is the finer and more subtle part of the *qi* that a person gets from water and grains. It runs in *jingmai* or *qi* channels and has the function of providing nutrition to the human body, hence the name nutritious *qi*. *Wei qi* is the stronger and faster *qi* from the *qi* of water and grains that runs outside *jingmai* or *qi* channels and functions as the kind of *qi* that protects, hence the name protective *qi*. In *Ling Shu* of *Huang Di Nei Jing*, "*Wei qi* is that with which the muscles are mildly warmed, the skin is nourished, the grain of the muscle is enriched and the opening and the closing action of *qi* is accomplished." These types of *qi* can be distinguished from *hunyuanqi* because they can only describe partial features of the organs in the human body, while *hunyuanqi* describes human holistic features.

Section I The *Hunyuan* Whole Entity of Humanity

Any existence in the universe is a *hunyuan* whole entity. Although the theory of *hunyuan* (mergence into oneness) and the theory of whole entity illustrate existences from two aspects, the two of them are actually hard to be separated. Therefore, when we specifically talk about the *hunyuan* whole entity of the human body, we must illustrate it with the theory of *hunyuan* and the theory of whole entity combined.

I. The *Hunyuanqi* of Humanity

The *hunyuanqi* of humanity is also called human *hunyuanqi*. It is the nature of a holistic human being, namely, the true being of *self*. Human *hunyuanqi* is a *hunyuan* whole entity formed with the mergence and change of the prenatal *qi* with the postnatal *qi* and the natural human being with the social human being. There are three forms of expressions of this *hunyuan* whole entity in the human body: The first is the concrete human body of flesh; the second is the magical mind and consciousness; and the third is the *qi* that fills and permeates the inside and outside of the human body and connects the human body and mind. They will be respectively stated as follows:

i. *Qi*

Qi is the fundamental existential form of human body *hunyuanqi*. Formless and imageless, it fills every tissue of the human body and spreads around this substantial entity that has form. It is the embodiment of the *hunyuan* whole entity of the human body.

i) Features of *Qi*
a. The feature of holography: All of a human being's life information is included in this person's human body *hunyuanqi*. It can merge and change with the genetic substance in

tissues and cells that have their own particular features and form entities having forms. It can also nourish *Yi Yuan Ti* (Mind Oneness Entity)—spiritual activities, through brain cells.

b. The feature of motion: Human body *hunyuanqi* can flow everywhere throughout the human body. The power for the flow can be the drive or the attraction of the activities of different tissues and organs (and even cells). It can also be the drive from mental activities.

c. The feature of gathering and adhering: It refers to that formless *qi* mostly gathers and adheres to and round substantial or concrete entities.

d. The feature of condensing and dispersing: *Qi* can gather or expand (including condensing and dispersing) according to the need of the body or structure and according to the instruction of mental activities.

e. The feature of energy transformation: *Qi* has the energy different from all known forms of energy and it is tentatively called *hunyuan* (merged-into-oneness) energy, which can transform into different forms of energy under certain conditions.

ii) Distribution and Function of *Qi*

Formless human body *hunyuanqi* is divided into three aspects according to the difference in its distribution and functions.

a. Body *hunyuanqi* (*quti hunyuanqi*): The part of *hunyuanqi* distributed in tissues throughout the human body (including tissues in the body, in the internal organs and in the brain) and used for the growth, multiplication and metabolism of cells themselves and also for the consumption of energy in the human body is called body *hunyuanqi*. It centers in the lower *dantian* behind the navel in the cavity and can be mobilized to every part of the human body under mental instructions. *Qigong* practice primarily enhances the functions of this part of *hunyuanqi*.

b. Internal organ true *hunyuanqi* (*zangzhen hunyuanqi*): The *hunyuanqi* that maintains the secretion function peculiar to the internal organs in the human body is called internal organ true *hunyuanqi*. On the one hand, internal organ true *hunyuanqi* guarantees the realization of all kinds of secretion function of "changing into somethingness from nothingness" in the internal organs so that different concrete substances can be secreted. On the other hand, it has a certain impact upon human spiritual activities. It centers in *Hun Yuan Qiao* (Merged-into-Oneness 'Acupoint') in the depth of gastric cavity and cannot be affected by familiar intelligence people's mental instructions.

c. *Yi Yuan Ti* (Mind Oneness Entity): We call the *hunyuanqi* formed with the *hunyuanqi* of nerve cells (essentially cerebral nerve cells) that is highly even in texture and has such special features as reflectiveness, memory, initiative and searching for information *Yi Yuan Ti*. Consciousness activity is the content and process of the activity in *Yi Yuan Ti*, and *Yi Yuan Ti* can permeate into each and every constituent part of the human body and has a commanding power over the life activities in the entire human body.

Generally speaking, *Yi Yuan Ti* belongs in the content of *shen* (human spirit). As it is also part of *hunyuanqi*, it is introduced here as well. There is only very little difference between body *hunyuanqi* and internal organ true *qi* and no essential difference exists between them. The difference lies just in that they are in different positions of the human body, thus having different functions. For example, when body *hunyuanqi* is used to supply metabolism in cells, the process in which different kinds of enzymes are produced in tissues and cells is also a process of generating somethingness out of nothingness. It is simply that this process in which enzymes are produced does not have as much independence as the process in which different kinds of substance are secreted in the internal organs.

Various enzymes produced in the cells in the tissues of the human body are used to satisfy the need in the tissues themselves in the metabolic process. Different kinds of substance secreted in the tissues and cells in the internal organs are not to directly satisfy the need of metabolism in the internal organs themselves but to satisfy the need of the life process in the whole human body. The *qi* that satisfies the metabolism of cells themselves in the internal organs also belongs in body *hunyuanqi*. It is hard to differentiate this *qi* from the *qi* used in supplying the secretion process.

As to the point that body *hunyuanqi* can be controlled by mental instructions while internal organ true *qi* cannot be controlled by mental instructions—this is also not absolute. Through *qigong* practice, the *hunyuanqi* of the internal organs can also follow the instruction of consciousness.

From this it can be more easily seen that there is no practical difference between body *hunyuanqi* and internal organ true *qi* and they are both part of human body *hunyuanqi*. It is just for the convenience of explanation and for the convenience of giving guidance to *qigong* practice that we give them different names according to their difference in distribution and function and explain them respectively.

ii. Form (Essence)

Form or the human body is the expression of human body *hunyuanqi* condensed into concrete substance. It is the manifestation of the time-and-space structure and function contained in human body *hunyuanqi* that mark the essential human holistic nature having been deposited and fixed by condensing into concrete substance. Once formless *hunyuanqi* gathers and is condensed into an existence with form, formless *hunyuanqi* will be "compressed" into the substantial body. Although there is still the *hunyuanqi* field of the concrete substance around, the density and scope of the *hunyuanqi* field are restricted by this concrete substance. It is already greatly different from the "free" state when it has not become condensed.

Since the human body is gathered into form with human body *hunyuanqi*, in the form of the human body there also exists the *quan xi* (See note on page 42) feature. This has already been introduced in the section on the concept of the human body whole entity in the previous chapter, hence it will not be repeated here. We will merely give a brief introduction here to the *quan xi* feature of essence (form).

Although essence stated in *qigong* theory includes the genetic substance in common cell nucleus, what will be mainly talked about here is the special essence of male and female reproductive cells.

It is well known that a human being develops from a zygote and the zygote is formed with a sperm and an ovum combined. The combination of a sperm and an ovum can grow into a human being. This shows that in the sperm and ovum all the information to develop into a human being is contained. But it is not that one sperm or one ovum contains all this necessary information. So do the sperm and ovum each contain half of the information? It has been scientifically proven that the sperm and ovum have 23 chromosomes respectively which form 23 pairs of chromosomes and also an intact cell when they become combined. Nevertheless, we still consider it better to say that the two of them complement each other, namely, it is only when the two of them combine that a complementary *hunyuan* whole entity can be formed.

For example, it has been proven in bio-genetics that difference in the gender of a zygote is determined by whether the gender gene (XY) in the chromosomes of the sperm is X or Y, but not by the gene (XX) in the chromosomes of the ovum. This shows that the gender gene in the X chromosome in an ovum and that in the X chromosome of a sperm are not the same; otherwise, two ova could also combine and form a new life. In this sense, it seems that the nature of the sperm takes a leading position in a zygote. This can also show to a certain extent that the chromosomes of the sperm contain the decisive information about the new life.

On the other hand, the ovum also plays a leading role in the development of a zygote. For example, when meiosis has not happened in the sperm, there are also 23 pairs of chromosomes, with X and Y chromosomes in the gender gene. Why can't a sperm develop into a new life by itself, but can develop into a new individuality only after it has split and becomes combined with an ovum? This shows that the 23 chromosomes in the ovum are absolutely indispensable for developing into a new life. Moreover, the cytoplasm of the ovum is also indispensable for generation of the new life. For example, in the experiment previously mentioned, the cell nucleus of the clawed frog's small intestine is transplanted into its ovum (note: into the cytoplasm of the ovum with its cell nucleus removed) and a tadpole can be produced. This has sufficiently proven that decisive factor for the growth and development of the new life exists in the cytoplasm of the ovum.

As to whether the factor of the sperm or the factor of the ovum is more important for the life of a newborn individuality, we think that it cannot be divided like this. They are both the complementary content of one *hunyuan* whole entity because both of them have the holistic information about human life in it in a certain sense and yet both the complementary halves of the holistic information. It is only the combination of them both that can form the holistic information structure of a *hunyuan* whole entity.

Nevertheless, we can still be sure that the roles that a sperm and an ovum play in the formation and development process of a newborn individuality are not the same. They are a complementary antagonistic unity of a whole entity. Both of them have the holistic information feature of life. Mature sperms and ova are the concentrated expression of the functional differences in the two genders of male and female and have the quintessence of male and female *hunyuanqi* deposited in them. Some readers may ask: Is there any essential difference between the *hunyuanqi* of a male and the *hunyuanqi* of a female? The answer is yes. The *hunyuanqi* of the male and that of the female seem to be the same, but there are opposing factors in their nature. The difference between the *hunyuanqi* of the male and the *hunyuanqi* of the female can be tested with certain means which will not be further talked about here.

It is because sperm and ova are both the quintessence of human *hunyuanqi* that they have extremely great impact on human health. When the essence of reproductive cells in the human body becomes abundant, on the one hand, the strength of the *hunyuanqi* field around the human body increases; on the other hand, the essence of reproductive cells can be turned into *qi* highly characteristic of the holistic time-and-space information to enhance the *quan xi* feature of human *hunyuanqi* and make one's vitality prosper accordingly. It is said in traditional Chinese medicine that the lower *dantian* is the source to generate *qi* and in *qigong* it is said that between the two kidneys is the *Hunyuan Shenshi* (Mergence-into-Oneness Miracle Chamber), and the reason just lies in this.

If one does not know the importance of maintaining the sufficient level of this essence and frequently makes it decrease, not only will the strength of this person's holistic information *hunyuanqi* field weaken, but a lot of human body *hunyuanqi* will swiftly gather into reproductive essence that has form, which decreases the nourishment of holistic information *hunyuanqi* in each part of the body and thereby weakens the human body's life functions. If the *quan xi hunyuanqi* field strengthened with abundant reproductive essence can be used to nurture the entire human body, the *quan xi* ability of the life of the human body will be enhanced.

iii. *Shen* (Spirit)

Shen is the most advanced function of human beings. It is the commander of human life activities and the mode and content of the activity in *Yi Yuan Ti*. *Yi Yuan Ti* is formed with the *hunyuanqi* of the nerve cells (particularly brain cells) which have special functions and structures. That is to say, *shen* is the special form of existence of *hunyuanqi*, namely, *hunyuanqi*'s motional mode and content. As for this, there are also similar statements in traditional *qigong*. Huang Yuanji once said,

"We should particularly know that *yuan shen* (primal spirit) has no trail and the most intelligent in *yuan qi* (primal *qi*) is *yuan shen*. ... From this we can deduce that in all the seeing, hearing, speaking activities and actions, in whatever we do in our daily life, nothing is without the function of *yuan shen*. Yet the conscious mental activity belongs to *shi shen* (perceptive spirit), while the sub-conscious mental activity belongs to *yuan shen*."(*Le Yu Tang Yu Lu*)

It is also said in *Bao Hun Yuan Xian Shu*, "In the entity of *hunyuan*, that which is pure with nothing else in it is *jing* (essence), that which makes *qi* and blood flow smoothly in the *qi* channels and blood vessels is *qi*, and that which is void, miraculous and with initiative in motion is *shen* (human spirit)." "*Yuan shen* is the utmost miraculous and sacred and it dominates all existences."

What kind of *hunyuanqi* is *Yi Yuan Ti* then? Physiological anatomy tells us that nerve cells differ from common biological cells. In their own metabolic process, not only are there the functions to absorb and release concrete substances (e.g., hydrated ions, molecules, organic molecules) as those in common biological cells, but they have also reinforced their functions of receiving and sending information conveyed with energy. This enables nerve cells to have more extensive channels and content in their contact with the outside world. And this function continuously improves with the evolution of the nervous system.

According to the theory of *hunyuan* whole entity, any concrete object is the condensed state of its *hunyuanqi* gathered in form, around which is its expanded *hunyuanqi* as well. On account of this, when the density of nerve cells has reached a certain extent, the *hunyuanqi* around each nerve cell will permeate into each other and connect and combine into a whole entity. This whole entity is affected by changes that take place in the nerve cells and can also react on the nerve cells.

When the evolutionary chain has progressed from animals to humans, the nervous system of humans becomes highly developed and its functions become exquisitely divided. This further strengthens the functions in nerve cells of receiving and sending energy and information. The whole entity formed with the combination of the *hunyuanqi* around the multitude of nerve cells is also greatly enhanced, and great changes take place in the whole entity's functions as well. This whole entity can not only reflect existences in the external world, but can also reflect all kinds of change in itself (as the result of the multitudinous aspects in *Yi Yuan Ti* that can reflect each other). Therefore the whole entity has its relative independence. This we no longer call *hunyuanqi* but *Yi Yuan Ti* or Mind Oneness Entity.

As *Yi Yuan Ti* has the feature of having the holistic information in it, as it has such features as having memory, initiative, being motional and able to reflect, to make choices, to condense and disperse (including gather and expand) and to penetrate (details of which will be provided in Chapter V The Theory of Consciousness), it has a commanding power over all the life activities in the human body and it determines the holistic characteristic and nature of human body *hunyuanqi*.

If we say that the essence with form (which refers to the human body) has the holistic information about human life deposited in it, it is only the holistic information about all the information regarding a natural human being having been fixed with the substance with form (human body). The holistic information in *Yi Yuan Ti*, on the other hand, includes all the information about the entire life activities of a human as a natural human being and a social human being. This is essentially the storage of formless information, a great deal of which has also been condensed into the substance that has form in the brain cells.

People's consciousness activity is the process in which the special kind of information regarding concept (vocabulary) is formed and actively applied. This is the fundamental difference from the *hunyuanqi* of animal brains. Although animals also have nerve activities and they also have their brain *hunyuanqi* with which they know and identify things, their nerve activities have not formed their integrated and independent activities. Instead, their nerve activities adhere to their entire life activities, or rather, animals' nerve activities are but part of their entire life activities. Yet human beings have formed independent spiritual activities on the basis of their nerve activities, and their life activities have become the means and processes to accomplish the instructions of their spiritual activities. This makes human life activities become purposeful planned actions with initiative under the instruction of their spiritual activities. Human body *hunyuanqi* can both change according to the laws of life activities and move and change according to spiritual instructions, which is hardly achievable for average animals.

The command of *shen* or the human spirit over life activities in familiar intelligence people's familiar intelligence state is chiefly accomplished through nerve activities. The end of the neurites (neurodendrites and neuraxons) of nerve cells is distributed everywhere in the human body, and *Yi Yuan Ti* envelops the entire nervous system. Therefore, any information from *Yi Yuan Ti* can be transmitted through nerve cells or directly transmitted to every part of the human body to command their activities.

131

iv. The Relationship between *Jing* (Form), *Qi* and *Shen* (Spirit)

Jing, *qi* and *shen* are three different manifestation forms of human body *hunyuanqi*. They have the common features of holistic information and gathering and expanding and so on. They can transform into each other under certain circumstances: *Qi* can fill and nourish form (essence) and can gather into form; form and essence with shape can transform into *qi*. In the human body, this process of transformation between somethingness and nothingness is going on all the time.

Zhineng Qigong practice especially emphasizes this connecting point because, when the transformation is taking place, change happens in both the form and function of *hunyuanqi*. The *hunyuan* (mergence-into-oneness) state when somethingness and nothingness are transforming into each other is not only an even state, but an even state that closely connects the change in form and *qi*. A subtle "stir" applied to it at this very moment can evidently influence the change. How do we influence it then? Consciousness can penetrate this state and guide its change. Therefore, no matter it is in *qigong* practice or in treating disease, using positive mental instructions to guide the process of change between somethingness and nothingness will achieve desired effect with the same or less effort and will gradually attain "the grand harmonious realm in which the void and the concrete are thoroughly connected."(*Hua Shu*)

Shen (human spirit) needs not only concrete nerve cells as something to adhere to, but also *qi* for nourishment. Since body *hunyuanqi* can be affected by the relevant part of the human body, the *qi* that nourishes *shen* must be the *qi* having been exquisitely refined in the human body, the *qi* that is highly characteristic of the holistic information about human life. This *qi* is essentially generated in the process of transformation between the somethingness and nothingness states of reproductive cells. In traditional *qigong*, "collecting medicine back to the stove" at "the moment of living *zi shi*"(which refers to sexual excitement) and not letting the essence *qi* (that can form the reproductive cells) excrete are called "returning the essence to nourish the brain", and the reason lies in this.

How do children nourish their *shen* in their childhood when the sexual function in their bodies is still not mature? Although children's sex organs are not mature, in their bodies there is another kind of function that can generate their essence *qi* (Notice that the essence *qi* mentioned here is not the reproduction essence of sperm or ovum, but the related internal secretion of *yuan jing* or primal essence). That the penis of a young boy becoming erected in his sleep is the sign of the existence of another kind of function that generates essence *qi*, and this is what Lao Zi said "Knowing nothing about the union of male and female, yet his sex organ is erected." Nevertheless, the function of *shen* at the time of childhood is after all not perfect. The perfection of *shen* and the maturity of *jing* are basically synchronous.

How do *Zhineng Qigong* practitioners who do not particularly practice lower *dantian* strengthen their functions of spirit then? We bring into full play the functions of mind guiding *qi* and mind gathering *qi* (which is another function of *shen* upon *qi*). We gather the primordial *hunyuanqi* in Nature to nourish the human spirit.

As for that *qi* can change according to mental instructions, it is said in *Wen Shi Zhen Jing*, "*Qi* is generated because of the function of the heart. It is like picturing a big fire in the mind and one feels warm after some time. It is like picturing the vast body of water and one feels cold after some time." It is also said, "Wherever the heart goes, *qi* follows. ... My one heart can transform into *qi* and into form. ..." This is a magical and profound truth that has not been recognized and mastered by people yet.

Although it is a universal phenomenon in traditional *qigong* that many practitioners take their own *jing* (body, form), *qi* and *shen* (spirit) as their target of self-cultivation, they do not know the cultivation of *hunyuan* ancestral *qi*. Nevertheless, quite a few practitioners who have achieved success in their practice already realize that "To cultivate the grand *Dao*, one cannot do without this truly void primordial *qi*." "This *qi* is actually the formless primordial *qi* in the void that generates the heaven, earth, humans and all other existences."(*Le Yu Tang Yu Lu*)

It is also said in *Yu Quan*, "The method of self-cultivation is not to guard the heart but to guard *qi*. *Qi* gathers at the beginning of *hunyuan*. Its birth comes before the heaven and earth. It establishes itself before *yin* and *yang* emerge. In closing and opening, an 'acupoint' of *Xuan Guan* is hewn out of *wuji*. This *qi* corresponds to *yin* and *yang* so it is called ancestral *qi*."

In *Zhineng Qigong*, practitioners directly practice the middle *dantian* and upper *dantian* instead of the lower *dantian* and they guide their *qigong* practice on account of this. They directly nourish their body and spirit with this prenatal *hunyuanqi* of Nature.

Shen being the commander of form and *qi* will be elaborated in Chapter V The Theory of Consciousness.

In sum, *jing*, *qi* and *shen* are all part of human *hunyuanqi*. From the perspective of phylogeny, formless *hunyuanqi* generates and brings about the human body that has form and the human body that has form brings about formless *Yi Yuan Ti*—*shen*. In the realistic life of human beings, the three of them depend on each other and transform into each other to ensure the normal continuance of the life activities peculiar to humans.

Moreover, the three of them exist in a mutually permeating and accommodating manner. For example, the substantial entity of the human body that has form is actually the condensed state of formless *hunyuanqi*. Not only can a *hunyuanqi* field be formed in and around the human body, but uncondensed *hunyuanqi* can also fill it. *Yi Yuan Ti* is the more exquisite form of *hunyuanqi*. It can permeate the *hunyuanqi* of the tissues in each part of the human body through the *hunyuanqi* of nerve cells and can also directly penetrate the *hunyuanqi* of the tissues and coexist with it in a mutually accommodating manner. The concrete human body entity contains both formless *qi* and *Yi Yuan Ti*. It is because humans have not completely grasped the laws of *qi* and *Yi Yuan Ti* that they are unable to consciously use *qi* and *Yi Yuan Ti* to serve their life activities.

II. Characteristics of the Distribution of Human *Hunyuanqi*

The characteristics of the distribution of human *hunyuanqi* talked about here refer to the distribution characteristics of the formless and imageless *hunyuanqi*, the part of *hunyuanqi* in the narrow sense.

i. The Agreement Feature of Human *Hunyuanqi* with the Human Body Structure that Has Form

Since human body *hunyuanqi* contains all the information about the time-and-space structure of the human body, and since this feature becomes fixed in the zygotic stage and the development of the embryo and fetus later on is the display of the information about this fixed structure, the distribution of *hunyuanqi* in the body of the adult agrees with this structure of the human body, namely, *hunyuanqi* fills every part of the human body and spreads around it. We should know that human *hunyuanqi* is a whole entity and, though it is influenced by the human body, it also has its independence to a certain extent.

As has been previously stated, the human body is an entity gathered into form with human body *hunyuanqi* in its development process of mergence and change. So the distribution of human body *hunyuanqi* agrees with the prenatal structure of the human body. For example, the *hunyuanqi* of the body, internal organs, facial features and four limbs accord in their distribution with their prenatal structural characteristics. If a person is born with four or six fingers on one hand, the distribution of the *hunyuanqi* of this person's fingers will be accordingly abnormal and agrees with the prenatal structure or form of the fingers. If a normal person has lost a finger in daily life, this cannot alter the distribution law that human body *hunyuanqi* forms according to the prenatal structure. In the position where the finger is gone, there is still the image of the finger's *hunyuanqi* distributed there. This has already been proven by biofield photography.

ii. Human Body *Hunyuanqi* Has a *Qi*-State Structure in the Human Shape Both Inside and Outside the Human Body that Has Form

This characteristic results from the fact that human *hunyuanqi* gathers into shape to different extent at different levels.

i)   There is a not very regular *qi*-state shape of human formed with human *hunyuanqi* relatively sparsely distributed outside and around the human body, and it can also be regarded as the human body's *hunyuanqi* field:

There is a layer of *qi* which seems to be distributed along the vertical axis of the human body outside the human skin within a distance of about 3cm to 5cm (at most no more than 10cm). It directly reflects the state of the human body. Generally speaking, people of strong physique have relatively sufficient *qi* in this layer. Outside this layer of *qi*, there is a layer of sparse *qi* (formed with the first layer of *qi* spilling out). It surrounds the entire human body and is related to the sufficiency of internal *qi*. Outside this layer, there is a layer of even sparser *qi* which seems to radiate outward. Consciousness activity has an impact upon these "three layers of *qi*". After *qigong* practice, the *qi*-state human outside the human body will be strengthened.

ii) The *qi*-state shape of human formed with *hunyuanqi* and distributed inside the human body:

There is even more *hunyuanqi* that fills the derma of the skin, the subcutaneous membranes, and the relatively deep tissues of the muscles and bones in the human body. As the distribution of the *hunyuanqi* agrees with the structure of the human body, it seems that there is also a *qi*-state shape of human within the human body that has form. When a certain level is reached in their *qigong* practice, *qigong* practitioners can perceive this phenomenon. *Fa Shen* (Dharmakaya) as stated in traditional *qigong* refers essentially to the *qi*-state shape of human inside and outside the human body.

Since human body *hunyuanqi* is at the command of mental activity, the shape and size of this *qi*-state human is unconsciously under the control of consciousness activity. With changes in mental activity, a myriad of changes in the phenomena of *qi* occur. This is usually referred to as the manifestation of the celestial being, Buddha, devil or ghost in traditional *qigong*. It is in fact different levels of change in form and shape caused by mental activity in human body *hunyuanqi*.

iii. The Unevenness Feature in the Distribution of Human Body *Hunyuanqi*

The distribution of human body *hunyuanqi* in the entire human body is uneven. Generally speaking, there is more *hunyuanqi* in positions where functions are greater (which refers to greatness in strength and more frequency in use in the relevant parts of the human body). In positions where tissues are not so densely packed, it is easy for *qi* to go through and gather. These are the main reasons for the unevenness in the distribution of human body *hunyuanqi*, which is shown in:

i) There are three intense *hunyuanqi* centers in the human body. *Quti hunyuanqi* (human body *hunyuanqi*) is concentrated behind the navel area. The center of *zangzhen hunyuanqi* (internal organ true *qi*) is in *Hun Yuan Qiao*, and *Yi Yuan Ti* centers in the brain. They will be respectively explained as follows:

a. The centre of human body *hunyuanqi* is in the correspondent body cavity behind the navel. As it has been previously pointed out, the duty of human body *hunyuanqi* is to supply the multiplication of different tissues and cells in the human body and to ensure the realization of the functions peculiar to different tissues. How can body *hunyuanqi* be concentrated in the area behind the navel?

a) The essential part of the *hunyuanqi* that a human being absorbs from the outside world is from food that has form. The digestion and absorption of food are mainly carried out in the small intestine, the position of which is equivalent to the navel and the place below the navel. The process in which the *hunyuanqi* of food is decomposed and turned into human *hunyuanqi* is firstly shown in this area. This actually concerns the generation and conveyance of human body *hunyuanqi* and will be elaborated in Section IV Generation of Human Body *Hunyuanqi*.

From the perspective of prenatal development, the taking in of nutrition and the sending out of waste when a new human life is in the form of a fetus are both carried out through the umbilical cord that conveys them from and to the placenta to exchange with the body of the mother. That is to say, the *hunyuanqi* that the fetus absorbs in the process of development is conveyed from the mother's body through the umbilical cord. Therefore, the navel area becomes the gate for the gathering as well as the going out and coming in of *hunyuanqi*.

b) Distant though adults' testicles and ovaries are from the navel, the testicles and ovaries as well as kidneys all originate from the urogenital ridge from the perspective of embryology. During the time when the internal organs of the fetus are being formed, the testicles or ovaries are on the inside of the kidneys, a position equivalent to the deep space behind the navel. Then the testicles and ovaries gradually move downward. The vein in the left testicle of the male adult is still connected to the vein in his left kidney in his adulthood and this shows their close relationship.

As the distribution of human *hunyuanqi* maintains its identical feature with the prenatal structure of the human body, the *hunyuanqi* that guarantees different functional activities of the testicles and ovaries gathers in the deep space behind the navel, a position which is called by the later generations *Ming Men*, or the gate of life, that faces the *Ming Men* acupoint on one's back. When the essence of reproductive cells that carries the entire life information is about to be generated and formed, *hunyuanqi* gathers and generates in this place.

It has been pointed out in Yuan Qing Zi's *Zhi Ming Pian*, "Between the two kidneys is *Hun Yuan Shen Shi* (Merged-into-Oneness Miracle Chamber)." It has also been pointed out in *Nan Jing*, "*Yuan qi* is the source for generating *qi* in humans. It is the motional *qi* between the kidneys. It is the foundation of the five *zang* internal organs and the six *fu* internal hollow organs, the root of the twelve regular meridians, the gate of breath, … with another name called God Guarding against Evils." And "*Ming Men* is the shelter for all spiritual activities and the place where the primarily generated *qi* resides."

From these statements we can not only know the place where the *hunyuanqi* that maintains life activities generates and gathers, but can also see that it completely agrees with what is advocated in the theory of *hunyuan* in this book. It is simply that the body *hunyuanqi* we talk about that is concentrated in the navel area is not only the motional *qi* between the kidneys which belongs to prenatal *qi* but also the postnatal *qi* of water and crops. Therefore, this is the place where prenatal *qi* and postnatal *qi* generate and gather.

c) The waist (particularly the backbone near the waist) is the central place that upholds the weight of the human body and is the column that maintains the balance of the human body. It is the critical position for the human body's movement. It concerns the movement of the entire human body and needs the nourishment of more *hunyuanqi*. This is also one of the reasons why *hunyuanqi* is gathered to the navel.

b. The center of internal organ true *qi* is in *Hun Yuan Qiao*. *Zangzhen hunyuanqi* essentially refers to the *hunyuanqi* of the heart, liver, spleen (equivalent to the pancreas and digestive glands in modern medicine), lungs and kidneys. How can the *qi* of the five internal organs, each of which has its own special functional abilities, be gathered in *Hun Yuan Qiao*? On the one hand, they have a certain identical feature. They all receive substances from the entire human body. For example, the heart receives blood and lymph that flow back from the whole body and meanwhile conveys them to the whole body; the lungs receive carbon dioxide produced by each part of the body and convey oxygen to the whole body; the kidneys receive blood as well as the waste produced in the metabolic activities from the whole body and convey many kinds of hormones produced by the kidneys to the whole body; the spleen (with the function of the digestive tract included) digests water and crops and conveys their essence to the whole body.

It has been scientifically proven that every internal organ produces certain endocrine hormones to supply the entire human body, and all this needs the five internal organs to carry out effective *hunhua* (mergence and change) so as to ensure the unity of each organ with the whole entity. As for the *hunhua* among the five organs themselves that forms *zangzhen hunyuanqi*—this will be expanded in Section IV of this chapter.

On the other hand, the *hunyuanqi* of the internal organs inevitably fills and spills out around the organs. *Hun Yuan Qiao* is at the center of the five internal organs being considered as a whole and the *hunyuanqi* that spills out of the five organs can gather here. What is even more important is the up and down movement of the diaphragm that connects the *hunyuanqi* around the five internal organs into one entity and makes it gather in *Hun Yuan Qiao*. As there is already no difference between prenatal *qi* and postnatal *qi* here but only the *hunyunqi* merged together through *hunhua*, it is more favorable for the *hunhua* motion of generating somethingness out of nothingness to take place.

c. The center of *Yi Yuan Ti* is in *Shen Ji* 'acupoint' which is in the center of the head. When the two Chinese characters *shen ji* are chanted, the place in the head where one feels a kind of motion is where *Shen Ji* 'acupoint' is located. *Yi Yuan Ti* is the especially exquisite *hunyuanqi* formed with the *hunyuanqi* of the brain cells. It is distributed everywhere inside the human body as well as outside it and shares a lot of similarities with primordial *hunyuanqi*. Yet it has a central point which seems to be the pivot and commencement point of the entire *Yi Yuan Ti*. At the level of familiar intelligence people, this function in the central point of *Yi Yuan Ti* of being the pivot and commencement point is distracted by so many different kinds of sensory information (or other intense information) that its concentrated strength is diverted (Details will be found in Chapter V The Theory of Consciousness) and its due functions are not exerted.

ii) The Distribution of Body *Hunyuanqi* Can Vary according to Changes in Consciousness

When consciousness concentrates on a particular part in the human body, *hunyuanqi* will gather toward it. In the course of time, *hunyuanqi* in this place will become more plentiful. For example, human hands are what we frequently use and they have the most chances to cooperate with human consciousness, therefore the *hunyuanqi* in our hands are relatively plentiful and the effect that they are at the command of consciousness is more evident.

It has been proven in modern biofield photography that when consciousness is concentrated on the hands, the biological light on the hands is apparently increased. Human sensory organs (e.g., the eyes, ears, nose, tongue and skin) are also the place where *hunyuanqi* is highly concentrated, and the degree of concentration is directly related to which of these organs is preferentially used.

iii) Greater Quantities of Body *Hunyuanqi* (*Quti Hunyuanqi*) Can Be Found in Such Places as the Skin and Membranes (*Muoluo*)

Moluo refers to the membranes on the surface of every tissue and organ inside the human body. As there are many tiny blood vessels and relevant *qi* on these membranes, they are called *moluo* which means membranes with tiny blood vessels. Since the *hunhua* process between the internal and external *hunyuanqi*, whether it is the internal *qi* going out or the external *qi* coming in, must be carried out with the help of the skin or the mucosa, and since the *hunhua* between different tissues and organs inside the human body must be carried out with the help of *moluo*, the skin and the membranes become the significant channels for the communication of internal and external *hunyuanqi*. In addition, in people's daily life, the outward feature of mental activity can lead the flow of *qi* to make it tend to expand outward instead of guiding it to the depth of the human body. Furthermore, the swift adjustment of body *hunyuanqi* in people, e.g., to swiftly make *qi* concentrate on a particular part of the human body, is carried out by conveying *qi* from *dantian* through relevant membranes under the guidance of consciousness.

It is shown in the *jingluo* research into traditional Chinese medicine theories by modern science that the most evident place where the phenomenon of *jingluo* is shown is in the skin and the subcutaneous tissues, as agrees with what has been stated about *moluo* in the theory of *hunyuan*. However, it should be pointed out that *moluo* we talk about still includes the membranes of the bones, muscles, blood vessels, nerves, body cavity, and internal organs as well as cells that are even smaller.

In summary, the distribution of human body *hunyuanqi* is affected by the structure and function of the human body and also by mental activity. *Qigong* practice is essentially about making use of the influence of consciousness upon *qi* to change human imperfection and make people achieve more advanced levels of health.

Section II The Evolutionary History of Human *Hunyuanqi* and Its Further Development

From the perspective of the evolutionary history of the human race, human *hunyuanqi* comes into existence through gradual evolution from low level organisms, which we will not talk in detail here. From the perspective of the evolutionary history of human individuals, human *hunyuanqi* develops from the zygote's *hunyuanqi* through the stages of the *hunyuanqi* of the embryo, fetus, baby, and finally to the *hunyuanqi* of adults. Each stage has a complicated mergence-and-change process.

I. Zygotic *Hunyuanqi*

According to the theory of *hunyuan* whole entity, a zygote is the substantial-entity expression form of zygotic *hunyuanqi*, while the zygotic *hunyuanqi* is the insubstantial-entity expression form of the zygote. They are one entity in two forms of expression and two forms of expression of one entity. Yet the zygote and the zygotic *hunyuanqi* cannot be completely equated with each other as in their process of formation there exist the fact of one coming into form earlier than the other and a relationship of cause and effect.

i. The Formation of a Zygote and Its *Hunyuanqi*

The *hunyuanqi* of a zygote is formed in the course when the sperm and the ovum are combined. It belongs in the prenatal of the prenatal of human *hunyuanqi*. According to the ancient Chinese, only truly formless *qi* is the prenatal of the prenatal[1]. (See footnote on next page)

According to modern medical science, a zygote comes from the direct combination of a sperm and an ovum. The reason why a sperm and an ovum can combine is because a sperm and an ovum both have only 23 chromosomes, each of which is half the number of chromosomes in a normal cell, and the combination of the two of them forms an intact cell. This viewpoint which has been proven through the observation of the fertilization process of tube babies seems to contradict our *hunyuanqi* theory. In fact, they are not contradictory because modern science observes and states from the macro-cosmos level while we state from the micro-cosmos level.

In the point of view of ancient *qigong* and that of modern science, it is both believed that any change in substances concerns two levels of change at the macro-cosmos level and the micro-cosmos level. It is believed in the theory of *hunhua* that, in the course of a substance turning into another substance, there is always a process in which the original substance disappears, a process that something disappears into nothingness, and a process in which something new is formed out of the nothingness (i.e., somethingness is born of nothingness). The formation of zygote is no exception. Let me try to explain further.

If the combination of a sperm and an ovum is only due to the fact that they each have 23 chromosomes, each amounting to half the number of the chromosomes in a normal cell, will the combination of two sperm or two ova be able to form the 23 pairs of chromosomes of a normal cell as well? In fact, it is absolutely impossible. This shows that sperm and ova have their own respective nature which makes it impossible for a sperm to combine with a sperm and an ovum with an ovum, or rather, they repel each other, whereas a sperm and an ovum are different in nature so they attract each other and have affinity to each other. It has been discovered in experiments related to protein electrophoresis that sperm gather at the positive electrode and ova gather at the negative electrode. It has also been observed that once a sperm and an ovum are combined, they do not combine with any other sperm or ovum. This shows that once they are combined, features such as the original affinity between a sperm and an ovum disappear and on this basis features of a zygote are established.

Here, between the disappearance of the features of a sperm and an ovum and the formation of the features of a zygote, there exists a certain relationship in space and time. Although this relationship in the precedence of time and in cause and effect cannot be displayed in the macrocosmic world (and it will also be incredible if it could be shown in the macrocosmic world), it is completely understandable in the microcosmic world. According to modern theoretical physics, when length measurement enters the realm of 10^{-35} m and time measurement enters the realm of 10^{-44} sec, change in substance cannot be divided into past, present and future. Therefore, in the process when a zygote is being formed, there is not only a process in which the features of the sperm and ovum disappear and a process in which the features of the zygote come

Note:

The *hunyuanqi* of the zygote is the prenatal of the prenatal, with the latter prenatal referring to the concrete entity of the zygote; the *hunyuanqi* that a zygote obtains from the maternal body is the postnatal of the prenatal.

into being, but also a very short period of time between the disappearance of the sperm and ovum and the generation of the zygote. At that moment, there seems to be no sperm or ovum, nor the existence of the zygote, and this is what was called by the ancient Chinese *kong* ("emptiness"). This argument is tenable. It is simply that this "emptiness" is not completely empty, but rather a kind of special material state that is formless and imageless. It is the result of the interaction between the sperm and the ovum as well as their interaction with the environment. This special material state is the *hunyuanqi* that is the prenatal of the prenatal, from which the zygote comes into existence. Therefore, we say that this *hunyuanqi* is not a simple sum of a sperm's *qi* and an ovum's *qi*, and the zygote is not simply a mechanical mixture of the sperm and the ovum.

According to the theory of *hunyuan* whole entity, the formation of a zygote's *hunyuanqi* has experienced two *hunhua* stages. One is the *hunhua* of the *hunyuanqi* of the sperm's and the ovum's chromosomes, which forms the *hunyuanqi* of an integrated cell nucleus. The other is the *hunhua* of the *hunyuanqi* of the cell nucleus with the *hunyuanqi* of the ovum's cytoplasm, which forms the *hunyuanqi* of an integrated zygote. Scientists have only noticed the combination of the chromosomes in the first step and have not noticed the significance of the second step. In actual fact, the mergence and change in the second step is also very important because, without the mergence and change with the cytoplasm of the ovum, the cell nucleus of a zygote alone could not accomplish the development of the zygote.

So far, the formation of a zygote's *hunyuanqi* seems to have been clearly illustrated. In fact it has not been completely illustrated because the *hunhua* process mentioned above needs to take place in a certain environment, i.e., *hunyuanqi* in the environment will also take part in the *hunhua* process and become part of the content in the zygote's *hunyuanqi*. This seems to be not easy to understand, and some specific examples will be cited as follows to illustrate this effect.

For example, there used to be a type of big burgundy-colored horses in the Japanese army that was used for training and inspection. They were about two feet taller than normal horses, and riding on this type of horses made riders look particularly mighty and lofty. However, horses of this type had no reproductive ability. It was said that in order to have this type of horses, when horses were mating, their surroundings were all covered with red cloth and ideal burgundy-colored horses were more easily available. This is using the influence of the environmental factor of red color on the zygote of the horse.

In addition, experiments and statistics in modern eugenics show that the color of the bedroom at the time when a man and a woman have sexual intercourse has a certain influence on the character of the baby conceived. If the bedroom is in warm colors, the child's character tends to be enthusiastic and extroverted; if the bedroom is in cool colors, the child's character tends to be placid and introverted. If melodious music is played in the bedroom, the child will have a sensitive ear for music. There are many such examples.

It is true that it is incorrect to excessively emphasize these effects because the formation process of the life of a human being is complicated, which can be fully shown from what will be talked about later. The purpose of talking about this is to make people realize this law that the environment in any *hunhua* process joins the mergence and change.

ii. Factors that Affect the Holistic Feature of Zygotic *Hunyuanqi*

Although a zygote of a human being is also a monoplast, it is already widely different from a lower level monoplast. A human zygote is formed with the combination of a sperm and an ovum in which all the evolution information from single-celled organisms to human beings is deposited and it contains the holistic information about all the time-and-space structure of humanity: For example, the information about the development process from a zygote to an adult as well as the information about the tissues, structures, functions of humanity and so on are already deposited in the zygote. So long as an appropriate environment is available to carry out the *hunhua* process with it, all the features of a human being can gradually be displayed. The holistic information about the time-and-space structure in the zygotic *hunyuanqi* is the potential form of all the features of a newborn individual. It plays a guiding role in the development after the zygote has come into form.

Readers might ask: Now that a zygote determines the formation and development of a new individual human being, how come that children of the same parents differ in their intelligence, health and looks? Is it that the sperm and ova from the same parents differ? This concerns not only the individual differences in the sperm and ova, but also the differences in the *hunhua* process.

i)   The Influence of the Sperm and Ovum on the Holistic Feature of Zygotic *Huyuanqi*

A mature sperm and a mature ovum are carriers of all the life information (including physiological, psychological and consciousness information) about the male and the female from whom they are produced. However, the life information in every person in different life stages is not always the same. For example, in the prime of youth, a person generally has exuberant vitality but still immature psychological activity and is relatively lacking in social contact. In the middle age, a person is relatively abundant in the information regarding all the three aspects of vitality, psychological activity and social contact activity. In late middle age and old age, a person is rather mature in psychology and consciousness yet declining in vitality. During these different periods, the life information deposited in the sperm and ova also differ. Even within the same age period, the health of the body can also affect the information composition of the sperm and ovum. For example, disease, especially the disease in the reproductive system, directly affects the healthy development of the sperm or ovum. Unwholesome emotion and consciousness can also be implanted into the *hunyuanqi* of the sperm and ovum.

In addition, the holistic feature of a zygote's *hunyuanqi* is also affected by the complementary induction of the sperm and ovum when they combine and mergence and change between them take place. As has been previously mentioned, the combination of a sperm and an ovum is not just a simple combination, and the functions of the sperm's *hunyuanqi* and the ovum's *hunyuanqi* in the formation process of the zygote are not the same, either. In a certain sense, the information about the time-and-space structure of a zygote is determined by the *hunyuanqi* of the sperm, while that the time-and-space structure with information can be fixed by a human form is determined by the *hunyuanqi* of the ovum. This is a coupling *hunhua* process of complementary induction. If the sperm's *hunyuanqi* is more abundant, the sperm will guide this process. On the contrary, if the ovum's *hunyuanqi* is more abundant, the ovum will guide this process. Whatever the situation is, the two of them inspire each other, induce and guide each other and complement each other in forming a unified whole entity in the *hunhua* process. If the sperm and the ovum are too identical in their genetic feature, the tendency of complementary induction will inevitably decrease and in the newborn individual prenatal defects might occur. This is the drawback in the reproduction between couples if they are from close affinitive ties of blood.

ii) The Influence of the *Hunhua* Process of the Sperm and the Ovum on the Holistic Feature of Zygotic *Hunyuanqi*

As has been mentioned above, the zygote's *hunyuanqi* is not only a simple combination of the sperm and ovum. The *hunyuanqi* in the environment also takes part in the mergence and change. The whole process of impregnation has been studied in physiology and it has been realized that, after the sperm enters the female's reproductive organs, a series of complicated changes takes place in the sperm as it progresses from the vagina to the oviductal ampulla, after which it can obtain enough energy to combine with the ovum. The conditions needed for these changes to happen can all be regarded as the *hunhua* environment.

In addition to this, this environment referred to in the theory of *hunyuan* whole entity is connected to the physical and mental state of the parents as the intercourse is being conducted and the degree of the sexual excitation reached. When the climax of sexual excitation is reached, all kinds of life functions in the human body are in a harmonious enhanced state. Particularly in the female, a chaotic state similar to a *qigong* state can be shown. At this moment, the *hunyuanqi* of the male and the female will experience a certain extent of *hunhua* and will form a kind of *hunyuanqi* field which instills into the sperm and ovum high quality integrate life information and lays a foundation and creates a condition for the *hunhua* between the sperm and the ovum to take place. In the climax of sexual excitation, not only the physiological and psychological conditions in the parents are in a state of excitement, the sensory nerves and the efferent nerves of the voluntary nerves are also all in a state of excitement, even the pair of the antagonistic sympathetic nerves and parasympathetic nerves which belongs to the autonomic nerves have also lost their antagonistic feature and show excitement simultaneously and function coordinately. All this will influence the *hunyuanqi* of the sperm and the ovum.

Readers may wonder: How do you explain the production of test tube babies then? We think that, though the formation process of such a baby's zygote is not influenced by the climax of sexual excitation in their parents, the father has also experienced a process of sexual excitation when the sperm is being obtained. Besides, when artificial insemination is being carried out, the process in which the sperm obtains energy outside the human body is also a kind of environment. However, from the perspective of the theory of *hunyuan* whole entity, there must be some imperfect and immoderate (compared with naturally inseminated babies) element in the *hunyuanqi* of an artificially inseminated baby. However, it may also have the advantages of being leisurely and peaceful, quiet and serene, pure of heart and having less desire.

In sum, the environment in which *hunhua* between the sperm and the ovum is being carried out is very important. This is very similar to the impact of initial conditions upon things stated in nonlinear physics. It has been sufficiently proven by many facts that in nonlinear physical changes any tiny little change in the initial conditions will have a great impact upon the result of the nonlinear matter in the future.

II. Embryonic *Hunyuanqi*

In modern medical science, the term embryo, including fetus, refers generally to the human body in the entire process from the zygote to when the fetus becomes mature in its development. It is roughly divided into three stages: First, it is the ovigerm stage, which refers to the first week after fertilization. Secondly, it is the embryonic stage, which refers to the second to the eighth week after fertilization. Thirdly, it is the fetal stage, which refers to the ninth to the 38th week after fertilization. The embryo's *hunyuanqi* we talk about here refers to that the embryo and its *hunyuanqi*, after having received the *hunyuanqi* from the maternal body, carry out mergence and change according to the specified time-and-space structure of embryonic *hunyuanqi* and begin to develop from the zygote to the germ layers of a multicellular organism, that they differentiate further and rudimentarily finish all the differentiation for the formation of a fetus in the first to the eighth week after fertilization. We call the *hunyuanqi* at this stage the embryo's *hunyuanqi*.

Strictly speaking, the first week after fertilization does not belong in the stage of embryonic *hunyuanqi* as in this stage the division of the zygote (which is called merogenesis) is only the increase in the number of cells, while the size of the zygote does not increase and the zygote relies mainly on its own ability to accomplish this whole merogenesis process. On the third day, the cell mass that consists of 12 to 16 cells are formed (which is called morula in medical science). On the fifth day, there are about 107 cells which look like bubbles (In medicine, this is called blastula). After the sixth day, it begins to be implanted in the mucous membrane of the womb, after which is the real beginning of the embryo. Before this, it is in fact still the zygotic *hunyuanqi*.

From the implantation of the blastula into the uterine wall to the formation of the umbilical cord in the sixth week, the embryo has gone through a complicated differentiation process. The zygote's *hunyuanqi* has become fully unfolded according to its feature of holistic time-and-space structure. Until the formation and the closure of the nerve canal, it is completely the growing in an open style. In every cell at this moment there exists the active all-dimensional information and the tendency of growth (the division and differentiation of cells) is very strong, thus the *hunyuanqi* field formed is also very strong. In the fourth week or so, the nerve canal is formed and the heart begins to beat. The tissue primordia in every part of the human body are differentiated and the primordial *hunyuanqi* of a human being is also rudimentarily founded and fixed. Since the embryo develops in a highly open style during this period, the embryo's *hunyuanqi* is very strong in its force of expanding outward. It not only fills the maternal body, making it become abundant in vitality (She who cannot adapt herself to this dashing force feels sick and even vomits) and not want to eat or drink, but also expands outside it. People having acquired the ability of telepathy in their *qigong* practice can perceive embryonic *hunyuanqi* from more than tens of meters away and its strength is no less than an ordinary *qigong* master. As the ectoderm closes in and forms the nerve canal, the embryo's *hunyuanqi* begins to be absorbed into the embryo.

This development process is, from the perspective of the theory of *hunyuan* whole entity, a very significant open-and-close process. It first opens and expands and then takes back and closes. It not only absorbs the mother's *qi* but also the appropriate *qi* in the environment that it meets, thus collecting more information for the formation of each part of the fetus and forming the centre of "the prenatal of the prenatal". This is a critically important point in fetal development. If we say that the embryo before the formation of the nerve canal already has the features of the lower organism and has replayed the evolution from the unicellular to the multicellular organism, the embryo after the formation of the nerve canal has the characteristics of higher animals. It begins to gather life force into the body of the individual and lays a foundation for the formation of an independent creature with advanced nerve activities.

The embryo's *hunyuanqi* is essentially the zygote's *hunyuanqi* that carries out *hunhua* by means of receiving the mother's *hunyuanqi* as well as the *hunyuanqi* outside the maternal body. This *hunyuanqi* has the characteristic of being strongly open and highly holistic. During this period, the *hunyuanqi* in the mother's body is also relatively abundant. It can be used to cure disease, to build up the mother's body, or even to attain better achievements in athletic competitions in this special state. This stage also has a significant impact on having a healthier child.

III. Fetal *Hunyuanqi*

The fetal stage begins from the eighth week after impregnation. The fetus' *hunyuanqi* refers to the holistic time-and-space structure of the fetal stage. To be specific, it refers to the *hunyuanqi* connected with the fetus' life activity, including the *hunyuanqi* of the fetus itself, the *hunyuanqi* of the afterbirth, the *hunyuanqi* of the umbilical cord and the *hunyuanqi* of the placenta. They are an integrate whole entity. We should not regard the fetus' *hunyuanqi* merely as the *hunyuanqi* of the fetus' body. The chief characteristic of the *hunyuanqi* during this period is the holisticness or unity of the placenta and the fetus.

Strictly speaking, fetal *hunyuanqi* actually begins from the fourth week after fertilization when the nerve canal is formed and closed, for the fetus already takes the outward expanding *hunyuanqi* inward to develop its own body. Since the nerve canal is formed with the inward subsidence of the ectoderm while the ectoderm will develop into the skin and the nerve canal will develop into the nervous system, there is an innate connection between the internal nervous system and the external skin. When the umbilical cord is formed in the sixth week, though the circulatory system of the fetus has not formally started, the formal channel with the placenta is after all established. In the eighth week, the gender of the new being can already be distinguished and it can be called a human fetus in the true sense of the word. After this, development continues, which makes different tissues and organs of the human body gradually grow mature for birth. General speaking, the nervous system and the heart are the earliest in development and they already have the function to live independently in the seventh month.

It is truly miraculous that in only a few months a zygote can develop into a mature fetus and the evolutionary history of organisms seems to have been replayed in it. This, however, is easier to understand if analyzed with the theory of *hunyuan* whole entity. There is a complicated *hunhua* process in every step of evolution in an organism, which makes the content of the holistic time-and-space structure of the organism more complicated. The *hunyuan* substantial entity of the organism is also to have more information deposited in it with the evolution of its *hunyuanqi*. The *hunyuanqi* having become deposited in the form of concrete substance then becomes the "radiating" source of the *hunyuanqi* field of the organism, while this *hunyuanqi* field is the guiding reason for the growth and development of the new individual.

Since human holistic information has already been deposited in the zygote—both the information about the time holistic structure and the space structure of a human being, mergence and change take place upon the existing foundation with every step of the zygote's development and will form a correspondent *hunyuan* whole entity, which therefore further strengthens and improves the *hunyuanqi* field and lays a foundation for further mergence and change. This process continues repetitively like this and makes the fetus accomplish the whole process of individual development very swiftly.

What is the feature of human fetal *hunyuanqi*? In brief, it is the unified chaotic state of the motion of growth of the natural life of the human body. During this period, different tissues and organs of the human body gradually come into form. Yet, except for the heart and the circulatory system which begin to perform their own duties, none of the functions of other systems and organs have started yet. Their main task is to perfect their own structures and functions, and the nutrition and the *hunyuanqi* needed are essentially obtained from the body of the mother. Therefore, digestion, absorption and excretion do not need to perform their functions yet.

In the theory of *hunyuan* whole entity, this period belongs to human prenatal chaotic state. The *hunyuan* whole entity of the human fetus results from the development of the entire zygote which includes the fetus, the placenta, the afterbirth, the umbilical cord and the amniotic fluid. The fetus is suspended in the amniotic fluid and is connected to the placenta with the umbilical cord. Around it is the afterbirth formed with the chorion (whose two layers consist of an outer layer formed by the trophoblast), the amnion and the decidua of the mother's womb. The fetus seems to be an independent "little heaven and earth" in the "big heaven and earth" of the mother's body. The state of the fetus is similar to that of the little chick having just come into form in an egg, with the difference merely in that the chick has no umbilical cord or placenta. The fetus displays a quite grand *qi* image of "merging in the universe".

The *hunhua* motion of the fetus' life is carried out at two levels. One is the mergence and change in different tissues inside the body of the fetus, i.e., the growth and development of the fetus itself; the other is the mergence and change outside the body of the fetus: *Hunhua* with the connected placenta through the umbilical cord. This process not only involves taking nutrition from and exchanging *hunyuanqi* with the mother's body, but is also a process in which quite complex mergence and change take place.

It has been proven in modern science that many kinds of hormones can be synthesized in the placenta and accomplishing this *hunhua* process also needs the help of the liver, gallbladder and adrenaline of the fetus, so the concept of the fetal-placental unit is put forward. It means that the fetus and the placenta work together to perform the function of bio-synthesis. Relying on either the fetus or the placenta alone cannot fulfill this task. From this it can be seen that the fetus' *hunyuanqi* is a *hunyuan* whole entity that includes the placenta.

In the fetal *hunyuanqi*'s *hunhua* process, if the *hunyuanqi* from the zygote is considered as the prenatal of the prenatal, the *hunyuanqi* absorbed from the mother's body can be considered as the postnatal of the prenatal. As the *hunyuanqi* from the maternal body also contains the holistic life information about the maternal body, it creates a convenient condition for the mergence-and-change process in the fetus.

The reason why we say that the fetus' *hunyuanqi* is in a prenatal chaotic state lies crucially in that, unlike what happens in the life activity after the baby is born, differentiation in the fetus' life activity in this period has not happened yet. The fetus' life activity is in a harmonious coordinated state of growth. Although the fetus already has the foundation for the activity of *jing*, *qi* and *shen*, the respective activities of *jing*, *qi* and *shen* have not started yet. The three of them serve the growth and development activity of the fetus together.

The *hunyuanqi* of course focuses on serving the fetus' growth and development process and there is no obvious body movement yet. The occasional movement of the fetus is just the adjustment of posture for the fetus' development. The activity of *shen* also mainly focuses on controlling heartbeat to convey blood so as to better serve the fetus' growth. The growth and development of different tissues and organs are carried out under the guidance of the unified time-and-space structure (i.e., the *hunyuanqi* field). Therefore the development of each part of the body, e.g., the trunk, four limbs, internal organs as well as other different organs, harmoniously and naturally goes on and displays a highly unified feature. This is also the critical point that ensures the swift growth and development of the fetus.

Since the fetus' spirit still have not truly begun its activity and its *shen* perception is still in a chaotic state, and since the environment of the fetus is the surrounding of amniotic fluid, this not only "rids" the body of the fetus from the influence of gravity, but the stimulation on each part of the fetus is also relatively even.

All these points stated above help create a condition for the even state of the fetus, which, therefore, makes the fetus' *hunyuanqi* seem to display a merging state between its *shen* (spirit) and *qi* and between its *shen* and *xing* (form or body). This, I'm afraid, is the reason why traditional *qigong* pursues to return to the prenatal state. However, this realm or state of mergence in a growing fetus differs fundamentally from the merging state of *shen* and *qi* and of *shen* and *xing* in *qigong* practice. In the fetus is the primal mergence in which different functions have neither divided nor become mature. It cannot be compared with the advanced-level mergence realm of *qigong* practice.

In sum, fetal *hunyuanqi* is a special *hunyuan* whole entity. It is quite similar to the celestial integral state of the universe described in ancient times. The earth of the human beings is suspended in the grand *qi*, and the body of the fetus is suspended in the fetus' *hunyuanqi*. The *hunyuanqi* of the fetus itself is part of the whole entity of fetal *hunyuanqi* and is the most crucial part, which can also be regarded as a whole entity itself. Although the fetus already has different tissues and organs, their respective functions have not displayed yet. The fetus' *jing*, *qi* and *shen* have not begun their own independent activities. All the life activities of the fetus are unified in its natural development.

The fetus connects itself to the outside world (which refers to the body of the mother) with the umbilical cord and through the placenta. This makes the navel area the fetus' *hunyuanqi* center. This center is in fact the postnatal center of the prenatal. It is formed on the basis of the center of the prenatal of the prenatal. On the other hand, important tissues with endocrine function, such as the kidney, the adrenal gland and the sex gland, all grow and develop near the navel during the fetal period and form a certain *hunyuanqi* field. Therefore, in traditional *qigong* theories, the navel and *mingmen* are regarded as the place where the prenatal *qi* is generated.

Shen hasn't actually begun its independent activity yet. During the fetal period, the functions of the entire nervous system have generally not really started (e.g., the efferent nerve and sensory nerve) except that trophic nerves adjust the bodily development of the fetus in a certain scope and to a certain extent. Although modern science has captured the EGG of the seven-month-old fetus, what is recorded is but the waves that periodically appear, which is equivalent to a normal person's sleeping state. Even if the fetus shows movement in the limbs and body, it is just the spontaneous response of lower spinal nerves. The function of *shen* in this period is chiefly shown in the control of cardinal autonomic nerves over heartbeat. Through the pulsation of the heart and blood vessels, macroscopic regular motion is formed in the fetus, which enhances the holistic feature of the fetus' life activity.

The process from the formation of the embryo to the development of *hunyuanqi* in the fetal stage is just to accomplish the natural development of the human fetus. During this process, the head enjoys priority in development. The fetus' head occupies a relatively large proportion of the fetus' body. In the third month, it makes up one third of the fetus' body; in the fifth month, it is one fourth of the fetus' body. The cranial nerve cells not only are sufficient in their amount of information, but also have the greatest number and density. The essence (*jing*) that belongs in genetic information also gathers the most here. The cranial nerve cells are the part that develops the earliest and the fastest in the fetus' body. The cerebrospinal fluid in the fetus is also more plentiful than that in the adult. This is perhaps the reason why the brain was called "the sea of essence and quintessence" by the ancient Chinese. As fetal *hunyuanqi* is still in a prenatal chaotic state, it is still impossible to talk about the distribution of its *jing*, *qi* and *shen*.

When each part of the fetus' body has become relatively consummate in development, or rather, when the fetus' *hunyuanqi* has developed to the stage when the mother's *hunyuanqi* cannot provide new mergence-and-change content to it, this marks the maturity of the fetus. The independent "small heaven and earth" in the "big heaven and earth" of the maternal body is no longer the perfect environment for the fetus' growth and development, so it is born into this world.

IV. The Young Baby's *Hunyuanqi*

After the baby is born, though it already possesses basically all the tissues and organs as a human being, the functions of these tissues and organs have not started. In fact, it does not possess all the characteristics of a natural human being yet. The various life functions still need a rather long period of time to be gradually realized and developed even after the start of the function of the internal organs when the umbilical cord is cut. Moreover, dramatic changes in the environment after birth require the baby to establish its new holistic life activity different from that of the fetus. This period lasts until infanthood and early childhood.

What requires particular attention is that in this period the consummation of the natural attribute of a human being and the consummation of the social attribute of a human being are closely related. To be more exact, this is a stage in which consciousness activities are established and united with life activities and, furthermore, guide life activities. Therefore, the holistic mode of the *hunyuanqi* in infanthood and babyhood is characteristic of the dynamic feature of being in continuous change based on the destroying of the holistic mode of the fetus. This change begins with the baby's first cry when its umbilical cord is cut.

i. Changes that Take Place after Birth

There is, first of all, a change of environment. The baby after birth has broken through the "imprisonment" of the small "celestial integral universe" of the afterbirth and enters the infinite universe of Nature. It has rid itself of the soaking of the amnion fluid and obtains the bathing of the air. What is more important, the cutting of the umbilical cord has not only given to Nature the amnion, villus, placenta, etc., that originate from the zygote and have once constituted a *hunyuan* whole entity with the fetus, but also cuts off the passage through which nutrition is provided and waste is removed. This brings about changes in the newborn baby's internal environment. One is the lack of oxygen; the other is the accumulation of metabolic waste in the blood and the tissue fluid. The density of carbon dioxide (CO_2) in the metabolic waste becomes increasingly higher. As carbon dioxide can bring about a greater degree of excitation in the central nervous system, particularly in the respiratory system, it inevitably stimulates the respiratory nerve center and bring it into the breathing action when the density of carbon dioxide has reached a certain degree in the blood. This is an important mark of the beginning of the newborn baby's independent life. The breath that begins henceforth will accompany the life activity in the baby's whole life.

When the lungs begin to perform their respiratory duty, the concrete waste products in the blood are either removed by kidneys or detoxified by the liver, thus making the liver and the kidneys begin to exercise their functions. Meanwhile, the digestive system begins to accept water and liquid diet such as milk that the baby can drink. The baby begins to accomplish all the metabolic tasks of absorbing nourishment and releasing waste by itself.

With the start of the functions of the internal organs, not only the tasks each organ fulfils evidently differ, but the supply of blood should also correspondently change. The mode of blood circulation in the fetal stage should completely change. First, this involves the establishment of a distinction between systemic circulation and pulmonary circulation to guarantee the difference between arterial blood and venous blood. Secondly, a change occurs in the volume of blood flow in different tissues and organs. In a word, the baby turns from that chaotic state with no differentiation in it into the state in which conditions are complicated and should be dealt with differently accordingly.

All these changes are the process of development from the prenatal to the postnatal state. The content and mode of the mergence and change of *hunyuanqi* are altered. In the fetal stage, the process of absorbing the maternal body's *hunyuanqi* by means of the placenta is carried out in the placenta outside the baby's body. What the baby absorbs in babyhood is no longer the mother's *hunyuanqi* with holistic human information, but the oxygen in the air, the *hunyuanqi* in such concrete things as food and drink, as well as the formless *hunyuanqi* in Nature. The newborn baby has begun its own life.

ii. The Differentiation of *Jing*, *Qi* and *Shen* and the Unification Based on the Differentiation

The newborn baby's independent life is shown in the direct connection of the objective existence of the baby itself with the natural world. The multiplicity in both the content and mode of connection breaks the prenatal chaos in the fetal stage and causes differentiation in the baby's *jing*, *qi* and *shen*. This is a process in which the baby's various functions are displayed. It is a process of progress instead of "loss of innocence" as was said in traditional *qigong*. This differentiation is the display of the multiplicity of holistic life, so it is also unified.

If we say that the prenatal chaotic state in the fetal period enables the fetus to replay billions of years of evolutionary process of the biological world in a short period of 40 weeks, then the differentiation and unification of *jing*, *qi* and *shen* in the period of infancy and babyhood enable the baby to accomplish the replay of the evolution from an advanced animal to a human being in merely several years. The replay in the fetus is mainly embodied in the change of shape, while function and information are deposited in the shape. The replay in the baby is mainly embodied in function and function reinforces the change in shape. Differentiation and unification in a baby's *jing*, *qi* and *shen* can be summarized into two stages.

i) Differentiation and Unification in the First Stage

The duration of this stage is from when the baby is newly born to when the baby is about three years old. This stage is related to the rudimentary differentiation and unification of the baby's *jing*, *qi* and *shen*.

The fetus departs the mother's body and is born as a baby into this world, and a dramatic change takes place in its environment. The prenatal chaotic state comes to an end and the baby enters its postnatal life. The newborn baby's *hunyuanqi* says goodbye to the task that the fetus' *hunyuanqi* is used completely to serve the development of the fetus' body and the new state in which *jing, qi* and *shen* are differentiated begins.

First of all, it is the differentiation and independence of *qi*. The newborn baby can no longer obtain from the mother's body the *hunyuanqi* already processed and having human characteristics. Instead, it begins to directly absorb lower level *hunyuanqi* from nature and makes it become the *hunyuanqi* that a human being needs through processing and *hunhua* or mergence and change. Meanwhile, the motions of opening and closing, gathering and expanding (even condensing and dispersing), going out and coming in as well as changing and transformation are carried out in the baby's *hunyuanqi* according to the laws peculiar to humanity, and in the baby a unique *qi* distribution system is formed. This will be elaborated upon in the section on the generation and *hunhua* of human *hunyuanqi* in this chapter.

Next is the differentiation and independence of *jing* (literally essence, referring to form, body). *Jing* as discussed here is *jing* in the broad sense. All the entities that have form (from cells to tissues and organs) belong in the category of *jing*. Unlike the fetal stage in which *hunyuanqi* is used merely for the multiplication of cells in the fetus in order to ensure the fetal body's development, the body of a newborn baby, on the one hand, carries out the multiplication of its own cells, and on the other hand, exercises its peculiar functions (such as stretching and contracting functions in the cells of the muscles, transmitting information function in the nerve cells as well as different functions in the various cells of the internal organs). That is to say, the way in which the human body uses *hunyuanqi* is more complex than mere multiplication. This is also the mark of the differentiation and independence of *xing* (form, body). It is true that the accomplishment of the dual tasks of the human body is carried out on the basis of the human body provided by the baby already having functions.

The differentiation and independence of *shen* is the root and foundation of the differentiation and independence of *jing, qi* and *shen*. It is established upon the differentiation and independence of form and is realized through the start of the various functions of nerve cells. Although the functions of the nerve cells are also the functional display of the human body having acquired its differentiation and independence, they are already different in essence from the general function based on the differentiation and independence of the human body. This is because nerve cells are highly differentiated cells having lost the reproductive ability. Therefore, all their consumption of *hunyuanqi* in their whole life takes place on the basis of using *hunyuanqi* in their peculiar functions. Vast numbers of nerve cells form a complicated grand whole entity. It has the functions of receiving and sending information and has formed a unique independent system upon the foundation of the human body having acquired differentiation and independence in *jing*, *qi* and *shen*.

As the nerve cells have stopped their own multiplication, their life force is entirely shown in their special mergence-and-change function as they interact with the outside world. This is the more advanced content and mode of *hunhua* than the mergence and change (*hunhua*) of the *hunyuanqi* of common biological cells which take their own multiplication as their main task. The display of this function is the holistic activity established upon the foundation of the activity of the great number of nerve cells. This is what we call the differentiated and independent *shen*. Its establishment is a typical process of differentiation and unification.

With the first cry of the newborn baby after its birth, the functions of the throat and vocal cords of the baby begin and the coordinated motion involving the brain and cranial nerves comes into existence. This is a coordinated activity that engages the activities of a lot of nerves, such as the nerves of the ribs and diaphragm that control the respiratory muscles, the vagus nerve that controls the internal organs, the laryngeal, hypoglossal, glossopharyngeal and accessory nerves that control the larynx and mouth, and also the facial nerves that control the facial muscles. For the first time, the holistic power of life activity is displayed. The activity of the autonomic nerves is also unfolded with the functional activity of different internal organs.

The functions peculiar to each sensory organ are strengthened in their interactions with the outside world. The senses of vision and hearing, the senses of touch and heat associated with the skin, etc., are all reinforced step by step. The state of *shen* being merely concentrated on the heart and blood vessels so that it helps form the regular motions of the entire body during the fetal period is gone forever. Diversified differentiation happens in the function of spirit. This differentiation is rather essential, without which the functions peculiar to human beings cannot be enhanced, without which closer connections between humans and nature cannot be established. However, like any other independent and integrate form of existence, human beings retain their holistic features as they interact with the outside world, or rather, things that interact with humans appear in the form of a whole entity. The two of them mean the same. Yet human sensory organs can only receive part of the information about existences in the outside world. Therefore the connections between different sensory organs must be strengthened in order to meet the demand of whole entity. This is a further requirement of different organs, each of which has its different functions.

Babies gradually accomplish the establishment of the holistic connections between different functions of different organs in their life and practice. For example, eyes make progress from being able to see only in one direction to seeing in different directions. Then association between vision and hearing is established—on hearing a sound, the baby's eyes track where the sound comes from. All this, of course, is still the function of form (*xing*) of relevant sensory organs.

But the association between them is the effect of *shen*.

Then what is going on with the motor functions of the baby's body that has form? Babies first free themselves of the motional mode involving their entire body's movement that accompanies them ever since they are born into this world (e.g., a baby's four limbs and trunk move simultaneously when it cries) and gradually learn simple actions using part of their body. Though this kind of movement in babies is a common life activity, it is already greatly different from the same kind of actions in higher animals. This is because, ever since babies begin to learn how to perform these actions, they have not only received help from adults, but are also influenced by their language. For example, when a baby is learning how to walk, usually the adult gives the baby the instruction to "walk" and at the same time encourages or holds the baby to step forward. After having made some progress, the baby will be able to step forward according to the adult's instruction. It is the same with babies who are learning how to pick up things with their hands. Usually the adult puts an object in front of the baby and encourages or instructs the baby to take it, and the baby follows this instruction. All this shows that the process in which babies learn to perform actions is affected by consciousness. It is the same in learning to develop the functions of sensory organs. Parents show toys to the baby for many times or stimulate the baby's sense of hearing with toys that can make sound to attract the baby's attention. During this process, the actions are always accompanied with language instructions.

Why does this "leaping-mode" development happen in the growth of the baby? Namely, why doesn't the baby first establish its sensori-motor functions and then forms its consciousness functions at a high level? Why is it that they learn to use their motor functions under the guidance of consciousness? This is because, on the one hand, babies' *hunyuanqi* contains the totality of the inherited human information and already has the conditions to accept language and to generate consciousness, which will become reality with *hunhua* once there is correspondent *hunyuanqi* in the external environment. On the other hand, babies have a very strong capacity to directly receive emotion information and consciousness activity. Adults' consciousness activities can affect the baby. Therefore, babies' motor and sensory functions are established at the very beginning upon the holistic foundation of being guided by consciousness. The establishment of functions then reinforces the development of the brain. The cerebrum of the baby between the age of one to three develops very fast. Neurodendrites and synapses, the linking mode of brain cells, also swiftly develop and consummate during this period. With the establishment of motion, sense and thinking, consciousness is already unconsciously combined with life activities.

ii) Differentiation and Unification in the Second Stage

The period of this stage is the time from the age of three to the age of six. It is about further differentiation and unification of *jing*, *qi* and *shen*. If we say that differentiation and development of babies in the first stage are carried out on the basis of prenatal chaos and that the result of these changes will be greatly affected by the environment (The significance of the influence of environment can be seen in the case that a baby in a flock of wolves has differentiated and developed into nonhuman, i.e., the wolf child), then the differentiation and development of babies in the second stage as described here are carried out on the basis of their directed *hunhua* with human beings. It is the continuation and further development of the first stage.

Independent *qi* begins to be differentiated and distributed according to different positions in the body. We know that the *qi* in the fetus' body is an integrate whole evenly distributed. After the baby is born, though the distribution of its *qi* has experienced rather great change, the holistic feature of the baby's *qi* is still fairly strong. The concentration of *qi* according to different positions in the body has not begun yet. With differentiations taking place in the body, in the internal organs and the spiritual activity, the integrate holistic *qi* begins to concentrate and differentiate correspondingly into three centers—body *hunyuanqi* (*quti hunyuanqi*), internal organ *hunyuanqi* (*zangzhen hunyuanqi*) and *Yi Yuan Ti* (Mind Oneness Entity).

Shen (spirit) begins to form its own independent activity, namely, the ability to think in images is formed and is developing toward logical thinking. Such thinking activities are already activities peculiar to the spiritual realm and are the beginning for *shen* to become independent of life activities and to command life activities. The form and structure of children's nerve cells at about the age of six already have no evident differences from those of adults' nerve cells. Their spiritual activities can already actively give orders to relevant parts of their bodies and make their activities purposeful activities. Meanwhile, functions of different sensory organs of children are already relatively consummate and their perception of things is becoming holistic.

What is worthy to be noted is that during this period the holistic feature of super intelligence and the holistic thinking in familiar intelligence are very similar. Therefore, during this period, as long as necessary guidance and instruction are given to young children (of about six years old), super intelligence will be shown in them. Since familiar intelligence people lack knowledge about this, they do not influence children in this way, so the young children's *shen* often progresses on the track of familiar intelligence.

In this stage, both *shen* and *qi* are experiencing comprehensive differentiation and the whole entity feature of life of the human body has also become highly developed. Not only is holistic thinking established, but the organic association between *yi* (mental activity), *qi* and form are also established. A familiar intelligence person's life mode of mind guiding *qi* and *qi* guiding body is preliminarily formed (The fact that babies cannot accurately catch toys with their hands in the babyhood is a display that babies still have not established the directed association between mental activity, *qi* and form). The command of mental activity over body and *qi* is shown not only in the fact that mental instructions can command the life motion of one's own body, but also in that emotions are able to influence physiological changes, which forms the holistic life activity under the guidance of *shen*.

In sum, although changes in the period of infanthood and babyhood are very complicated, they can be summarized into three points: First, it is the *hunhua* (mergence and change) between prenatal *qi* and postnatal *qi* at the level of *qi*. Secondly, it is the formation of function and the transformation between *qi* and form at the level of form. Thirdly, it is the *hunhua* between internal information and external information at the level of *shen*. Furthermore, *shen* has permeated into form and *qi* and has experienced mergence and change with them. The whole entity of form, *qi* and *shen* guided by *shen* is preliminarily formed.

V. The Adult's *Hunyuanqi*

The human *hunyuanqi* talked about in Section I in this chapter refers actually to adults' *hunyuanqi*. Here we will see how adults' *hunyuanqi* develops from babies' *hunyuanqi*. What is the difference between adult *hunyuanqi* and the *hunyuanqi* formed in late babyhood with form, *qi* and *shen* in a whole entity commanded by *shen*? The difference lies in that the holistic *hunyuanqi* of a baby and a child is still not the concentrated display of all the functions of a human being. This is because organs in the genital system have not exercised the functions peculiar to them yet while these functions are the important content in the most fundamental life activities of a human being.

Generative organs and sexual functions begin their activities on the basis that human body, *qi* and spirit have developed to a fully matured and relatively consummate state. Sexual activity of mature generative organs is the mark that the holistic life activity of a human being has entered its prime state. It is only at this moment that the content of *hunyuanqi* of a human being is realistic and entirely integrate and can be condensed inward to form genetic substance that has the ability of reproduction. This *hunyuanqi* which is expressed both as a human body with form and as formless *hunyuanqi* and contains the entire life information about a human being pushes the life activity of a human being onto a summit on the one hand and, on the other hand, pushes it down to the depth of a ravine because all animals have this procreate ability as well. Moreover, extensive human communications in society as well as multiple forms of labor enhance further *hunhua* of human *hunyuanqi*.

In the theory of human body *hunyuanqi* that we have talked about, we have seen three different situations at the level of cells, the most fundamental unit in the human body. One is that all the function of the cell is concentrated on the reproduction of offspring, and this is the sperm and the ovum. The second is that the cell has both the function of self-reproduction and the general life function, and this is the cell of the body. The third is that the cell has no reproductive ability but only special life function, and this is the nerve cell. These three kinds of cells have different positions in the evolutionary history of organisms. The reproductive cell is the characteristic that can be found in the monoplast of lower organisms. The body cell is the characteristic found in functional differentiation in biological cells. The nerve cell is the characteristic achieved when biological cells have developed into the level of being highly differentiated.

When a human being has grown mature, these three kinds of cells concurrently exist in the human body. Reproductive cells are the foundation for physical development (reproduction), body cells are the foundation for life, and nerve cells are the foundation for generating human spirit. The three of them influence and serve each other, complement, contrast and realize each other.

If *jing*, *qi* and *shen* are concentrated on *jing* (essence), *jing* will be abundant and flourish. If one follows the impulse caused by relative sufficiency in *jing*, one will just be an average person and waste essence. If one resists such impulse and practice with appropriate life-cultivation methods, one can achieve *dan* in one's body and enjoy a long life. If *jing*, *qi* and *shen* are concentrated on the body or form, the human body will be strong. To strive and outshine others will waste one's energy. To deal with life naturally as currents go with the stream, one will have a long life. If *jing*, *qi* and *shen* are concentrated on *shen*, *shen* will be plenty and affluent. Being haunted with selfish thinking will exhaust it, while serving the public and being unselfish will increase wisdom and intelligence and prolong life.

The human beings who have developed up to now, though having decisively mounted the throne of being the most intelligent of all creatures, are still distant from reaching their true self-awakened and self-knowing nature with perfect self-mastery. One reason for this is that human beings have not long separated themselves from the world of animals and have not completely rid themselves of the trace of animal impulse. The other reason is that the degree of human beings' civilization, whether it is material civilization or spiritual civilization, is still relatively low, particularly that human beings still don't truly know themselves. Therefore, people inevitably show their blindness, partiality and limitation in their long history of development, which lead to the partiality and stubbornness at the level of their epistemology, thus making human beings' magically brilliant innate nature with perfect conscience, self-control and self-mastery, this special nature of the human race, unable to be truly realized.

The overall feature of human beings today is the display of the distorted or deviated human nature because human beings so far still do not own their true selves to realize their own nature. Instead, their innately magically brilliant nature (of knowing fully one's true self, knowing both the inside and outside, being magically capable to do all the right things all in the right ways) is occupied and veiled by material concerns. The holistic characteristics of *hunyuanqi* previously stated are a true reflection of this state in human beings. For example, human *hunyuanqi* can merge and change according to natural laws (at the level of cells, for instance) and can also merge and change according to consciousness instructions. However, human consciousness activity has lost its innately magically brilliant nature and is "muddied" because of the desire to possess. This makes human *hunyuanqi* which is actually part of nature *hunyuanqi* lose its natural feature. It is true that human beings' *hunyuanqi* at the present stage is the natural display of the currently distorted human nature which is originally brilliant and enlightened. This will be comprehensively illustrated in The Theory of Consciousness (Chapter V) and The Theory of Morality (Chapter VI) in this book.

VI. *Lingtong Hunyuanqi* (or *Hunyuanlingtong Qi*)

This is a special kind of *hunyuanqi* that is gradually formed through *Zhineng Qigong* practice. *Lingtong Hunyuanqi* not only is the genuine unification of human *jing*, *qi* and *shen* but can also actively communicate with the *hunyuanqi* in Nature. It enables humans to enter the truly free realm in which they can mobilize and command *hunyuanqi* in the natural world as they wish. *Lingtong Hunyuanqi* is the *hunyuanqi* guided with *shen yi* (high-level mental activity) and has many magical functions: There is nothing it cannot penetrate, nothing it cannot smash, nothing it cannot dispel, nothing it cannot grow, nothing poisonous it cannot detoxify... *Lingtong Hunyuanqi*'s outstandingly miraculous effects seem difficult for people to comprehend, yet the miraculous effectiveness has been preliminarily shown in *Zhineng Qigong*'s practice.

For example, it can heal fracture in an instant; it can dispel tumor (e.g., lipoma, sebaceous cyst, fibroma) in a blink; it can make spur vanish in a split second, and these have become common occurrences in *Zhineng Qigong* practice. It can reduce the weight of human body by three to five or even up to 10 kilograms within several minutes; it can increase the height of human body by two to four centimeters (with the most effective being 11cm so far). These effects have been proven in many experiments.

It can revive dead bones. For example, learner Pan Hong once suffered from a tumor of the femur. The bone tissue with bone cancer in her body was taken out, inactivated, and then put back into her body in an operation. After three months of *Zhineng Qigong* practice, the dead bone in her body came back to life. Now the metal object which was used to fix her bone in the operation has been taken out.

Lingtong Hunyuanqi can remove metallic objects from the human body or make them disappear. For example, learner Zhang Yushan had two metallic shell fragments in him that were blasted into his head in the Korean War. They had stayed there for 38 years. He began practicing *Zhineng Qigong* in 1989. One of the shell fragments came out of his mouth, the other disappeared.

Lington Hunyuanqi can make bone regenerate. In 1993, among learners in the 15th session classes which lasted for 50 days in the Restoration Department, seven patients once had some part of their bones cut away because of cancer (part of the skull or ribs) before they came to the restoration center. After *qigong* practice, the bones which had once been cut off grew out and became healthy again. Among 14 patients in the 16th session classes who once had their bones cut off, new bones have also grown out after their *qigong* practice.

In addition, great achievements have also been attained in the experiments into increasing production in different trades, such as agriculture, industry, forestry, husbandry and fishery. Crops ripen earlier and yield greater harvest, and quality and quantity of goods in industrial production are respectively improved and increased. Experiments have taken place in many fields, including jiemycin, aspergillusoryzas, rifampin (RFP), gibberellins, winemaking, cotton yarn processing, iron and steel processing.

Lingtong Hunyuanqi can increase compressive strength in the architectural material of precast concrete panels. It can make the worn out machine tools under numerical control become normal in operation. It can improve the yield of fruit trees. Examples like these are numerous.

These are just the preliminary applications of *hunyuanqi* and it is shown that *hunyuanqi* brings about miraculous effect wherever it is applied.

It is true that this is just the beginning of the experiments on the function and effectiveness of *hunyuanqi*. It is just the effects of *hunyuanqi* having been actively applied in the initial stage and is far from the entire effects of truly miraculously functional *Lingtong Hunyuanqi*. From this we can know that, how powerful it will be once truly miraculously functional *Lingtong Hunyuanqi* is successfully accomplished through practice.

Section III Motions of Human *Hunyuanqi*

The motions of *hunyuanqi*—opening and closing, going out and coming in, gathering and expanding (including condensing and dispersing) as well as change and transform, have been expounded in the theory of *hunyuanqi* in Chapter II. Here more specific illustrations about motions of *hunyuanqi* will be given in light of human *hunyuanqi*.

The motions of human *hunyuanqi* include two parts: The motions of opening and closing, going out and coming in, gathering and expanding (with condensing and dispersing included), and the motion of *hunhua* (mergence and change, here change particularly refers to transform). The two parts are inseparably interrelated. Opening, closing, going out, coming in, gathering and expanding (or condensing and dispersing) are the preparation process for mergence and change (transform), and mergence and change (transform) is the target of opening and closing, going out and coming in and gathering and expanding (even condensing and dispersing). Mergence and change without the former three pairs of motions and the three pairs of motions without mergence and change are both unimaginable.

I. Motions of Opening and Closing, Going out and Coming in and Gathering and Expanding (Including Condensing and Dispersing) in Human *Hunyuanqi*

The motions of opening and closing, going out and coming in and gathering and expanding (plus condensing and dispersing) in a human being as a *hunyuan* whole entity are carried out at two levels. They are carried out at the boundary (i.e., external boundary) where the body of the human being interacts with the outside world and also in the tissues inside the human body. Although there are such differences as *xing* (form, body), *qi* and *shen* (spirit) in human *hunyuanqi*, the motions of *shen* and *qi* are usually embodied in *xing*. In view of this, we plan to talk about the motions of human *hunyuanqi* from the level of *xing*.

i. Motions of Opening and Closing, Going out and Coming in and Gathering and Expanding (Including Condensing and Dispersing) at the External Boundary

The boundary where the human body interacts with the external world refers to the skin all over the human body and the relevant membranes with which the human body are in contact with the outside world. To analyse with this standard, in the scope of mucous membranes, not only should the respiratory tract and the mouth cavity be included, but even the mucous membranes of the trachea and the bronchus, the epithelial tissues in the pulmonary alveolus, the mucous membranes of the esophagus, of the stomach and of the small and large intestines should also be included. It is because the whole alimentary canal is "enclosed" in the human body by the throat and the anus, etc., which seems to make them become the cavity of the human body, that people even forget they are the boundary in contact with the outside world.

The motion modes of human *hunyuanqi* on the boundary that has form are mainly *kai he* (opening and closing) and *chu ru* (going out and coming in). According to the theory of *hunyuan* whole entity, on the boundary that has form, there are not only pores, sweat pores and adipose gland pores for the going out and coming in of concrete substances, but also acupoints, *cou li* (the grain in the skin and muscle) as well as even more subtle tiny *qi* holes for the going out and coming in of formless *qi*. For the opening and closing as well as the going out and coming in of substances with form: The skin can release concrete substances through opening, such as sweat (which contains inorganic salt and organic compound), fat, water vapor, gas with special smell as well as various secretions of the mucous membranes. It can also close and take in relevant substances. A lot of traditional Chinese medicines for external application are absorbed through the skin. The skin and the mucous membranes of the respiratory tract have the function of absorbing oxygen and releasing carbon dioxide; the mucous membranes of the stomach and intestines take in water and various nutritious substances, and their secretions contain various contents. All this is accomplished in the motion modes of *kai he* (opening and closing) and *chu ru* (going out and coming in).

The going-out and coming-in process of these concrete substances has already become known in modern science. However, modern science has known nothing about the more important content in the theory of *hunyuan* whole entity, namely, the process and nature that formless *qi* goes out and comes into the human body with the opening and closing of the human body. Therefore, the explanation of it is relatively difficult to comprehend.

It has been pointed out in the theory of *hunyuanqi* in Chapter II that around the concrete existence there is its relatively sparse *hunyuanqi* distributed. This amount of *hunyuanqi* actually spreads out of the concrete existence through the opening action of the "*qi* pathways" on the external boundary of the existence. It is the same around the human body: The *hunyuanqi* field is both the frontier for the connection and *hunhua* between the concrete existence (here it refers to the human body) and the external *hunyuanqi* and also the buffer between the inside and outside. This *hunyuanqi* field is directly affected by the opening and closing and the going out and coming in of the boundary of the concrete existence. Different *hunyuanqi* has different laws and methods to adjust the balance between the two of them, namely, the concrete existence and its *hunyuanqi* field. In the human body, the adjustment of opening and closing and going out and coming in is carried out through breath, tissue metabolism and mental instruction (*yi nian*).

In normal life activities of the human body, when a person is breathing out, *couli*, pores, sweat pores, acupoints and *qi* pathways on the skin and the mucous membranes open, the internal *qi* goes out, the *hunyuanqi* field around the human body expands outward and its strength is enhanced. Sensitive people can feel the opening of pores, sweat pores and acupoints in the skin and there is a feeling of *qi* flowing out or pouring out (like steam). These feelings can be more apparent in the part of the human body with big acupoints, such as the finger tips, the center of the palm, the top of the head, *yintang*, *tanzhong*, *mingmen*, the navel, *huiyin* and *yongquan*. When a person is breathing in, pores, sweat pores, acupoints, *couli* and *qi* pathways on the skin and the mucous membranes close inward, external *qi* comes in and the scope of the *hunyuanqi* field around the human body shrinks and its strength is reduced. Sensitive people can feel that, in the part of the human body mentioned above, there is *qi* flowing in. But, generally speaking, the feeling is not as obvious as that when *qi* is flowing out.

The opening and closing and the going out and coming in brought about by breath talked about here belong to the movement generated by "inhaling and blowing", that is, breathing mobilizes the going out and coming in of grand *qi*, thereby the opening and closing and the going out and coming in of the boundary of the *hunyuan* entity is brought into being (as in the operation of bellows).

In addition, the increase and decrease in the amount of *hunyuanqi* formed in the human body can give impetus to *kai he* and *chu ru* at the boundary of the *hunyuan* entity based on changes in *hunyuanqi*'s "strength of pressure". For example, when a person is abundant in *jing*, *qi* and *shen*, the boundary opens and expands and *hunyuanqi* goes out. Otherwise, the boundary closes in and *hunyuanqi* comes back.

As far as human spiritual factors are concerned, when a person's spirit focuses on the outside of the person's body, it leads *qi* out; when a person's spirit guards the inside of the person's body, it leads *qi* back. Emotions of fright, anger and sorrow make *qi* go out and lost; the psychologies of thinking and fear gather *qi* inside. Yet when a person is too absorbed in thinking, the flow of *qi* becomes sluggish. There are many such examples. It is simply that familiar intelligence people still do not know nor have they grasped the great guiding power of mental activity (*shenyi*) over the opening and closing, the going out and coming in and the gathering and expanding (including condensing and dispersing) of *qi*.

In fact, consciousness can cause changes in motions of *qi* not only by means of emotions but also by giving active instructions. There is no exception whether it is on the external boundary or at the internal level of the human body. It is also the same with the *hunhua* motion to be talked about later. The principles of application of consciousness in *Zhineng Qigong* are formulated because of this.

However, opening and going out, closing and coming in do not always happen at the same time because there is a difference between pressing and guiding in the dynamic factors for opening and closing. For example, pressing open with internal high "pressure" of *qi* and guiding to open with mental instruction can both cause a person's *qi* to flow out. Yet it is not the same with closing. It is only with internal guidance that *qi* outside the human body can enter the human body, while external pressure is likely to cause closure thus *qi* cannot enter. From this it can be known that closing and coming in of *qi* are contradictory to some extent.

Opening and closing talked about here are both relative. Opening is not complete opening and closing is not complete closing. Otherwise, if the two of them go to the extreme, normal life cannot be maintained. Although we do not emphasize *ju san* (gathering and expanding or even condensing and dispersing) here, in fact, in *kai he* (opening and closing) and *chu ru* (going out and coming in) there must be the accompaniment of gathering and expanding or condensing and dispersing to a certain extent. For example, the process of opening and going out is the process in which human *hunyuanqi* expands outward; the process of closing and coming in is the process in which human *hunyuanqi* gathers inward.

ii. Motions of Opening and Closing, Going out and Coming in and Gathering and Expanding (Including Condensing and Dispersing) at the Tissue Level inside the Human Body

The human body is a complicated multi-level super system. Modern science divides it into some systems which are further divided into organs, and organs are further divided into different tissues and eventually cells. From the perspective of the theory of *hunyuan* whole entity, though it is not necessary to divide the composition of the human body as detailed as this, stratums are also emphasized in *hunyuan* whole entity, and, in the integrate whole entity, there are also its constituent parts. Although parts in the whole entity are determined by the whole entity, they also have their own independence. In the concrete entities of these parts, there are also their own boundaries. For example, on the boundaries of any tissue and organ in the human body there are membranes. Not only the internal organs have their membranes, different tissues in the human body also have their own membranes: Bones have periosteums, muscles have sarcolemmas, nerves have nerve membranes, blood vessels have blood vessel membranes, and even cells have their cell membranes. Nucleoli and mitochondria, etc., in the cells also have their membranes.

The membrane is the very boundary of the relevant part of the human body and its surrounding environment. It is the place where the motions of *kai he* (opening and closing), *chu ru* (going out and coming in) and *ju san* (gathering and expanding, including condensing and dispersing) of the *hunyuanqi* of this part are carried out. The motive power of motions at this level comes mainly from changes in *hunyuanqi*'s strength brought about by the metabolism in tissues. Let me give specific illustration about this with the instance of cells. Although cells in the human body differ in numerous ways and bio-chemical reactions in each cell also differ from each other, all cells separate themselves from the surrounding environment with cell membranes to ensure their own independence. Cells interact with the external world through cell membranes, thus forming the *kai he*, *chu ru* and *ju san* motions at the level of cell membranes. When the *hunyuanqi* pressure in cells is high, the membranes of cells expand and the *qi* pathways on the membranes open. The *qi* inside the membranes takes substances out to join the *hunyuanqi* of the whole body. Meanwhile, the *hunyuanqi* needed is also absorbed from the surrounding environment and enters the membranes with the closing movement. Motions as such are going on repetitively like this.

II. Motion of *Hunhua* in Human *Hunyuanqi*

The motion of *hunhua* is carried out on the basis of the motions of opening and closing (*kai he*), going out and coming in (*chu ru*) and gathering and expanding or/and condensing and dispersing (*ju san*), and there are also two stratums that are at the external boundary and at the internal tissue level in the mergence and change. Some of the *hunhua* processes are realized by means of concrete substances (here it refers to macromolecules, such as amino acids and monosaccharides) and some are realized directly with formless *qi* (including simple elements). The former is usually carried out in part of the human body while the latter is mostly carried out in the whole entity of the human body.

In a certain state, humans can live on mergence and change with formless *hunyuanqi*. *Bi Gu* (literally means evading eating crops, which actually means eating nothing at all) in *qigong* practice is just an instance of this. Among people who do not practice *qigong*, there are also some who can maintain their lives without eating food but only through drinking a little amount of water. The author of this book has met three ladies who have not eaten any food for many years. They have not practiced any *gongfu* but can also live as normal humans and have given birth to children. They just look thinner and weaker in their figure than other people. Evidently these people maintain their vitality by depending on *hunhua* with formless *hunyuanqi*. This function in fact exists in every human being. It is simply that the strengths of this function in people differ. Familiar intelligence people are accustomed to absorbing life materials from concrete substances and have formed "biased" adherence in their consciousness which repels the above-mentioned function and makes such function seem to disappear.

It has been talked about in the theory of *hunhua* in Chapter II that the *hunhua* process of concrete existences is a merging and changing process of the subject with the external object. This process includes the split of complicated integrate substance, which forms the simple *hunyuanqi* of smaller substantial entities that have lost the original complicated substantial entity's holistic features and have acquired the features of smaller individualities. Then the simple *hunyuanqi* of smaller substantial entities merges and changes with the subject, enters the substantial entity of the subject and becomes part of the substantial entity of the complicated subject. This mergence and change process of concrete substances in the human body is carried out in two stratums, i.e., the external stratum and the internal tissue stratum.

i. *Hunhua* at the External Boundary

The *hunhua* talked about here that is carried out at the external boundary refers to the mergence and change of the *hunyuanqi* field formed near the external boundary. It refers mainly to the *hunhua* of food (the *hunyuanqi* in concrete form) carried out by the stomach and the mucous membranes of the stomach and small intestine. This is the first step in the *hunhua* of concrete substances.

Hunhua carried out in the stomach and the small intestine with concrete substance is first of all a process of making the concrete substance that has form lose its innate *hunyuan* holistic features. This is realized through grinding the concrete substance that once has form and adding secretions from the mucosa of the stomach and small intestine and from other digestive glands to mix with it. This together with the functioning of the *hunyuanqi* field formed in the cavity of the digestive system makes the integrate *hunyuanqi* of the original food decompose into small units.

We know that substances at different levels in Nature have different levels of basic units (which are called fundamental bricks). For example, electrons, protons and neutrons are the fundamental bricks in the stratum of elements; elements are the fundamental bricks of compounds; amino acids are the fundamental bricks of cells with life; cells are the fundamental bricks of complex living organisms. Since all living organisms are established on the basis of cells, the mergence and change going on at the level of concrete substance in living organisms must meet the requirement of fundamental brick at the level of cell. The whole entity in any stratum is formed by integrating the fundamental bricks of this stratum into a particular time-and-space structure. The integration cannot happen if the requirement of the fundamental bricks is not met.

The decomposition of the macromolecules in food is not just simple decomposition but a preliminary *hunhua* with human body *hunyuanqi*. From the perspective of modern biochemistry, the digestive process in the stomach and small intestine is realized with the function of various enzymes that make such concrete substance as protein, sugar and fat decompose. These enzymes are gathered into form under the effect of human body *hunyuanqi*. Therefore, this digestive process is a process in which human *hunyuanqi* merges into object and make the concrete substance of object not only conform to the standard of life fundamental bricks, but also acquire the identical feature as the whole entity of the human being. Isn't it so? Milk directly injected into the muscles of the human body can cause fever in the person who has taken the injection as no effective *hunhua* has taken place in the absorbing process; milk that enters the digestive tract of a person, after having experienced mergence and change, can be safely absorbed. This evidence sufficiently shows the importance of the mergence-and-change process.

In addition, the process in which the mucous membranes of the stomach and small intestine take in nutrition is also an effective *hunhua* process. The objective concrete substance at this moment has lost its original holistic features and has become part of the human body. It is still not a direct participant in the life activities yet.

ii. *Hunhua* at the Level of the Internal Tissue

No matter it is the nutrients absorbed from the stomach and small intestine or the oxygen taken in from the lungs, it is only when they are conveyed by blood to the internal tissues and experience mergence and change at the level of cells that the *hunyuanqi* taken in from the outside world can take part in the metabolism of human life activities. It has already been realized in modern science that metabolism at the level of cells includes energy metabolism (which mainly refers to the metabolism of sugar and fat that provides energy) and material metabolism (which mainly refers to synthesis and decomposition of protein), and it is believed that these processes (esp. energy metabolism) are already clearly known.

In fact they are not clearly known because this is a very complicated mergence-and-change process. Take the metabolism of common cells in the human body for instance. No matter it is the process of energy metabolism or material metabolism, it is a very complicated changing process of material structure and energy transfer under the effect of enzymes. And in the generation and adjustment of various enzymes that participate in this process, there is a process in which enzymes are produced and destroyed. Besides, metabolism at the level of cells is affected not only by the genetic factors in the cells themselves, but also by hormones in the whole body and even by the nervous system. In sum, in energy metabolism there is the generation of substance (the mergence and change from nothingness into somethingness) and in material metabolism there is also the involvement of energy. In all this, there are still many questions (e.g., the functional mechanism of enzymes as well as energy transfer and change in protein synthesis) that are far from having been made clear in modern science. This is because none of these processes are just simple change but the whole entity level of *hunhua* or mergence and change.

This holistic *hunhua* is also a long chain reaction. For example, in the process of energy metabolism (which is called "tricarboxylic cycle" in modern medicine), over ten kinds of enzymes are needed, and the production of these enzymes needs other kinds of enzymes since the accomplishment of any bio-chemical change in the body of an organism needs the involvement of enzyme. This is a long and complicated catalytic chain of cause and effect involving the application and generation of enzymes.

In the theory of *hunyuan* whole entity, it is not required to go into extreme details about this since the relationship between enzymes in any stratum and the substances that they affect is basically a kind of antagonistic and complementary relationship of mergence and change. The function of hormonal adjustment and the receptor (leptocyte) is also in antagonistic and complementary mergence and change like this. The adjustment of nerves, to a certain extent, also needs the functioning of hormones. The theory of *hunhua* studies the mergence-and-change process in the stratum of cells, and the purpose is to exercise the function of human spirit (*shen*) and make the guiding effect of human spirit permeate into the level of cells. This is to further expand the sphere of human beings' consciousness functions and to enable human beings to comprehensively and completely command their own life motions.

To come back to the metabolism in the human body, energy metabolism consumes the *qi* of concrete substance and provides the energy for human beings' life activities, whereas material metabolism consumes a certain amount of energy and integrates the *hunyuanqi* of concrete substance into the human body's own time-and-space structure to make the concrete substance's *hunyuanqi* the embodiment of the substantial material state of human time-and-space structure. The mergence-and-change motion in this mode of metabolism is exclusively going on in tissues in every part of the human body. It is simply that the energy and substance produced in the *hunhua* of body cells are used by the body cells themselves, while in the *hunhua* of energy metabolism and substance metabolism in the cells of the internal organs, some substances generated, except for being used by the internal organ cells themselves, are also used by relevant tissues in the whole body, e.g., hormones secreted by the endocrine gland and the secretion of digestive juice in the digestive gland and so on are all like this. *Hunhua* of the two kinds of metabolism of substance and energy in the nerve cells, aside from supplying the nerve cells' own need, should provide all the information for the thinking activity and provide the energy that is needed in the motion of the thinking activity. This is to say that mergence and change going on in body cells, internal organs cells and nerve cells are not completely the same, and this is the reason why we divide human body *hunyuanqi* into body *hunyuanqi*, internal organ *hunyuanqi*, and *Yi Yuan Ti*.

In addition to this, consciousness activity has it own special content of mergence and change. Although *Yi Yuan Ti* is the *hunyuanqi* of the brain cells, it differs fundamentally from *hunyuanqi* at the level of life activity when independent consciousness activity is formed. Consciousness activity becomes the conceptual activity in the realm of thinking, and the divided state with consciousness activity and life activity "on the ends of two poles" is formed. Yet consciousness activity and life activity are also a unified whole entity, with consciousness acting as the commander of life. The process of their unification is their process of *hunhua* or mergence (and change). (Details about this can be found in Section IV Formation Process of Human *Hunyuanqi* in Human Life Activity).

The *hunhua* motions in a human being, aside from those carried out by the human being as subject to absorb things from the external world, also include those to interact with external existences: There is a *hunhua* process in both exerting functions on external existences and in receiving external stimulations. To exert influence upon external things, to transmit *yi* (mental information) and *qi* into object with or without the help of the human body and cause the object to change—this is a process to impress things in nature with the mark of the human being and is undoubtedly a kind of *hunhua*. In the process of receiving information when a person is cognizing external things, there is also a *hunhua* process with them. In actively searching for external existences, there is of course mergence and change in it which is similar to the mergence and change to exert functions upon external things. Even in passively receiving information, so long as attention can be roused, be it intentional attention or unintentional attention, consciousness can be implanted into the object, which forms an information exchange between the mental activity (*yi nian*) and the objective substance and imprints a certain mark in the realm of the subject's consciousness. This is a special kind of *hunhua* or mergence and change.

Section IV Formation Process of Human *Hunyuanqi* in Human Life Activity

All human life activities, such as metabolism of cells, motions and senses in organs and tissues as well as spiritual activities peculiar to humanity, are the embodiment of the mergence and change of human *hunyuanqi* at different levels. Human *hunyuanqi* is, on the one hand, consumed in the process of human life activities, and on the other hand, merges with *hunyuanqi* from the external world and changes it into human *hunyuanqi* to maintain the balance of human life activities.

The generating process of human *hunyuanqi* that frequently goes on in humans, similar to its formation process in the evolutionary history of the human race and the evolutionary history of human individuals, also experiences a changing process of firstly forming life level *hunyuanqi* and then evolving into human *hunyuanqi*. These two processes are to be introduced respectively in this section.

I. Generation Process of Life Level *Hunyuanqi*

The generation process of life level *hunyuanqi* in humans is a *hunhua* process in which human *hunyuanqi* merges with and transforms external *hunyuanqi*. This mergence and change is divided into *hunhua* in the abdominal cavity and *hunhua* in the tissues of the human body.

i. *Hunyuanqi* Generated in the Mergence and Change in the Abdominal Cavity
i) Mergence and Change in the Abdominal Cavity

The digestive process of various nutrients in the stomach and small intestine is a process in which human *hunyuanqi* merges with them and changes them. During this process, not only different kinds of nutrients are turned from their original complex structures into simple "fundamental bricks", but the "fundamental bricks" are meanwhile made to integrate and transform with human *hunyuanqi* to form *hunyuanqi* at the level of life. This will be expounded as follows:

Hunhua of different kinds of nutrients in the stomach and small intestine goes on in two aspects.

a. Mergence and Change of Food with Digestive Juices
In the digestive tract of an adult, six to eight liters of digestive juices are secreted everyday. In the digestive juices, there are various kinds of digestive enzymes. These enzymes act respectively upon different nutrients, such as protein, fat and amylum, and decompose them into "fundamental bricks". This is not a simple digestive process, but a process in which human *hunyuanqi* gathers, merges with and changes the food. This is because digestive juices come

from the body fluid that is processed by cells of the digestive organs and secreted by these cells, and this is a complex process. For example, digestion of protein needs pepsin in the stomach and trypsin and chymotrypsin in the small intestine. Pepsin comes from pepsinogen activated by gastric acid (dilute hydrochloric acid) and the secretion of pepsinogen and hydrochloric acid involves a chain of biochemical processes. Trypsin comes from trypsinogen activated by enterokinase and the secretion of pancreatic juice needs not only the adjustment of such body fluid as secretin, but also the adjustment of the vagus nerve, and the vagus nerve exercises its function by releasing acetylcholine. The biochemical reactions going on in the human body, like those going on in all other organisms, are accomplished under the effect of enzymes. Digestion of various nutrients needs enzymes and, in the production process of enzymes, other enzymes are also required. We know that all these substances with form are the condensed state of formless *hunyuanqi*. Therefore, the digestive process itself is a process of mobilizing human body *hunyuanqi* to generate and produce substances, hence it is a process for *hunyuanqi* to gather as well.

The digestive juices and the nutrients interact upon each other and make the complex structures of the nutrients decompose. For example, amylum turns into monosaccharide, fat becomes glycerol and fatty acid, and protein is changed into amino acid. In all, they are turned into tiny small structures (fundamental bricks) that can be absorbed and used by the human body. During this process, the fundamental bricks generated are assimilated by human *hunyuanqi* due to the influence of the enzymes in the human body.

b. Mergence and Change of Food and Fundamental Bricks with the *Hunyuanqi* Field

The position in the human body where food goes through mergence and change and becomes assimilated is right the place where human body *hunyuanqi* and the internal organ *hunyuanqi* are concentratedly distributed, thus the *hunyuanqi* and the *hunyuanqi* field here are relatively strong. The powerful *hunyuanqi* field has a great promotive effect on facilitating the digestion of food and the integration and transformation of fundamental bricks to make them have an identical feature as the human body. The effect of prenatal kidney *qi* is particularly significant in this enormous promotive effect. It, on the one hand, enhances the digestive function of the digestive system (In traditional Chinese medicine, there are the methods of nourishing the kidneys to nurture the stomach and giving *mingmen* more *qi* to invigorate the spleen) and, on the other hand, directly endows the *qi* freshly produced from *hunhua* with the *quan xi* nature of life.

Readers may wonder why the prenatal kidney *qi* can have such a great effect. It is said in *Nan Jing*, "*Yuan qi* is the source for generating *qi* in humans. It is the motional *qi* between the kidneys. It is the foundation of the five *zang* internal organs and the six *fu* internal hollow organs, the root of the twelve regular meridians, the gate of breath, ... with another name called God Guarding against Evils." The kidney *qi* stated here is in fact equivalent to the *hunyuanqi* field of

169

mingmen. "Between the two kidneys is the *Hunyuan* Maracle Chamber" stated by the ancient Chinese refers to this. In order to deepen our understanding about this, further illustration is given in light of modern medical science.

It has already been proven in modern science that kidneys and the adrenal gland can secrete many kinds of hormones directly related to holistic life activities. So the *hunyuanqi* field of kidneys will inevitably influence the generation and production of *hunyuanqi* in the human body. What is more important is that, in embryology, testes (and ovaries) not only come from the same origin as the kidneys, but are once positioned between them. With further careful inspection, it can be found that the cells that form the essence of testes (or ovaries) are not the cells of the three germ layers having gone through repeated divisions, but are from the cells with reproductive capacity in the yolk sac in its early stage. Since these cells have not been repeatedly divided, they still maintain the reproductive function with all the life information in it. The *hunhuanqi* field thus formed with such concrete substance inevitably retains its innate feature. This is perhaps the reason why it is said in traditional Chinese medicine that "the kidney is the fundamental of the prenatal". It is this prenatal holistic information function of kidney *qi* that endows the *hunyuanqi* freshly formed with human life feature.

ii) *Hunyuanqi* Formed in the Mergence and Change in the Abdominal Cavity
a. The Formation of Formless *Hunyuanqi*

We know that every substance has its *hunyuanqi* that marks its own holistic features. When a complex substance breaks down into simple substances, part of the *hunyuanqi* it once has merges into the simple substances and part of the *hunyuanqi* once in the complex substance is released, and the release of the combined energy talked about in modern science is only partial display of this part of the released *hunyuanqi*. This part of *qi* is released under the *hunhua* effect of human *hunyuanqi*. It has already changed its original feature, has become assimilated and can be directly used by the life activity in the human body, so it belongs in the category of body *hunyuanqi*.

Generally speaking, this part of *qi* is stored in the mesentery, epiploon and retinophore, peritoneum, abdominal cavity and membranes. It fills the abdominal cavity and permeates the stomach and intestines. It participates in the turning-into-*qi* change in food and can be mobilized with mental instruction to relevant parts of the human body to enhance the sensory and motor functions and so on. This *qi* is similar to *wei qi* (guarding *qi* or protective *qi*) stated in the theory of traditional Chinese medicine ("Guarding *qi* is the strengthful *qi* in water and grains. Bold, swift and smooth in motion, it is unable to enter blood vessels, so it flows in the skin and muscles, immerses among *huangmo*, the thin membranes between the five internal organs, and is distributed in the chest and the stomach." "Guarding *qi* is used in warming up flesh, filling up the skin, nourishing the grain of skin and the texture of the subcutaneous flesh, and carrying out the opening and closing motions of *qi*."). Mobilizing *dantian qi* to cure disease in some *qigong* practice uses mainly this part of *qi*. Hard-form *qigong* also uses mainly this part of *qi* to strengthen the functions of skin, muscles, etc..

b. The Formation of Fundamental Bricks of Nutrition

This refers to the exquisite substances formed through digestion, such as amino acids, glucose and fructose, fatty acids and vitamins. These substances that are characteristic of the feature of fundamental bricks have already experienced *hunhua* under the effect of human *hunyuanqi* on the one hand, and on the other hand, they can be absorbed and enter the circulatory system and can be conveyed into tissues in every part of the human body to carry out *hunhua* with the human body. Being similar to *ying qi* or nurturing *qi* stated in the theory of traditional Chinese medicine, this part of nutritious substances still cannot be directly called *hunyuanqi*.

ii. *Hunyuanqi* Formed in the Mergence and Change in Tissues of Each Part of the Human Body

The mergence and change going on in tissues in every part of the human body, on the one hand, carry out synthesis *hunhua* of the fundamental bricks of nutrition and form different kinds of tissues or macromolecule compounds in the human body. For example, amino acids form various proteins, fatty acids and glycerol form triglycerides in cells, glucose forms glycogen, etc.. On the other hand, synthesis and decomposition are also going on in ribonucleic acid (RNA) and deoxyribonucleic acid (DNA) in cells, and oxidation is going on in glucose and fatty acid.

The energy released in these processes is merely a kind of energy display when *hunyuanqi* has released from concrete substance, and such series of reactions are accomplished under the effect of various enzymes. There are as many as more than 3,000 kinds of enzymes known in the human body. Yet there are still a lot of reaction processes unknown. This shows that metabolism going on in tissues in each part of the human body is a rather complicated mergence-and-change process. Although all metabolism goes on at the level of concrete substance, the mergence-and-change process between somethingness and nothingness accompanies the entire metabolic process. There are both the process of concrete substance turning into *hunyuanqi* and the process of *hunyuanqi* condensed into concrete substance and also the process of combination and decomposition between substances with forms. The *hunyuanqi* produced from these processes is used both for the formation of tissues in part of the human body and for the consumption in various bodily motions, such as various muscle motions. This *hunhua* is generally the same whether it is in the body cells or in the cells of the internal organs. This is the reason why we say that there is body *hunyuanqi* in the internal organs.

Since *hunhua* in internal organ cells, aside from accomplishing the metabolism of internal structures of the cells themselves, secretes such substances as hormones that are not used by the internal organ cells, since in familiar intelligence people the internal organ cells only accept the adjustment by autonomic nerves and hormones and usually do not accept the command of cranial cells, we call the *hunyuanqi* of internal organs *zangzhen hunyuanqi* (internal organ true *hunyuanqi*) in order to show its difference from the *hunyuanqi* of the body (*quti hunyuanqi*).

II. Formation of Human *Hunyuanqi*

The *hunyuanqi* formed in the mergence and change in tissues in every part of the human body mentioned above can only be called *hunyuanqi* at the level of life since it has not become combined with consciousness activity yet. It is only when consciousness activity merges and combines with life level *hunyuanqi* that such *hunyuanqi* can be called human *hunyuanqi*. We know that *Yi Yuan Ti* is a kind of very fine and exquisite *hunyuanqi*. It can penetrate all levels of *hunyuanqi* and can therefore merge with life level *hunyuanqi* and change it into human *hunyuanqi*. This makes this special kind of *hunyuanqi* of *Yi Yuan Ti* permeate and command life *hunyuanqi* and imprint life level *hunyuanqi* with the mark of consciousness.

Hunhua of consciousness with life level *hunyuanqi* is carried out through three channels.

i. *Hunhua* through Nerve Cells

It has been pointed out in physiology that the nervous system is the commander of human life activities, and the commanding function is realized mainly through nerve cells all over the human body that transmit information. From the perspective of modern science, this process is a connection between the substantial entities of nerve cells and the substantial entities of tissues in different parts of the human body, and the medium is the nervotransmitters that convey information. It is still not clear in modern science how the instructions that belong to consciousness activities are turned into the information in nerve cells. The theory of consciousness in the greater theory of *hunyuan* whole entity gives a brief exposition about this motional process. It results essentially from that *Yi Yuan Ti* which is the special kind of *hunyuanqi* guides nerve cell *hunyuanqi* to *hunhua* (merge and change) with tissue *hunyuanqi* in different parts of the body. The various chemical substances which are the transmission medium produced in this process are but the substances condensed into form in accompany with the *hunhua* process. According to the theory of *hunyuan* whole entity, there must be reactions between the nerve cells that do not rely on chemical medium or electrical transmission. As to how to design experiments to explore this, it is the task of physiologists.

ii. *Hunhua* through Secretions from Nerve Cells

The secretions from the nerve cells talked about here refer not to the hormones secreted by the hypothalamus and the pituitary (as these hormones are the content of internal organ true *hunyuanqi* instead of the content of *Yi Yuan Ti*), but to the changes between somethingness and nothingness which are brought about by the changes in *Yi Yuan Ti* in the process of consciousness activities. Some substances produced in the brain cells can be distributed all over the body through the circulatory system, or through "neurons in contact with cerebrospinal fluid" to make the secretions of the brain merge and change with cerebrospinal fluid so as to influence the *hunhua* of the nerves with the entire human body. The neurons in contact with cerebrospinal fluid include the ependyma cells in the paraventricular organs on both sides of the third ventricle, the photoreceptor cells of the pineal complex, and the fibers formed under the apparatus that pass through the mesencephalic aqueduct and the fourth ventricle and reach directly the spinal cord.

iii. *Yi Yuan Ti* Directly Carries out *Hunhua* with *Hunyuanqi* in the Whole Body

This kind of *hunhua* in familiar intelligence people is carried out not consciously but naturally. It is primarily shown in that familiar any mental activity that occurs can cause the flow of *qi* in the process of different kinds of human activities. It is secondly shown in the influence of emotions on such changes as gathering and expanding (including condensing and dispersing) and rising and falling of *qi*.

Through the three modes of *hunhua* mentioned above, life level *hunyuanqi* is endowed with content of consciousness and thus becomes the *hunyuanqi* characteristic of holistic life features of human beings—human *hunyuanqi*. This mergence-and-change process is gradually realized in the development process from the late period of fetus to the period of infanthood and babyhood, in the process of both the development of the human body and its functions and the formation of consciousness. It is because of such a close relationship between consciousness activity and human body *hunyuanqi* that, once consciousness activity occurs in calm *Yi Yuan Ti* (particularly instructive consciousness activity), changes in the entire human body will take place. Functions of Consciousness (Section IV) to be talked about in The Theory of Consciousness in the next chapter as well as *Dao De* and Its Impact upon the Human Body's Life Activity (Section II) in the chapter that follows it in The Theory of Morality are both unfolded on this basis.

III. Re-Recognition of Humans as a Whole Entity of the Trinity of *Jing*, *Qi* and *Shen*

We have talked about in The Theory of *Hunyuan* (Chapter II), The Theory of Whole Entity (Chapter III) and as well in this chapter that a human being is a whole entity of the trinity of *jing*, *qi* and *shen*. It is particularly emphasized in this chapter that *jing*, *qi* and *shen* are the different forms of expression of human *hunyuanqi*, and human *hunyuanqi* is the manifestation of human holistic life feature. Although so many words have already been used to describe this whole entity, readers might still have some perplexity with regard to this whole entity. When studying the theories step by step, readers feel them well organized with the key points systematically explained; after having put the book aside and trying to associate these theories with realistic life, readers might feel at once at loss as in a dense fog. This is because modern science to which we have become accustomed has just touched upon the substantial or concrete structural characteristics of humans in its study of human beings and has made a lot of progress. In the study of consciousness activities, though such subjects as psychology, brain science, neuropsychology and physiopsychology have been established, still little about consciousness activities is known, thus is formed the deviated tendency of taking the study of the human body as the study of human life.

In people's common sense, existence of consciousness activity can be shown in people's actions (including language), so it is difficult for people to deny that it exists. In the case of *qi*, however, it is rather different. *Qi* is formless and imageless and resides in the activities of the human body and the human spirit. Not only is it hard for familiar intelligence people to perceive its existence, but it is also hard for modern apparatus to detect its trace before it is transmuted into specific energies. This is because *qi* is the *hunhua* state with all kinds of energies merged in oneness. We say that *qi* is the holistic existence with all the content of the human body and the human spirit in it. Yet as people have not established the mode of super intelligence to sense *qi*'s existence, it is very hard to comprehend.

So some of our readers might ask: Now that we can't understand *qi*, why do you still talk about it? Why don't you directly talk about the concept of body-and-heart whole entity instead? You can use the achievements in modern science, and it is easier for people to understand, study and grasp. That is really "killing two birds with one stone" and why don't you follow this advice? It is because without touching upon *qi*, the concept of human body-and-heart whole entity cannot be established, nor can the true human nature be reflected and revealed. To take an overview of the statements of our predecessors, be it the theory of soul in theology, the theory of the four elements in Buddhism, the theory that "The human is a machine" according to some physicists, or philosophers' dualism of matter and spirit, … none of them can accurately and profoundly illustrate the nature of human life, nor can they reveal the laws for human beings to develop into more advanced stage. Even the concept of *jing*, *qi* and *shen* as a whole entity in traditional *qigong* and traditional Chinese medicine has not fulfilled the task of facilitating human beings to develop into more advanced level. Why is it so? You may wonder. It is because, in the development up to the present, human beings have not actually formed the genuine whole entity of the trinity of *jing*, *qi* and *shen*, nor have they sensed the existence of *qi*.

We know that the nature which tells human beings apart from animals is that human beings' consciousness activities established on the basis of their life activities can control their life activities with initiative under the guidance of their consciousness. This shows that human beings' consciousness activity has acquired its independence. However, in the early stage of human beings' existence, though people already differed in nature from the chaotic state of animals in which nerve activities and life activities were combined, human consciousness activities were still subordinate to their life activities. For human beings today, human consciousness has acquired great independence—it is like an unbridled horse that can gallop freely in every realm in the universe and like a magic bottomless bag that can take into it all other existences in the universe. Human beings' consciousness seems to have become King having mounted the throne over the universe. However, this "god" of human consciousness does not know to consciously give orders to its own flesh but gives it free rein instead. As a result, human beings who presume to dominate the universe have not even controlled themselves.

From the perspective of the theory of *hunyuan* whole entity, *Yi Yuan Ti* differs from life level *hunyuanqi*. In fact, *Yi Yuan Ti* overrides the *hunyuanqi* at the level of life activities. Although consciousness carries out mergence and change or *hunhua* with life level *hunyuanqi* as stated above, consciousness and life level *hunyuanqi* are still far from reaching genuine unity. This is because a) human beings still have not recognized the channel, the method and the paramount significance to actively command the *hunyuanqi* at the level of life, therefore they still cannot command it with initiative; b) *Yi Yuan Ti*'s activity has not formed an integrated and unified state of mind and human consciousness is distracted by the states of mind based on different sensory organs.

Since human sensory organs all receive partial features of existences in a particular aspect, people who want to know the holistic feature of things have to make a comprehensive analysis of information received with different sensory organs before they finally abstract an integrate concept from it. Such thinking processes are the process in which holistic *Yi Yuan Ti* takes possession of different states of mind related to sensory organs, thus it is also a process in which *Yi Yuan Ti*'s integrate state of mind is divided by them. In such a state of mind, it is impossible to take real command over different organs' activities which belong to the level of life activities. This is the present state of human beings' lives today. This undoubtedly provides food for thought for dualism in philosophy. The whole entity of the trinity of *jing*, *qi*, *shen* in traditional *qigong* and traditional Chinese medicine is also a description of this state and is therefore incomplete because human consciousness has not truly permeated into the life activities in different parts of the human body.

Then how can the whole entity with *jing*, *qi* and *shen* in genuine unity be realized? The answer lies in establishing *Yi Yuan Ti*'s integrate and holistic state of mind, and this needs and depends on developing and tapping super intelligence. This will not only enable people to perceive the real state of the genuine unity of *jing*, *qi* and *shen* in their own body and get rid of their puzzlement about the mysteries of life, but will also make human life activities united in *Yi Yuan Ti* so that human beings will enter a truly free, self-awakened, and self-commanding realm and thus advance rapidly into a more advanced stage. Yet all this, instead of being included in this chapter, is what will be elaborated in The Theory of Consciousness (Chapter V).

Chapter V The Theory of Consciousness

According to the theory of *hunyuan* whole entity, consciousness activity is a special motional mode of substance. It is the activity state of the special substance of *Yi Yuan Ti* and the content and process of motion in it. Consciousness activity comes from the reflection of all kinds of relevant information in the internal and external environment of the human body's life process (such as social information, natural information and life information) and, through certain channels, it controls human life motion and movement.

What is to be talked about in this chapter, such as the nature of consciousness, the laws of motion of consciousness and the functions of consciousness, is relatively novel. They seem to be quite different from the knowledge of consciousness in people's common sense, particularly in philosophy. In order to make it easier to understand these statements on consciousness, let us leave the theory of *hunyuan* whole entity for a while and begin with the philosophical proposition of the materiality of consciousness.

Section I Philosophical Demonstration of Materiality of Consciousness

I. Materiality of Consciousness according to Great Philosophers of Dialectical Materialism

In Marxism, dialectical materialism is applied to the process of cognition. It is believed in Marxism that cognition begins from sense perception to perception and then to image, which is at last abstracted into concept and enters the realm of consciousness. It is illustrated from the evolutionary history of the human race and the evolutionary history of human individuals that consciousness is the product of the human brain.

Friedrich Engels said, "Our consciousness and thinking, no matter how supersensory they seem to be, are after all the product of the material flesh organ of the human brain. Substance is not the product of spirit while spirit is but the most advanced product of substance. This is of course pure materialism."(*Ludwig Feuerbach and the End of Classical German Philosophy* in *Selected Works of Marx and Engels*, Vol. 4, p. 223)

When talking about the motions of matter, Engels said, "The motion of matter is not merely crude mechanical motion, mere change of place, it is heat and light, electric and magnetic stress, chemical combination and dissociation, life and, finally, consciousness."(*Dialectics of Nature* in *The Complete Works of Marx and Engels*, Vol. 20, p. 376)

This is equal to dividing the motional mode of matter in the universe into seven levels, i.e., displacement, light, electricity, magnetism, chemistry, life and consciousness. The seven levels can be summarized into three larger domains: the domain of the motion of inanimate matter, the domain of life and the domain of consciousness (Modern science has conducted a lot of research into the first domain, but very little into the latter two domains. Thus we know rather little about them). This shows that Engels considered consciousness as a motional mode of substance.

According to dialectical materialism, motions are all motions of substances. There is no motion that is not the motion of substance and there is no substance that has no motion in it. Now that consciousness is a mode of motion, it is inevitably a mode of existence of substance.

Some people say that the motional mode of substance cannot be equated with the mode of existence of substance. We say that this is incorrect. An explicit answer to this has already been given by Engels, "Motion in the most general sense, conceived as the mode of existence, the inherent attribute of matter, comprehends all changes and processes occurring in the universe, from mere change of place right to thinking."(*The Complete Works of Marx and Engels*, Vol. 20, p. 408), and "Matter without motion and motion without matter are equally unimaginable."(*The Complete Works of Marx and Engels*, Vol. 20, p. 65)

Vladimir Lenin also pointed out in his book *Materialism and Empirio-Criticism*, "Things psychological and consciousness, etc., are the most advanced product of matter (namely, things physical). They are the function of this particularly complicated mass of substance called the human brain." He also cited what Plekhanov had said, "Consciousness is the internal state of substance." As to this, it was further pointed out in a textbook of philosophy for college students *Dialectical Materialism and Historical Materialism* in the ex-Soviet Union in 1950s, "Consciousness is the internal state of the substance that has the special structure of the human brain." Mao Zedong further explicitly pointed out in *Outline of Dialectics and Materialism*, "To see this something of consciousness with the viewpoint of complete materialism, consciousness is nothing but a motion form of matter."

Up to now, though the scientific community is still not clear about what form of motion the motion of consciousness actually is, a justifiable explanation can be obtained about it through deducing from achievements in modern science. It has been scientifically proven that in any concrete substance, so long as its internal state exists, there is a field form existence outside it correspondent to it. Concrete substance and field form substance always exist in a dual-structural pattern. The great masters mentioned above regard consciousness as the "product" and the "internal state" of the human cerebrum. Therefore, it can be deduced and imagined that consciousness is the relevant motion of substance that exists with the concrete substance of the cerebrum as a dual-structural, field-like form of existence.

II. Scientific Experiments Have Proven that Consciousness Has the Material Characteristics Confirmed in Science

i. Consciousness Activity Can Demonstrate the Characteristics of Energy

Quite a lot of experiments in recent years in super intelligence have proven that activity of consciousness can cause such changes as transformation and displacement in objects. For example, the experiments in teleportation conducted by Wang Bin and Wang Qiang (two children with super intelligence) monitored by Prof. Lin Shuhuang in *Beijing* Normal University were very rigorously designed and met the requirement of having only one sample for each experiment, which eliminated the possibility of cheating. The samples having been successively teleported in their experiments included magnets, insects, M3 screw nuts, matchsticks, M6 screw nuts, and mechanic watches. They even slowed down the electronic watch and made it behind in time. To make objects with body and weight change in place or form, material force must be applied; otherwise, it would be impossible, and this is already a final conclusion in science. Now that consciousness activity can make displacement and transformation happen, this proves that consciousness activity has strength in it and the material attribute of consciousness is thereby confirmed. Success in these experiments provides scientific grounds for what we say in our daily life "spiritual force" and "exercising subjective dynamic function". In fact, Marx pointed out a long time ago, "Consciousness is also a kind of force."(*The Complete Works of Marx and Engels*, Vol. 3, p. 496) Engels also mentioned, "Thinking is a form of energy and a function of the brain."(*The Complete Works of Marx and Engels*, Vol. 3, p. 368) If this force of consciousness comes not from substance, it can only be attributed to the power of God or gods. This is just as what Lenin said by citing Joseph Dietzgen's words, "The force of non-existent substance does not exist... If idealistic scientists of natural science believe that force is nonmaterial existence, at this point they are not scientists of natural science, but ... the men who catch sight of ghosts."(*The Selected Works of Lenin*, Vol. 2, pp. 272-273)

ii. Consciousness Can Display Material Functions

In the experiments conducted in *Zhineng Qigong* in recent years, different kinds of mental activities or instructions can not only be used to grow or kill cancerous cells, but can also be applied to production in industry, agriculture, forestry, husbandry and fishery. In the two academic symposia held at the beginning and the end of 1992 by *Zhineng Qigong* Research Group of China *Qigong* Science Research Association, nearly a hundred of research papers were respectively submitted in which achievements in experiments carried out by *Zhineng Qigong* organizations or individuals in cooperation with local researchers of related research organizations in different places in China were reported. In these papers, reports on experiments in increasing the yield of crops constituted the major part. For example, the yield of wheat increased by 4.8 percent to 13.5 percent; the yield of rice increased by 4.6 percent to 26 percent;

the yield of corn increased by 3 to 28.5 percent; the yield of cap fungus increased by 30 to 45 percent; the yield of Chinese cabbage increased by 32.87 to 35 percent, the yield of pumpkins increased by 26.7 percent. Other crops such as cotton, soybeans, tomatoes and onions all successfully increased in their yield.

What is more amazing is *Zhineng Qigong* external *qi* experiment conducted by 73-year-old Prof. Li Xiaofang of Beijing University in the county of Chi Feng in Inner Mongolia: She brought about the unprecedented achievement of getting in two yields within one year, something normally impossible at the altitude of Chi Feng in China. In the experiment in Shijiazhuang Railway Institute, the activity of mental instruction helped to increase the anti-pressure strength of pre-cast concrete panels. There are numerous additional examples like these. All these achievements have convincingly proven that mental activity has special material features; otherwise, these material achievements would have been impossible.

III. Scientific Experiments Have Proven that Consciousness Has the Material Characteristics Recognized in Philosophy

It was pointed out by Lenin that "The concept of substance in epistemology, as we have talked about, refers to nothing but the objective substantiality which does not rely on human consciousness yet is reflected by human consciousness."(*The Selected Works of Lenin*, Vol. 2, p. 267) He also pointed out that "Substance is a philosophical domain which marks objective substantiality. This objective substantiality is what is perceived with human senses. It exists independent of our senses and is copied, photographed and reflected by our senses."(*The Selected Works of Lenin*, Vol. 2, p. 128)

A great many experiments over many years have proven that consciousness conforms to this definition of substance.

At the conference for changing personnel upon completion of a term of office held by *Beijing Qigong* Research Association in 1981, a *qigong* master disagreed that there exists telepathy ability in *qigong*. Mr. Wang Zhongping, who at that time was working in the *qigong* research association, tried to convince him with an experiment. Wang, touching a washbasin with his hand, asked this *qigong* master to let the turning action around this basin occur in his mind and Wang said he could perceive in which way the *qigong* master mentally turned around it. The *qigong* master wanted to have a try. He first thought of turning around the basin one circle clockwise, then a circle counter-clockwise before at last he turned half a circle clockwise and half counter-clockwise. It turned out that Wang Zhongping judged all correctly. In this experiment, this person's consciousness activity was independent of Wang Zhongping's consciousness; it was not affected yet was perceived by Wang's consciousness.

In 1988, a meeting was held for the establishment of the *Qigong* Research Association in Northeast Electric Power Institute in the city of Ji Lin. Wang Youcheng was invited there to give a *qigong* performance. He said to the audience that he could know what they were thinking and this aroused great interest among the teaching staff and the students. The president of the institute came to the front. "Let me have a try," he said, and he thought of a person's name. He asked Mr. Wang what the name was. Wang wrote the person's name on a piece of paper first, then he asked the president to speak it out, after which Wang displayed the name he wrote on the paper to the teachers and the students. He was completely correct. The vice president went onto the front stage and thought of a mathematical formula in a foreign language. Wang Youcheng, who did not know any foreign language, drew it on a piece of paper according to what he perceived, and it was also correct. The vice president then thought of attending a scientific conference in Belgium, and Wang again gave a correct answer. Wang Youcheng's performance was acclaimed with bursts of applause.

In 1980, a three-year-old child's performance was broadcasted on Beijing TV. Touching his father who was doing calculations in his mind, the child could perceive the thinking activity going on in his father's mind and could tell the answer.

From 1980 to 1981, we performed many such experiments: A crowd of people all simultaneously concentrated their mind on thinking of a Chinese character. The children with super intelligence could tell what Chinese character these people were thinking about.

The perceptibility of consciousness proven in these experiments shows that human consciousness conforms to the definition of substance pointed out by Lenin: "The concept of substance represents nothing but the objective substantiality we perceive with our senses."(*The Selected Works of Lenin*, Vol. 2, p. 273)

IV. To Acknowledge Materiality of Consciousness Does not Violate the Philosophical Proposition of Primariness of Substance and Secondariness of Spirit

Lenin once pointed out, "The opposition between substance and consciousness has its absolute meaning only in a rather limited sphere. Here, it has its absolute meaning only in the sphere concerning the fundamental question of acknowledging which is primary and which is secondary in epistemology. Beyond this sphere, the opposition between substance and consciousness is undoubtedly relative."(*The Selected Works of Lenin*, Vol. 2, pp. 147-148) Mao Zedong also pointed out in *Outline of Dialectics and Materialism*, "The opposition between substance and consciousness is meaningful only in the realm of epistemology. If they are regarded as being opposing to each other beyond the realm of epistemology, it is undoubtedly going against dialectics." This is a fundamental question in philosophy. In terms of this fundamental question, we certainly believe that the universe is material and that substance determines spirit. Denying this principle is idealism which is sharply against materialism.

We admit that substance determines spirit and that consciousness (spirit) is also substance. How can the two points be unified? The proposition that substance is primary and spirit is secondary and that substance determines spirit in the epistemology of dialectical materialism is stated from the general and universal sense of the universe. We say that consciousness is also substance and the substance we describe here is not substance in the general and universal sense but a special type of substance. There is a difference of universal proposition and particular proposition between them.

The earth was already a material world before the appearance of human beings on it. But human consciousness (spirit) couldn't exist at that time. It is only when human beings appeared with the evolution of species on the earth that consciousness as the special function of the human brain came into existence with it. Therefore, consciousness is a special motional mode of matter established on the basis of a relatively high-level and complicated material stratum. It is determined by the substance of the human brain. Therefore, to say that consciousness is also substance does not contradict the statement that substance is primary and spirit is secondary or that substance determines spirit. This agrees with what Lenin illustrated when he cited Joseph Dietzgen "Substance is the boundary of spirit and spirit cannot go beyond the boundary of substance. Spirit is the product of substance, but substance is something bigger than the product of spirit."(*The Selected Works of Lenin*, Vol. 2, pp. 249-250)

What is the difference between the acknowledgement of this material feature of consciousness and the viewpoint of the material feature of spirit in vulgar materialism?

Vulgar materialism is a philosophical school of materialism that appeared in the 1930s when the new Hegel School disintegrated. It is believed in vulgar materialism that everything in the universe is material and that spirit is also material. It played an active role at that time in opposing the idealistic view which believes that all is spiritual. Nevertheless, it is believed in vulgar materialism that this substance of spirit is secreted by the substance of the human brain and that the human brain secretes spirit in the same way as the liver secretes bile. This, however, vulgarizes, simplifies and absolutizes the existential form of substance. It seems that the existence of substance can only be in the form of concrete entity and that it does not have any special form of existence. As to this point, even Lugwig Feuerbach, who was a materialist at that time, could not agree, saying that if they were materialists, he himself would rather not be a materialist.

We say that consciousness is material and we mean that consciousness is a special form of substance that differs from the general existential form of substance, namely, it is a special mode of motion of substance established upon the foundation of the special material structure of the human cerebral cortex. It is born in the human brain, belongs to the human brain and cannot separate for even a moment from the human brain. Consciousness is not concrete substance. All these are viewpoints drastically different from vulgar materialism.

However, we should also realize that vulgar materialism should still be categorized in the domain of materialism. In the fundamental point of the material feature of consciousness, it is correct. When criticizing the mistake in Hegel's idealism, Feuerbach dropped from it the viewpoint of development which is the nucleus of dialectics. Engels criticized Feuerbach, saying that he threw the baby out with the bath water. Therefore, when we criticize vulgar materialism, we should protect its correct fundamental point that consciousness is material and avoid "throwing the baby out".

V. Difference between Acknowledging Materiality of Consciousness and Monism and Dualism in Idealism
i. Difference from Idealistic Monism

In idealistic monism, it is believed that spirit is spiritual and substance is also spiritual. Its argument is set forth on the basis of spirit and spirit is believed to be incognizable. We believe that substance is material and that spirit (consciousness) is also material. Our argument is set forth on the basis of substance and we believe that substance is cognizable. In the statements of many viewpoints, there seem to be some similarities between Buddhist idealist monism and our theory. Yet in fact, we have already changed its fundamental point. Consciousness in Buddhism is believed to be fundamental. It comes from time immemorial and has no birth or decease, hence it is incognizable. Consciousness, according to our theory, does not come from time immemorial. It is gradually established with the appearance of the human cerebral cortex and also with human social practice. The generation and motion of consciousness have special laws of its own and these laws are cognizable.

ii. Fundamental Difference from Dualism

In dualism, the universe is believed to be both material and spiritual and the two of them simultaneously exist. The material world is the realistic world, the world that is on this bank, whereas spirit is the thing-in-itself, the world on the bank beyond. The two of them are independent of each other and irrelated to each other, and between them lies a wide gap that is impassable. Human beings can only know the superficial phenomena of this dualistic world, but they cannot know its fundamental content and nature. In *Zhineng Qigong*, it is believed that both consciousness and material are objective existences and that consciousness has its material feature. This cannot be equated to dualism because in *Zhineng Qigong* it is also believed that consciousness is the product of the advanced substance of the human brain (and society as well) and is closely related to substance. Consciousness and material can combine with each other and merge into each other. Their content, nature and laws of motion can all be cognized.

VI. Great Significance in Acknowledging the Material Feature of Consciousness

What is the significance of acknowledging the material nature of consciousness? The significance, we think, can be evaluated from at least two aspects. One is that it concerns whether the research in *qigong* science can develop forward along the right track; the other is that it concerns whether the world outlook and methodology of dialectical materialism in Marxism can be realized in the realm of life science and the nature of God can be cognized.

In Section IV Functions of Consciousness in this chapter, we are going to talk about the enormous functions and tremendous power that consciousness can exercise on people's life activities and on some substances in the objective world. And this will bring about a question: Where does the force that cause changes in these substances come from? There can only be two answers. One is that the power that causes changes in these substances comes from another substance and so we must acknowledge the materiality of consciousness. The other is that the power that causes these substances to change comes not from substance, that is to say, a nonmaterial force brings about changes in them. What is this nonmaterial power? Where does it come from? The second answer will inevitably lead to two results:

a) Agnosticism. It is not known where the power comes from nor what the power actually is. Yet it is believed in materialism that all things are cognizable. Some of the things may not be perceived for the time being, but they will not remain incognizable forever and agnosticism is incorrect. Denying the materiality of consciousness will make research into questions about consciousness impossible, and this will inevitably lead one who denies the materiality of consciousness to agnosticism.

The study of consciousness in psychology at the present is just the study of consciousness through people's behavior, which can only study the relationship between consciousness and behavior instead of studying the true laws of consciousness itself. With such kind of research method used in studying consciousness, the true colors of things sometimes cannot be discovered. For example, a Chinese saying goes that "words spoken are the voice of the heart." In fact, sometimes the words spoken may probably not be the voice of the heart. A hidden secret agent of the opposing side, for instance, behaves and talks actively and positively but thinks otherwise. Denying materiality of consciousness and thinking that consciousness is incognizable will make it relatively difficult to reveal the nature of things like this. To admit the material nature of consciousness and study the innate laws of consciousness itself, and particularly to use super special intelligence in perceiving people's inward mental activities, will make disclosing the nature of such things easier.

b) Pantheism. People with super intelligence can move things with their consciousness, e.g., they can move cups or watches, etc., without leaving any trace of how these objects change their places. What force is functioning in this? It is the force of consciousness. Consciousness can cause displacement in objects, which proves that consciousness is also material because only material force can make objects move. If materiality of consciousness is denied, the displacement of objects that happens with no external forces acted upon them can only be attributed to the power of gods, and this will fall into the realm of pantheism. People who have opinions like this will become those "who catch glimpses of ghosts" as what Lenin said. Some people who in fact want to maintain the purity of dialectical materialism and deny the material nature of consciousness turn themselves, however, into followers of pantheism as a result. Dialectics is thus merciless: You go beyond truth by just one step and you are on the opposite side of truth.

Our task now is to implement dialectical materialism in *qigong* science. Should dialectical materialism be carried through to all realms of research or should it be given up halfway? We must carry it through to all academic fronts and this means we must implement dialectical materialism in the realm of the human spirit in order to disclose the laws of motion of this special substance of consciousness and to realize the true nature of the motion of human life so as to drive theology out of every corner of the natural world. Engels said in *Ludwig Feuerbach and the End of Classical German Philosophy*, "After people synthesized alizarin, God's last haven was demolished." This statement was very powerful and very significant in doing away with theological thinking in the West at that time since people who worship God believe that all substances in this world are created by God. Now that humans have synthesized the new substance alizarin, the doctrine that God creates all comes to an end and the sanctuary of God is dismantled.

However, this sanctuary torn down is not God's last refuge. He has another refuge in the realm of consciousness. This is because many people now still believe in the existence of God, including some scientists. For example, many Western scientists worship God, saying that inspirations visit them when they say prayers to Him and that this is helpful for their creations and inventions. In fact, being absorbed in saying prayers is actually a kind of *qigong* practice. Inspirations that sometimes occur in prayers are the effect of *qigong* practice. If we, through *qigong* science research, realize and grasp the materiality and the laws of motion in the realm of consciousness, the thought of dialectical materialism will take command of the realm of consciousness and the last refuge of God will be truly demolished. This will be a great contribution of the cause of *qigong* science to the philosophy of Marxism.

VII. What Kind of Material Motion Is Consciousness

All the content about consciousness will be expounded in the following sections in this chapter in light of the theory of *hunyuan* whole entity. Before these theories are expounded, it is relatively difficult to explain clearly the materiality of consciousness from a theoretical perspective. In order to make it easier for readers to understand, a brief illustration will be given here with the aid of knowledge in modern science.

In the theory of *hunyuan* stated previously, we divided substances in the universe into three levels with the aid of modern scientific knowledge. The first is the substance that has the three features of mass, energy and information and exists essentially in the form of mass, and this is concrete substance. The second is the substance that has energy and information and is shown essentially in the form of energy, and this is field substance. The third is the substance that manifests essentially in the form of information while both energy and mass are not evident, and consciousness belongs to this level. Consciousness is a substance that exists with energy and mass in a hidden state and information in an evident state. It is a substance that still cannot be received with any instrument. However, its existence can be cognized and confirmed in *qigong*.

Why do we say that consciousness is a substance that exists with energy and mass in a hidden state and information in an evident state? To illustrate this point, we need to trace back how consciousness is formed. An existence has many characteristics such as mass, energy and information. When it is received by human sensory organs, its compound information will form in *Yi Yuan Ti* an image of this substance. The formation of this image results from the imprint of the energy and the information of this substance in *Yi Yuan Ti* with the mass of this substance having been extracted during this process. The image continues to be processed in the cerebrum and is abstracted into concept with the energy of the substance also extracted, while what is retained is only information. Therefore, we say that consciousness activity is actually the information activity in the human brain and that consciousness is the substance with information in the evident state.

Some people say, "Isn't consciousness activity the activity of the brain cells?" We respond that the activity of consciousness and the activity of brain cells are related yet they are not the same. Consciousness activity is the activity that takes place on the basis of the activity of the brain cells. Brain cell activity includes physical change and chemical change, yet consciousness activity cannot be equated to the physical and chemical changes of the brain cells. As Engels pointed out a long time ago, "One day we shall certainly 'reduce' thinking experimentally to molecular and chemical motions in the brain; but does that include all the nature of thinking?"(*The Complete Works of Marx and Engels*. Vol. 20, p. 591)

This shows that the physical and chemical changes taking place in the brain cells are not equal to the activities of consciousness. This is just like that there are both strong and weak electric currents moving to and fro in the transmission antennas of radio and TV stations, while outside the antennas electromagnetic waves are formed and transmitted to far-away places. There is a close relationship between the electromagnetic waves and the electric currents in the antennas, yet we cannot say that the electromagnetic waves are just the electrical currents. They are not things at the same level. Consciousness and the activity in the brain cells are not things at the same level, either. The electromagnetic waves are based on the electric currents but they are not the same thing. Consciousness is based on the activity of the brain cells, yet they are not the same, either.

As to how the functions of the cranial nerve cells help to form consciousness activities, this cannot be clearly explained in just a few words. Let us start from the fundamental points of the entity, nature and functions of *Yi Yuan Ti*—Mind Oneness Entity.

Section II *Yi Yuan Ti* (Mind Oneness Entity)

Yi Yuan Ti (Mind Oneness Entity) is a special kind of *hunyuanqi* formed with mergence of the brain cell *hunyuanqi* having developed to a certain stage. According to the theory of *hunyuan* whole entity, any concrete object is the condensed-in-form state of the *hunyuanqi* of the object, around which is its *hunyuanqi* sparsely distributed. One feature of nerve cells is that, aside from the function of absorbing and releasing concrete substances in their own metabolic process, they also have the function of receiving and sending information conveyed by energy. This enables nerve cells to not only receive and reflect external information extensively, but also send information and act upon objective existences, which makes the connection between the subject of a human being and the outside world have new channels and content. When nerve cells gather to a certain extent and form the brain, the *hunyuanqi* around the cranial nerve cells merges into a whole entity and can integrate the partial information received by every nerve cell into holistic information. We call the *hunyuanqi* of the brain with this kind of reflective feature *Nao Yuan Ti* (Brain Oneness Entity). When *Nao Yuan Ti* can further abstract and form concept and think in terms of concept, we call it *Yi Yuan Ti* (Mind Oneness Entity), and this function continuously improves with the evolution of the nervous system.

Yi Yuan Ti contains all the life information of the human body. It has the functions of receiving all kinds of information and reflecting existences, of saving, synthesizing, analyzing and extracting information and sending instructions (also a form of information). When it is not disturbed by the objective world, it maintains its original even state similar to "being solitarily still" as said by the ancient Chinese. When it is disturbed by various information, reflection and effect that conform to object will occur in it and this is similar to what was described by the ancient Chinese—"The moment it (which refers to *Yi Yuan Ti*) feels it (which refers to information), the inside (*Yi Yuan Ti*) and outside (information of object) becomes immediately and thoroughly connected". *Yi Yuan Ti* is both the subjective world and the objective existence, which can be perceived by consciousness in an advanced *qigong* state. It is the internal foundation for the generation of consciousness activity.

I. *Yi Yuan Ti*'s Features

Yi Yuan Ti is a formless and imageless material state that people with familiar intelligence cannot see or feel. It is extremely even inside, without any difference at all. It exists in a mutually accommodating manner with the substantial entities of tissues of the human brain. It is centered in the brain (the center of the cerebrum) and is distributed all over the inside and outside of it. *Yi Yuan Ti* is also connected with the entire human body and spreads outside it. *Yi Yuan Ti* is the *hunyuan* (merged-into-oneness) state of the holistic functions of all the nerve cells of the nervous system in the human body and their *hunyuanqi*. The physical, chemical and biological changes of the nerve cells (essentially the brain cells) are the foundation upon which *Yi Yuan Ti* is established. It also has a certain independence and can react upon the nerve cells. The relationship between nerve cells and *Yi Yuan Ti* can be analogized to a live conductor and its surrounding electric field or to a magnet and the magnetic field around it. Yet there is also a great difference between them. *Yi Yuan Ti* has the characteristics in its entity and attribute in the following aspects:

i. The Feature of Evenness and Exquisite Subtleness

As it has been stated above, *Yi Yuan Ti* is formed with the cranial nerve cells' *hunyuanqi* merging together. It should be made clear that the nerve cells' *hunyuanqi* talked about here refers to the *hunyuanqi* that embodies the function of the nerve cells of receiving and sending energy and information. We know that, though body cells and internal organ cells differ in some aspects in their functions, they are the same in that none of them have surpassed the metabolism of concrete substance, including energy metabolism and material metabolism. Nerve cells, however, have developed further into "information metabolism", such as receiving, conveying and sending information, the functions which embody the going out and the coming in of information similar to yet in fact different from common material metabolism. This is the function peculiar to nerve cells and is also their essential function. Both material metabolism and energy metabolism in the cranial nerve cells serve information metabolism. *Yi Yuan Ti* formed with the mergence of the *hunyuanqi* that embodies this part of function of nerve cells has been elevated to a level higher than body *hunyuanqi* and internal organ *hunyuanqi* whether it is in the degree of exquisiteness or at the level of function and is rather similar to primordial *hunyuanqi* of the universe.

Moreover, special features in the structure of the cranial nerve cells help bring about *Yi Yuan Ti*'s evenness and exquisite subtleness features: a) It is already known in modern science that cells, instead of being hard and solid, are gelatinous units enclosed in liquid crystal state cell membranes and the "soft" feature of the nerve cells, particularly the cranial nerve cells, is even more outstanding. This helps to enable the very large number of nerve cells to closely gather together, which makes the *hunyuanqi* of the cranial nerve cells permeate into each other and merge with each other to the utmost extent and reach an infinitely even state. b) The feature of the brain cells that do not further divide or reproduce reduces their disturbance to *Yi Yuan Ti* and guarantees the formation of *Yi Yuan Ti*'s infinitely even state. c) Cells with numerous synapses combine into one entity. All this lays a foundation for the formation of *Yi Yuan Ti*'s characteristics in entity and attribute in all aspects.

ii. The Feature of Reflectiveness

Yi Yuan Ti's infinite evenness characteristic in entity and nature determines its reflective characteristic when interacting with things in the outside world. *Yi Yuan Ti* can truthfully reflect differences in object, which is similar to a mirror reflecting the shape and color, etc., of objective existences. It is just that a mirror merely reflects the optical features of objects and the reflection is of one stratum (of the reflected objects) and on one stratum (of the mirror), whereas *Yi Yuan Ti* can reflect all the features of things. Moreover, *Yi Yuan Ti* has innumerable "stratums" which cannot only reflect the state of object but also the internal state of *Yi Yuan Ti* itself. Reflecting existences in *Yi Yuan Ti* can be realized through changes in the functional state of nerve cells which induce changes in the *hunyuanqi* of nerve cells. It can also be realized through *Yi Yuan Ti*'s directly receiving the holistic information about objective existences to make the reflection shown in *Yi Yuan Ti*.

The higher the degree of *Yi Yuan Ti*'s void, initiative, brightness and clearness, the greater *Yi Yuan Ti*'s sensitivity in reflection. Improving *Yi Yuan Ti*'s degree of voidness and initiative as well as brightness and clearness, apart from relying on enhancing the nurturing of *qi* to it, is determined by the improvement of its degree of regularity in motion and the elimination of various distractions. Generally speaking, enhancing regular motion can reduce the disturbance of various "noise" while reducing distraction helps the smooth continuance of regular motion. The two of them thus complement and complete each other. Traditional *qigong* usually begins with calm-form *qigong* which reduces disturbance. *Zhineng Qigong* begins with the enhancement of regular motion and movement.

Once *Yi Yuan Ti* is formed, all the information about life activity of the human body can be reflected in *Yi Yuan Ti*, and this forms *Yi Yuan Ti*'s feature of holistic information about life. All kinds of information in the natural world can also be reflected in *Yi Yuan Ti*.

iii. The Feature of Dynamic Function

Yi Yuan Ti's dynamic function is established upon the human body's life activity, particularly on the basis of the activity of nerve cells. Various life activities are not only reflected in the multitudinous "stratums" of *Yi Yuan Ti*, but are also holistically synthesized and form a dynamic *hunyuan* whole entity. That is to say, in every part and every "stratum" of human *Yi Yuan Ti*, there exist the features of both active and passive activities.

i) The Features of Condensing and Dispersing. The feature of condensing can be shown in concentration in spirit and attention in psychology. It can gather *qi* and condense it into energy and even form a concrete substance. The feature of dispersing can be shown in indifference in spirit and distraction in psychology. It can dissolve energy and concrete substance.

ii) The Feature of Motion. There are two kinds of motion in *Yi Yuan Ti*. One is the motion of fundamental information in the "microcosmic" realm. This is a motion going on faster than the speed of light in an unconscious state, such as the integration of various information in the process in which a concept is being formed. We know that the human cerebrum is a complicated information bank with approximately 14 billion neurons, each of which has 10^{10} base-pairs in its DNA. From this it can be deduced that compound information in the cerebrum can reach an almost infinite amount. Therefore, integration and selection of information are carried out against a very vast and complicated background and cannot be accomplished without a speed faster than light.

The other kind of motion in *Yi Yuan Ti* is the motion of concept information in the "macrocosmic" realm, and familiar intelligence people's thinking activity belongs to motion of this type. This kind of motion is at variable speed—it can both be faster and slower than light. The information motion of concept is the consciousness activity that can be noticed by the "thinkers" themselves. When it is in an activity state at a controllable low speed, "thinkers" can concentrate their mind and enhance their ability of mental activity. If it is in a free activity state at a very high speed, stopping its activity with a lot of mental effort is hardly effective. For example, when distraction occurs in a practitioner's *qigong* practice, it is usually effective if the practitioner guides the mental activity in a relaxed manner. If the practitioner pays special attention to the distraction and tries to control it with a great effort, the effect is usually quite the opposite: Distractions occur unceasingly one after another. This is a bit similar to the feature of antimatter as deduced in physics which "accelerates when meeting resistance."

Reasons for motion in *Yi Yuan Ti* do not have the difference of being either active or passive as the reasons for motion in life activities. This is because it is hard to differentiate active motion and passive motion in *Yi Yuan Ti*. Passive motion in daily life is the motion brought into being by the influence of object. We know that it is difficult for an object to bring energy in its effect into *Yi Yuan Ti*. What is shown in *Yi Yuan Ti* is but the reflection of the time-and-space structure of the object. The series of change thus caused in *Yi Yuan Ti* all belong to the domain of reflection in different "stratums". The motion caused by the reflected information (i.e., the time-and-space structure) displayed in *Yi Yuan Ti* results from the functioning of *Yi Yuan Ti*'s condensing and dispersing feature and has no difference in essence from *Yi Yuan Ti*'s active motion. This is because the active motion in *Yi Yuan Ti* and the active motion in daily life are brought about by different reasons. The active motion in daily life refers to the activity of *Yi Yuan Ti* giving an instruction without the premise of having received stimulation from the outside world, while any mental activity (including initiative instruction) formed in *Yi Yuan Ti* itself results from the "competition" within *Yi Yuan Ti*'s vast amount of information. In the process of information "competition", there is not such a thing called being active. It is only when a series of information has been integrated into a set pattern that a conceptual instruction in the realm of consciousness occurs. As to which information will take the advantageous position among the multitude of information, the reasons are multi-faceted according to familiar intelligence. If analyzed according to the theory of *hunyuan* whole entity, it is determined by the time-and-space structure of the *hunyuan* whole entity of the specific thing to be done having formed in *Yi Yuan Ti*. In this sense, an active instruction also comes from the reflection formed in *Yi Yuan Ti* and has no difference in essence from the passive motion mentioned above. It is simply that the passive motion is the motion caused by reflecting the time-and-space structure of the realistic existence, which is essentially the motion caused by reflecting the time-and-space structure of the thing being considered in the mind. Passive and active motions in *Yi Yuan Ti* merely differ in this subtle point.

Concepts, when having formed in *Yi Yuan Ti*, belong to the domain of the activity of consciousness, and consciousness activity has its feature of motion at will which is shown in the drive feature, direction feature, penetration feature, condensing and dispersing feature, etc..

II. *Yi Yuan Ti*'s Functions

Yi Yuan Ti's functions are chiefly shown in the following aspects of receiving information, processing, storing, extracting information and sending information.

i. Receiving Information

This is the specific display of *Yi Yuan Ti*'s feature of reflectiveness. It is the process in which material information in the objective world is internalized in the subjective world. There are two patterns.

i) To Receive Information and Reflect Objective Existences with the Function of Normal-Intelligence Perception of the Sensory Organs

Human sensory organs (the eyes, ears, nose, tongue, and skin) can sense stimulations of physical and chemical features of objective substances respectively. The eyes are the visual organ that can receive optical stimulation as well as related information, such as luster, color, light and shade, and shape. The ears are the auditory organ that senses the acoustic feature of substances as well as stimulation of related information, such as audio frequency, tone quality and intensity of sound. Each part of the human body is a sensory organ of perception, of the sense of pain and the sense of temperature and can receive such stimulation as temperature, hardness, strength and texture of substances. The nose is the organ of the sense of smell and it receives stimulation of the smell of volatile substances. The tongue is the organ of taste and it receives stimulation of different kinds of tastes.

When different kinds of stimulation mentioned above act upon receptors of relevant sensory organs, they can cause excitation in the afferent nerves, which further causes responses in the brain. These responses can all bring about relevant changes in *Yi Yuan Ti*. If only one stimulation is received, it will produce sense; if related compound information is received, it will produce perception. When "all" the information about an existence has entered *Yi Yuan Ti*, an image of this existence can be formed. Once an image is formed, the entire image can be called to the mind according to part of the characteristics of the image.

Every existence has the different feature of its own. The information they send out acts on different sensory organs, causes different nerve impulses and is accompanied with different physiological changes. This makes it seem that the integrated existence is disintegrated by different sensory organs. However, once these kinds of information received by different sensory organs enter *Yi Yuan Ti*, the original feature of the information will be restored. The different types of information will then combine according to their original state, so their original integrate state will be displayed.

ii) To Reflect Objective Existences with the Function of Super Intelligence Perception

This is a method regarded as reflecting things with super extraordinary abilities or paranormal abilities. *Yi Yuan Ti* can receive and reflect information transmitted to it by the sensory organs and other organs. It can also directly receive and reflect information about objective existences. *Yi Yuan Ti* can directly feel the holistic feature of space, the holistic feature of time and the holistic feature of both time and space of things. Therefore, paranormal abilities are relatively quick in cognizing existences. It can draw correct conclusions in an instant. It cognizes things, as far as the nature of cognition is concerned, not according to familiar intelligence people's logical calculation, but according to direct reflection.

ii. Processing, Storing and Extracting Information
i) Processing Information

As it has been stated above, the sensory organs receive different information about each aspect of things. After the information is reflected in *Yi Yuan Ti*, various partial information of each aspect of things can gather together and the original integrated look of things will be restored.

This is a process in which an image of a thing is formed. The image is formed naturally through repetitive reflection, so it is only processing at a preliminary stage. Such an image formed in *Yi Yuan Ti*, though not having included all the holistic information about an objective existence, can already reflect the general picture of the existence. It can be regarded as the coupling reflection in *Yi Yuan Ti* of object and it takes up a certain *Yi Yuan Ti*'s "space".

If we call this process of forming reflection in *Yi Yuan Ti* "processing of the first time", in *Yi Yuan Ti* there is also "processing of the second time", that is, contracting the integrate reflection which has complicated information in it into a piece of simple information, i.e., a particular symbol—a word to take its place. This is a process of abstracting images of things and becoming separated from specific information. This is the process of how concepts are formed in familiar intelligence people's consciousness activity. In modern psychology and philosophy, this abstraction is considered as the result of analysis, synthesis, deduction and sublation of phenomena, and concepts or judgments thus obtained are considered as having grasped the nature of existences.

Once an image is abstracted and is endowed with a word with meaning, the word thus formed then leaves the "stratum" where image resides and enters into another "stratum" and establishes other systematic connections. This is the second signal system. If we call *Yi Yuan Ti*'s space occupied by image "the First Reflection Space," we might likewise call *Yi Yuan Ti*'s space occupied by the second signal system "the Second Reflection Space". Symbols in the second space can be obtained through abstracting reflections in the first space, or by deducing from the basic composition of the system of the second space, or by directly receiving concepts from the outside world transferred through *Yi Yuan Ti*'s "First Reflection Space".

It needs to be pointed out that the information processing process in *Yi Yuan Ti* cannot be equated to the mechanism of a computer. Instead of being a simple binary-system calculation which gives answers of yes or no, this processing process is active multi-channel, multi-level information collection, a process in which creation can occur and new things can be known.

ii) Storing Information

The storing information function of *Yi Yuan Ti* is similar to the retaining information function of *hunyuanqi* in general. All information that enters *Yi Yuan Ti* can be retained. As both *Yi Yuan Ti* and information take up no time or space, the information that *Yi Yuan Ti* can store is infinite. Not only the information that is noticed can be received and retained, even the information that is not noticed yet having entered *Yi Yuan Ti* can also be stored. It is simply that the former becomes dominantly stored information and can be extracted, while the latter is implicitly stored information and is difficult to be extracted at will.

It is only when information has entered and has caused motion in *Yi Yuan Ti*, which makes *Yi Yuan Ti* gather to the information and makes the amount of information increase, that *Yi Yuan Ti* can clearly reflect it. When evident information change has occurred in *Yi Yuan Ti*, *Yi Yuan Ti*'s condensing-into-form function will be strengthened and causes a series of concrete substance change in the brain cells. Once information change in *Yi Yuan Ti* turns into change of concrete substance, the change will be fixed and the information will be stored in the concrete substance. The information in *Yi Yuan Ti* and the concrete substance in the brain then form a dialectical unity and a new calm state is displayed.

iii) Extracting Information

The *hunyuanqi* of the brain neurons also contains the information stored in the brain neurons. Thus such content of information also becomes part of the content of *Yi Yuan Ti*. When consciousness needs to extract the information, due to the searching and driving-force features of consciousness, it will actively gather to the point where the information is stored in *Yi Yuan Ti* and the information in the concrete substance is thus clearly shown.

It should be pointed out here that, though all kinds of information can be received by *Yi Yuan Ti* and stored in it, it is not necessarily easy to extract them because a lot of information is unconsciously received and stored. Information received and stored like this has caused relatively little change in the entire *Yi Yuan Ti*, so it is difficult for the recalling process to discover and extract it.

Efficiency of information receiving, reflecting, storing and extracting mentioned above is determined by *Yi Yuan Ti*'s sensitivity, namely, the state of how much of *Yi Yuan Ti* is stirred and mobilized, and also by the degree of how orderly the information in *Yi Yuan Ti* is stored.

iii. Sending Information to Act on Objective Existences

This is a process in which information in the subjective world is externalized and becomes combined with objective existences. All the information in the subjective world of humans results from receiving and processing external information. Since the modes of receiving information differ in familiar intelligence and super intelligence, the modes of sending information are accordingly different.

i) Sending Familiar Intelligence Information

This is a process in which people with familiar intelligence send information to objective existences with different organs, including sensory organs: a) According to *qigong* science, human sensory organs, when receiving stimulation from the outside world, do not receive it passively. When attention is paid to the target of perception, information has already been sent to it. Although this has not changed the macroscopic state of the target, it increases the amount of stimulation to the organs and meanwhile imprints consciousness on the target being perceived. b) Information is sent through the afferent nerves which act upon the bones and the muscles that accordingly act upon the target of perception. The two situations mentioned above exist in everybody.

ii) Sending Super Intelligence Information

This is a process in which only holistic information of consciousness activity is used to act upon objective existences. For example, "the start of mental activity" designed in modern scientific experiment to study paranormal abilities and the therapy using purely mental activity in *qigong* external *qi* treatment both belong to sending super extraordinary information. It should be pointed out that sending super extraordinary information is a functional ability in everybody. It is simply that people differ in the strength of the information they send.

Here is an additional point. The two modes of information transmission differ widely in their effects upon objective existences. When people with familiar intelligence send information, they rely on the organs with only partial (instead of holistic) features to interact with objective existences. This is the interaction with the partial attribute of objective existences. This function involves mostly physical and chemical features, and the relationship established with the objective existences is the *hunyuan* whole entity at the physical and chemical levels. In sending super intelligence information, *Yi Yuan Ti* directly sends holistic information to interact with objective existences. This is to interact with the holistic attribute of objective existences. This function involves the time-and-space structures of *hunyuan* entities. The relationship established with the objective existences is the *hunyuan* whole entity that transcends the physical and chemical levels.

III. *Yi Yuan Ti*'s Formation and Change
i. Formation of *Yi Yuan Ti*

Yi Yuan Ti comes into being as the result of the continuous evolution and development of primordial *hunyuanqi* in the universe. It is a special motional mode of substance displayed when the cerebral cortex with special structure and function emerges from evolution. It is the *hunyuan* or merged-into-oneness state on the basis of highly divided functions of nerve cells.

From the perspective of the evolutionary history of the human race, *Yi Yuan Ti* evolves into being on the basis that the world of animals has developed into the stage of having relatively advanced brains. Many lower vertebrates, though they can also reflect the objective world in a certain sphere, have not acquired the ability to abstract and summarize so as to form concepts. Therefore, they cannot establish the second signal system, nor can they reflect the state in their own realm of consciousness. Although these animals also have a certain ability of thinking, such as low level analysis and synthesis, *Yi Yuan Ti* which has relative independence cannot form in them because of the simple connection between the brain cells in their cerebral cortex. This state in the brain of animals might as well be called "*Nao Yuan Ti*" (Brain Oneness Entity) to be distinguished from *Yi Yuan Ti* (Mind Oneness Entity).

It is only when the universe has developed to the stage of humans that the distinction in the functions of different tissues and organs (such as the hands and speech organs) becomes more exquisite and subtle and their forms and structures become further enhanced, that the cerebral cortex becomes highly developed and evident changes happen in the number and composition of the brain cells. What is particularly significant is the connection extensively formed between brain cells (i.e., dendrites and synapses) which enormously enriches the amount of information in the brain neutrons. When this change reaches a certain critical value, it becomes possible for the brain cells to connect into an integrated and holistic entity and enables the *hunyuanqi* that fills both the inside and outside of the brain cells to accomplish mixing and mergence and change to the utmost extent. This, therefore, forms the *hunyuan* entity even in texture and centered in the human brain—*Yi Yuan Ti*. The emergence of *Yi Yuan Ti* improves the functions of the nervous system to a new level and lays a foundation for the establishment of a human being's self.

It is also the same in the development process of the human fetus. When the brain cells have grown rather mature in their development, namely, around the seventh month in a woman's pregnancy, *Yi Yuan Ti* in the fetus accordingly appears. As it has been stated in "Fetal *Hunyuanqi*", the fetus in the womb is in a development stage and every tissue and organ in the body of the fetus is growing according to their own laws. This determines that the stimulation that the fetus in the womb receives from the internal and external environment is comparatively even. Moreover, the brain cells of the fetus are still in a relatively low functional state and they have not established functional connections with each part of the fetus' body yet. So the fetus' brain is essentially in a calm state and does not have specific reflections of different kinds of existences. The baby basically maintains this state after birth before it begins to breathe.

It should be pointed out that *Yi Yuan Ti* at this moment already contains all the life information about the human body. Since *Yi Yuan Ti* at this moment only has the information regarding the life of an individual human and also the human species, and since the information has not been unfolded in the life of the baby after birth and *Yi Yuan Ti*'s functions are still far from being displayed, *Yi Yuan Ti* is in the most primordial even state with neither perception or knowledge nor distinction or differentiation in it. Therefore, we call it *Chu Shi Yi Yuan Ti* or Original Pure Mind Oneness Entity (equivalent to amala in Buddhism, which is also called the ninth consciousness).

It is true that the "fundamentally blank and pure" nature (amala) of Pure Original *Yi Yuan Ti* is subject to a certain influence even in the fetal period. For example, the life activity in each part of the fetus' body, due to genetic factors and its actual state, has already become the specified content in *Yi Yuan Ti* even when Pure Original *Yi Yuan Ti* has just come into existence. This, however, is the very characteristic of the theory of *hunyuan* whole entity and Pure Original *Yi Yuan Ti* and a point of difference from amala in Buddhism.

In the three months after Pure Original *Yi Yuan Ti*'s formation in the fetus, change to a certain extent in the fetus' *Yi Yuan Ti* indeed takes place, e.g., receiving various information from both the inside and outside of the fetus' body. Yet preliminary consciousness activity is not quickly established as after the baby is born, and the key reason for this lies in that the information that the fetus' *Yi Yuan Ti* receives is not intensified. There is neither confirmation of object with the sensory organs of familiar intelligence, nor combination of language information with inform-ation regarding concrete substance, nor connection of *hunyuan* holistic super intelligence information with the fetus' *Yi Yuan Ti*. Therefore, during the three months before the baby's birth, the fetus' Pure Original *Yi Yuan Ti* is basically in a chaos state. If this situation of having no intensification of information mentioned above can be effectively changed, the fetus' *Yi Yuan Ti* will definitely change accordingly.

ii. Changes in *Yi Yuan Ti*

After a baby is born into this world, its independent life is gradually established and the functions of each part of its body are successively exercised in its contact with Nature. All this will imprint in the baby's cerebrum a certain "mark". So changes accordingly take place in *Yi Yuan Ti* and, in the even Pure Original *Yi Yuan Ti* with no distinction between the subjective and the objective world, Self *Yi Yuan Ti*, a partial subjective world (Atma-graha) and series of consciousness activities are formed.

i) Formation of Self *Yi Yuan Ti*

Self talked about here differs from what Sigmund Freud called ego. In Freud's opinion, ego is the coordinator of id (impulse) and super-ego (rationality). Self *Yi Yuan Ti* talked about here is an integrated unified entity of the reference system of both subject and object formed when the objective world (including society, nature and one's own life activity) is internalized into Pure Original *Yi Yuan Ti*. It is not only a kind of feeling of one's self, but also a real existence with a certain power in it. This power is called consciousness strength. It is an "information integration force" based on material strength and life vitality and is formed in the process of the human body's development. Further elaboration will be provided as follows:

After a baby is born, its external environment becomes different from the state with equal and even force around the body of the fetus floating in the amniotic fluid in the womb. Part of the baby's body contacts the rigidity of material things (e.g., the bed) and part of it is in contact with the air, so the stimulations that each part of the baby's body receives become different. What's more, the start of every internal organ's functions changes the fetus' even internal environment. All this imprints "marks" in Pure Original *Yi Yuan Ti*.

With the activity of different sensory organs, various information from the outside world, both natural and social, inputs continuously into Pure Original *Yi Yuan Ti*, and the objective world is gradually internalized into Pure Original *Yi Yuan Ti*, i.e., the subjective world. The reflections in this subjective world begin to become complicated: There is both the information about one's own life activity which is the information about the subject (though the person himself or herself still does not know it) as well as the information in the surrounding environment, including the information about the natural environment and about the people who are around. This is the most fundamental content of the objective world internalized into *Yi Yuan Ti*. It is the first stage of Self *Yi Yuan Ti*. Activity in this stage is still in the reflection stage. *Yi Yuan Ti*'s commanding power over life activity is still not exercised. The relationship between *Yi Yuan Ti* and life activity may well be described as agreement in reflection, that is, *Yi Yuan Ti* reflects life activity as it really is.

Since different kinds of information in the environment are more intensely internalized into *Yi Yuan Ti* and the information reflected in *Yi Yuan Ti* is the reflection of broader environment different from the reflection of one's own life information, difference between the reflection of subject and the reflection of the objective environment is formed in the overall reflection in *Yi Yuan Ti*. Realistic environment and genuine human relationship in the objective world are shown in *Yi Yuan Ti*'s reflections. However, *Yi Yuan Ti* at this moment still does not have the ability to distinguish the difference between subject and object. In the reflections in the subjective world, the reflection of the subject's life activity is among the reflection of the environment. They are, in terms of *Yi Yuan Ti*, both the reflections of object.

Once the relationship between subject and object is established in *Yi Yuan Ti*'s reflections, these earliest reflections having entered *Yi Yuan Ti* will become the baby's Reference System by means of which the baby knows existences. In other words, all the consciousness activities established after this go on against this background and it becomes the fundamental background against which a human being knows the world. The self (familiar intelligence self) that an individual human recognizes about oneself is established on this basis. Strictly speaking, it is similar to *yi gen* or indriya ("the root of consciousness") in Buddhism. The baby now also has some contact with society and already begins to establish relationship with people, but it has not yet established the standards to distinguish between right and wrong and between good and evil. Nevertheless, the baby at this moment has already become both a natural human being and a social human being and is the root and foundation upon which an entire human being's changes take place. The content in *Yi Yuan Ti* now is already different from Pure Original *Yi Yuan Ti* because activities of the subjective world have begun in it and the "fundamentally blank and pure" feature of Pure Original *Yi Yuan Ti* is destroyed. We call this state of the subjective world Self *Yi Yuan Ti* (equivalent to the eighth consciousness or Alaya-vijnana stated in *Cittamatra*). Although Self *Yi Yuan Ti* has the ability to receive and display the differences in such factors as right and wrong and also good and evil in the objective world, all this has not been unfolded yet. The agreement between the human being and Nature as well as between the human being and society now takes the leading place.

The stage of Self *Yi Yuan Ti* is in continuous change. It is a transitional period from Pure Original *Yi Yuan Ti* to Partial Identity *Yi Yuan Ti*. In theory, boundaries between these three stages of *Yi Yuan Ti* seem to be fairly clear. But in reality, it is hard to draw clear lines between them. In the strict sense, Self *Yi Yuan Ti* steps on the track of Partial Identity *Yi Yuan Ti* the moment it is formed. Although Pure Original *Yi Yuan Ti* already has various functions, particularly super intelligence functions, the super intelligence in a baby's *Yi Yuan Ti* cannot be reinforced and developed since the people around the baby do not have super intelligence and cannot communicate with the baby using super intelligence information. Thus only familiar intelligence reflections are formed in the baby's Reference System. But viewed from the overall situation, what is reflected in *Yi Yuan Ti* agrees with one's life activity and the natural environment.

ii) Partial Identity *Yi Yuan Ti* (Familiar Intelligence *Yi Yuan Ti*)

With the baby's growth and development, functions of the human body's different tissues and organs, particularly functions of the nervous system, become gradually enhanced, so the ability of human self-mastery and self-control becomes increasingly strong, the difference between the human being and the objective world becomes increasingly evident and the distance between the two of them becomes increasingly wide. Existences in the outside world are just the essential condition or environment that guarantees the living of the individual human. When this reality is reflected in one's *Yi Yuan Ti*, it causes changes in one's Self *Yi Yuan Ti*'s reflections—the reflection of me as subject gradually grows more outstanding and the reflection of the environment begins to become subordinate to me the subject. The entire background of the Reference System is gradually taken up by the subject of me. As this subject of me is established on the basis of familiar intelligence and is partial, the relationship established upon this basis afterwards with the outside world is also partial, further development of human life activities guided by this and one's character, temperament and concept of morality accordingly formed are inevitably partial. We call the *Yi Yuan Ti* in this state Partial Identity *Yi Yuan Ti*. It is the internal reason and also foundation of people's consciousness activities, which is equivalent to Manas or the seventh consciousness stated in *Cittamatra*.

It should be pointed out that, be it the partial subjective world or Self *Yi Yuan Ti*, they are both developed on the basis of Pure Original *Yi Yuan Ti*, or rather, they are formed by adding particular information to Pure Original *Yi Yuan Ti* and making it become fixed to a certain mode. From a development point of view, this evolution is fine as it makes some of Pure Original *Yi Yuan Ti*'s functions objectively displayed. However, the various kinds of information that form the Reference System in Pure Original *Yi Yuan Ti* and the various kinds of modes established against this background are all based on familiar intelligence functions of different human organs, and all this is but *Yi Yuan Ti*'s familiar-intelligence functions. Since this part of functions is reinforced, *Yi Yuan Ti*'s paranormal abilities are depressed, which makes people's cognition about the objective world stay at this level. Since human beings obstinately believe that the universe and nature are merely like this, they cannot know deeper levels of the cosmos, the levels perceived with *Yi Yuan Ti*'s paranormal abilities (super intelligence). On account of this, people are just confined in a small narrow sphere and draw many incorrect conclusions instead of gaining true knowledge at advanced levels (and this is equiva-lent to what is referred to as "ignorance" in Buddhism).

From *Yi Yuan Ti*'s formation and change stated above, we can see that Pure Original *Yi Yuan Ti*, Self *Yi Yuan Ti* as well as Atma-graha formed afterwards all gradually come into being in the process of human growth and life practice. They are not "everlasting" as some people claim. The so-called "ignorance" is a special phenomenon in a particular stage in the evolutionary history of the human race and is brought about by anthropic factors. When Self *Yi Yuan Ti* changes into Partial Identity *Yi Yuan Ti*, self in the Reference System is differentiated: First, it is *Yi Yuan Ti*'s functional activity that becomes separated from and independent of life activity. Then, it is the successive establishment of self needed by one's psychology, self needed by the objective situation as well as self needed by one's rationality. After that, it is the division of the self to observe, the self being observed, the ideal self and the realistic self. On the basis of changes in *Yi Yuan Ti*'s Reference System, partiality in notions occurs, such as the notions of private ownership, theology and mechanical materialism. Once these notions are formed, they react upon *Yi Yuan Ti*'s Reference System, which further intensifies and solidifies the Reference System's partiality.

iii) *Yuanman Yi Yuan Ti* (Complete Mind Oneness Entity)

Through *qigong* practice, on the one hand, the information function (the functions related to receiving, processing, storing, extracting and sending information) in the familiar intelligence state will be enhanced, which enables *Yi Yuan Ti* to receive and store more information; on the other hand, potential function of receiving and sending super intelligence information will be tapped and developed, which adds new content to *Yi Yuan Ti*'s Reference System so that one can comprehensively and consummately deal with the relationship between oneself and Nature.

Since the partial phenomenon that familiar intelligence information alone occupies the Reference System is conquered, we call the *Yi Yuan Ti* in this state *Yuanman Yi Yuan Ti* or Complete Mind Oneness Entity.

iv) *Hunhua Yi Yuan Ti* (Developed Integrate Mind Oneness Entity)

After *Yuanman Yi Yuan Ti* is established, the super extraordinary information in the super intelligence state in *Yi Yuan Ti* can not only be sent to and act upon the natural world but can also be sent to and act upon oneself so as to change one's state of being fixed with familiar intelligence information and to make one enter the state of the level of super extraordinary information substance. Since the super extraordinary information in one's *Yi Yuan Ti* has taken command of one's entire human body and makes the body and *Yi Yuan Ti* identical in nature, we call *Yi Yuan Ti* in this state *Hunhua Yi Yuan Ti* or Developed Integrate Mind Oneness Entity (i.e., further developed *Yi Yuan Ti* making the human body integrated and identical with Mind Oneness Entity in the most extraordinary super intelligence information substance state) and the newly formed *hunhua* state in which *Yi Yuan Ti* and the human body are unified *Yi Shi Hun Yuan* (Consciousness Integrated in Oneness).

The display of *Yi Yuan Ti*'s change at different levels stated above is, in fact, not the change of *Yi Yuan Ti*'s original entity, but the change of the Reference System in it. Therefore, in order to truly understand the change in *Yi Yuan Ti*, an exploration must be made into *Yi Yuan Ti*'s Reference System.

IV. *Yi Yuan Ti*'s Reference System

"Frame of reference", which is also called "reference system" or "reference substance", is a term in physics that refers to an object or a system chosen as the standard to measure the position of another object and to describe the state of its movement. The term "reference system" as used to refer to the Reference System of *Yi Yuan Ti* has a great difference from the meaning of this term in physics. *Yi Yuan Ti*'s Reference System is the overall time-and-space structure of various features of the objective world internalized into *Yi Yuan Ti* through the human body's various perceptive functions. It is the internal prescriptive and systematic mode of consciousness activities, the measurement mode in the human subjective world to estimate and evaluate various existences, and the foundation upon which a person cognizes and judges existences and gives guidance to his or her own conduct. All human thinking activities are carried out against the background of *Yi Yuan Ti*'s Reference System.

i. Formation of *Yi Yuan Ti*'s Reference System

Yi Yuan Ti's Reference System is not formed once and for all. Instead, it is gradually formed with *Yi Yuan Ti*'s development and change. As it has been previously stated, Pure Original *Yi*

Yuan Ti has its "fundamentally blank and pure" feature. However, this "blank and pure" feature is destroyed the moment it is formed because of its feature of receiving information. This is because, once *Yi Yuan Ti* is formed, its entity and attribute will exercise their functions and various information in the natural world will be inevitably received by *Yi Yuan Ti*, which makes *Yi Yuan Ti* become a holistic information bank of all the information existent in the universe, including the information already in one's own life. It is simply that the information received in *Yi Yuan Ti*, having not been reinforced or activated by Pure Original *Yi Yuan Ti*, is in a natural hidden state, which is similar to the void that contains the information about all existences yet is unable to unfold them. The multiplicity of all existences in the universe cannot be reflected in Pure Original *Yi Yuan Ti* in a displaying or unfolding manner, either, so it is in *Yi Yuan Ti* in a chaotic state.

With the information about specific existences (as stated in Self *Yi Yuan Ti*, for example) repeatedly coming into *Yi Yuan Ti* accompanied with a certain amount of energy, part of the information in *Yi Yuan Ti* is reinforced, activated and manifested. Although *Yi Yuan Ti* at this moment is still unaware of the information, nor can it make use of it out of its own choice, the information about natural events, people, language and consciousness and so on gradually internalizes into *Yi Yuan Ti* from general to specific and from simple to complex and becomes part of the human subjective world. When the information internalized into *Yi Yuan Ti* has reached a certain extent, *hunyuanqi*'s feature of being able to condense into form will automatically take effect, which makes the information and *hunyuanqi* gather into concrete substance, facilitating change in the cranial nerve cells, and make the information internalized into *Yi Yuan Ti* become fixed.

As the information settled begins to form a fixed pattern and enters *Yi Yuan Ti* in the form of *hunyuanqi*, they become "newly added" content in *Yi Yuan Ti*, which *Yi Yuan Ti* already has. The content of the information becomes not only the mode with which objective existences are cognized, but also the background against which all kinds of information are to be reflected.

Yi Yuan Ti's Reference System is the reflection of the objective world, including human life activity, in *Yi Yuan Ti*. It is the result of the objective world internalized in the subjective world. In the Reference System, there exists not only the most fundamental information about the natural world but also the content of the interrelationship and interaction between all existences as well as people's state and status in this above-mentioned background. It is true that all these things are not completely the same in different people's Reference Systems. There can even be subtle differences that are difficult to describe due to everybody's different experiences and the accordingly different information internalized into their *Yi Yuan Ti*.

We know that the content of the Reference System gradually changes with the internalization of various information and further condensing and settlement of information. In the life process of people with familiar intelligence, information received with different sensory organs is all familiar intelligence information having partial rather than holistic features. Although familiar intelligence information varies in a great many ways, it does not exceed the scope of familiar intelligence. So any increase in it still belongs to change in quantity. It is because of this that the great variety of knowledge in every subject in modern science established upon familiar intelligence does not contradict the fundamental point in the Reference System. The input of various kinds of knowledge is but the further extension of the sphere of the Reference System already formed. In this changing process, the more extensive and fixed the content of partial information in the Reference System, the more depressed is *Yi Yuan Ti*'s innate function of receiving super intelligence information and consequently the harder it is to show the super intelligence. Since super intelligence information in familiar intelligence people's Pure Original *Yi Yuan Ti* cannot be reinforced and activated, and since it is even more difficult for it to become condensed inward and settled, there is no super intelligence information in their Reference System. This is the reason why familiar intelligence people cannot exercise super intelligence and why cognizers in general who have super cognitive abilities (e.g., reading with the ears) have to perceive through the familiar intelligence Reference System.

ii. Relationship between the Reference System and *Yi Yuan Ti*

As stated above, *Yi Yuan Ti* is *hunyuanqi* having special features and functions formed with the cranial nerve cells' *hunyuanqi* that is able to interact with information. *Yi Yuan Ti*'s Reference System is the overall time-and-space frame and structure established on the basis of the information about the objective world that has become internalized into *Yi Yuan Ti* by means of *Yi Yuan Ti*'s information receiving function (through the sensory and perceptive functions of the human body). In this sense, it seems to be reasonable to say that *Yi Yuan Ti* is the owner and *Yi Yuan Ti*'s Reference System is the guest of the human body. However, *Yi Yuan Ti*'s Reference System is not merely a guest because content of the Reference System of *Yi Yuan Ti* can be condensed into concrete substance and deposited in the nerve cells, and the *hunyuan* entity of the cranial nerve cells with the Reference System's content deposited in it can be integrated into the Mind Oneness Entity (*Yi Yuan Ti*) and becomes the content that occupies *Yi Yuan Ti*, which makes all the features of *Yi Yuan Ti* accessible to the influence of the Reference System. If we analogize *Yi Yuan Ti*'s original entity and attribute to a piece of white cloth, the *Yi Yuan Ti* with the Reference System already having a certain content formed in it is equivalent to a piece of dyed cloth.

Generally speaking, functions of a familiar intelligence person's *Yi Yuan Ti* are exercised with the participation of the Reference System. *Yi Yuan Ti*'s innate right of freely exercising its own functions seems to have been usurped by the Reference System to a certain extent. In this sense, the Reference System seems to have become the master while *Yi Yuan Ti* seems to have become the servant. In a certain sense, *Yi Yuan Ti* is affected by the Reference System in a manner similar to "frequency lock" in synergetics. It is because of this that "Changes in *Yi Yuan Ti*" (III. ii) take place. This is also the reason for "taking the wrong master for the real owner" to be further illustrated in "Mistaken Identity of One's Internal 'Master' in Section V in this chapter.

Yi Yuan Ti's changes having been previously stated, such as Self *Yi Yuan Ti*, Partial Identity *Yi Yuan Ti* and Complete *Yi Yuan Ti*, are in fact not the changes of the different characteristics of the entity and attribute of *Yi Yuan Ti*, but the adding of the content of the Reference System to the originally fundamentally blank and pure *Yi Yuan Ti* and the different changes taking place in the Reference System. For human beings who have kept developing till today, the Reference System in *Yi Yuan Ti*, though having extremely greatly expanded its scope with the development of modern science, has not gone beyond the realm of partiality, which restricts *Yi Yuan Ti* from freely exercising its authority. Nevertheless, the innate original entity and attribute that *Yi Yuan Ti* possesses have not disappeared. *Yi Yuan Ti* will receive more information through *qigong* practice so as to break through the confinement of the old Reference System and enter into a new realm.

iii. Relationship between *Yi Yuan Ti*'s Reference System and Consciousness Activity

Consciousness activity will be comprehensively talked about in Section III in this chapter. Here only a brief introduction to the relationship between consciousness activity and the Reference System will be given. Consciousness activity goes on against the background of the Reference System. The content of consciousness activity can become part of the Reference System under certain circumstances. The Reference System prescribes the mode of consciousness activity. Under the effect of consciousness activity, the mode of the Reference System can also be changed. This is the dialectical relationship between the Reference System and consciousness activity.

iv. Relationship between the Reference System and One's Own Life Activity

In the previous chapter of Human *Hunyuanqi*, the whole entity of *jing*, *qi* and *shen* in unity has been comprehensively introduced and the commanding effect of *shen* (*Yi Yuan Ti*) upon the human body's life activity has been expounded. When we become clear about the function of *Yi Yuan Ti*'s Reference System and then look back upon *Yi Yuan Ti*'s commanding effect on the human body's life activity, we find that the commanding power is in fact to a great extent exercised by the Reference System.

We know that the Reference System does contain a vast quantity of information about the life of the human body so that it can make life activity proceed healthily under its command. However, *Yi Yuan Ti*'s Reference System established in human beings up to now is still partial and biased instead of holistic, so it confines human life activity to the sphere prescribed by the Reference System and makes it unable to obtain greater freedom. We know that human life activity, like any other kind of life activity, is inevitably restricted by the law of nature. Nevertheless, human beings have already formed *Yi Yuan Ti* that has special functions and they can, under the command of *Yi Yuan Ti*, break through the confinement of the law of nature and enter a new free realm. It is simply that the Reference System having formed restricts *Yi Yuan Ti*'s functions, and with the Reference System in control of life activity, human life activity is confined to a small narrow natural scope. Although human beings can gradually develop forward like other organisms in a continuous *hunhua* process, merging and changing with the *hunyuanqi* in the outside world, they cannot jump out of the circle of natural development.

For example, with regard to the natural development laws of existences, the laws that people have concluded to be perfectly justified and unchanging (this is what is called "primal ignorance" in Buddhism) – as a matter of fact, a great part of the laws of human life motion and movement can be changed. Take for example the fact that human beings must maintain their own metabolism by eating food. To maintain metabolism by means of eating food seems to have become an absolutely unchangeable conclusion. Not all human beings are overwhelmingly like this in their lives, however. According to the theory of *hunyuan* whole entity, the metabolic process in humans can be accomplished by means of substantial (concrete) substance that has form and can also be accomplished by means of insubstantial (intangible) substance without form. In human beings' natural development process, life activity is maintained mainly in the former pattern. Therefore what is impressed in human *Yi Yuan Ti*'s Reference System is that food is the material premise that guarantees life activity, without which life cannot continue to exist. In modern medicine, it is believed that if a person does not eat food for over seven days, irreversible harm or even death will come as a result. However, this conclusion is shattered by the practice of *qigong*. So long as the truth of *qigong* is clearly explained to people to make changes happen in their old Reference System and to let them realize that humans can also live by relying on the support of *qi* instead of eating food, and they then practice eating nothing (*Bi Gu*) for more than ten days, their health will not be affected at all. What is more amazing is that some people have lived for many decades so far without eating anything. They not only live their lives as normal as others, but females among them can also give birth to children. From this we can see what an immense impact *Yi Yuan Ti*'s Reference System has on human body's life activities.

V. Appearance of *Yi Yuan Ti* Makes the Universe a Conscious Universe that Knows Its Self

It is pointed out in the theory of *hunyuan* whole entity that all existences in the universe today evolve from primordial *hunyuanqi* along two tracks. One is the division from an overall entity into small units, a track that the further the division, the tinier the unit: primordial *hunyuanqi*

→metagalaxies→clusters of galaxies→galaxies→stars→the earth→all existences→ humans. The other is a track that the further specific existences evolve, the more complex they become: primal *hunyuan* state of primordial *hunyuanqi*→the gravitation field and radiation field →particles→elements→compounds→organisms: plants and animals→humans. Although the two evolutionary tracks mentioned above evolve in different ways, the evolution and development of universe *hunyuanqi* is the same, and the development process goes on naturally with every existence functioning as part of nature.

Ever since the appearance of human beings, since *Yi Yuan Ti* already has the independent feature in relation to life activity and has the function of cognizing the attribute of nature, since *Yi Yuan Ti* can not only know nature, but can also create nature with initiative, human beings' activity in the world of nature becomes different from completely natural existences. Humans with *Yi Yuan Ti* begin to know and reform nature purposefully according to plans and make the world of nature a conscious natural world that knows its self. The natural world can no longer change totally naturally according to natural laws as it once did in the past.

Humans as part of the world of nature are a tiny part of the whole entity of the natural world. However, they are able to know infinite mysteries in nature and make them serve humanity. A great many philosophers in both ancient and modern times felt puzzled at this. Some of them attributed this phenomenon directly to gods. The reason for this is because they did not know the special features of the human body, and particularly because they did not know *Yi Yuan Ti* (Mind Oneness Entity).

It is pointed out in the theory of hierarchical structure of substance that "The more advanced the stratum of the substance, the less the abundance of the material system of this stratum in the universe, and the greater the multiplicity of its structures and functions."(*Journal of Dialectics of Nature*, Issue 83: *On Formation of the Hierarchical Structure of Substance*). Astronomy has told us that, after the Big Bang, only part of the substance formed hydrogen clouds, of which only a small part became stars. In the evolution of stars, only one percent or so of the hydrogen and helium turned into heavy elements. The heavy elements that gathered into the planets were even fewer, and the substance that formed life was even less than this. Life substance developed and evolved, and an extremely small amount of it formed the human brain.

We know that there are currently 108 known elements and more than seven million known molecules in Nature, and by estimation one billion organisms have once lived on the earth. Humans are the favored children in innumerable biological competitions during such a long and complicated evolutionary process. The reason why human beings can tower over all other existences in nature is because *Yi Yuan Ti* which is peculiar to human beings contains all kinds of information about all existences in nature. It is, on the one hand, the information deposited in the subject (humans) in the evolution of the human species and, on the other hand, the information that *Yi Yuan Ti* receives from the objective world. In this great quantity of information, there is both the information about primordial *hunyuanqi* and the information about all existences in the universe.

It is pointed out in the anthropic principle that it is the features peculiar to humans that determine the laws of change in the universe since the Big Bang to be inevitably like this. It is pointed out in the theory of *hunyuan* whole entity that information can gather energy to further form various kinds of concrete substance. Before the appearance of human beings, it was natural information that exercised its functions and brought about changes in the natural world. Ever since human beings came into existence, it is essentially the familiar intelligence in *Yi Yuan Ti* that exercises its functions and upon this basis all branches of science are established, which makes the world of nature enter the knowing-itself stage. In other words, ever since the appearance of human beings, the universe is no longer a natural universe but a conscious universe.

However, familiar intelligence is partial and limited. Therefore, the self-knowledge of the universe is still not thorough or complete. With the tapping and application of super intelligence, all *Yi Yuan Ti*'s functions will be commenced. This will not only make the natural world become a natural world that centers on and serves human beings, but will also make humans become truly free self-awakened and self-knowing humans with perfect self-mastery. When human beings truly understand the mysteries of their own lives as well as their consciousness, the universe will have attained genuine self-recognition and self-awareness.

Section III Consciousness and Consciousness Activity

Yi Yuan Ti is the manifestation form of universe *hunyuanqi* having developed to its most advanced stage, whereas consciousness activity is the content and process of motion in *Yi Yuan Ti*. Consciousness in the theory of *hunyuan* whole entity is regarded not only as a special motional mode of substance, but also as a special objective existence, and its content and laws are expounded in this section with respect to these two aspects. The meaning of consciousness will be first introduced as follows:

I. Meaning of Consciousness

The term consciousness appeared in China after the introduction of Buddhism. Before this, consciousness was called *xin* (heart) and *shen* (spirit) in ancient China and they were used in a general sense to refer to advanced nerve activities. Now the term consciousness is adopted in many subjects in science, yet different branches have their own understanding about consciousness. In order to understand accurately the concept of consciousness in *Zhineng Qigong* Science, let me first introduce views of consciousness in other subjects.

i. Concept of Consciousness in Different Subjects
i) In Philosophy

In idealism, consciousness is considered as the expression of the soul and also the function of the spirit that is incognizable. In dialectical materialism, consciousness is considered as the thinking activity using concepts. It is the sum of all the psychological activities in a human being's subjective world, including will, missing something or somebody as well as cognition. It is the product of both the cerebral cortex and society. It is antagonistic to objective existence as well as a reflection of it.

ii) In Psychology

Consciousness in psychology is considered as the sum of all psychological activities. Such activities refer to sense, perception, emotion, sentiment, aspiration, temperament, thinking, memory, ability and so on. All these psychological activities take place against the background of consciousness. Atkinson and Hilgard, et al., define consciousness in *Introduction to Psychology* as this: "We are conscious when we notice the events outside and inside ourselves, when we think of our past experiences, engage ourselves in solving problems (translated from Chinese into English) and deliberately select one course of action over others in response to environmental circumstances and personal goals. In short, consciousness involves (1) monitoring ourselves and our environment so that percepts, memories, and thoughts are represented in awareness, and (2) controlling ourselves and our environment so that we are able to initiate and terminate behavioral and cognitive activities (Kihlstrom, 1984)" They also categorize consciousness activities into four levels of consciousness, unconscious processes, preconscious memories and subconsciousness.

Philip Zimbardo, an American professor of psychology in Stanford University, defines consciousness in his book *Psychology and Life* (1988) as follows: "Consciousness is the general term for *awareness*—including awareness of ourselves as distinct from other organisms and objects. Ordinary waking consciousness includes the stream of immediate experience comprising our perceptions, thoughts, feelings, and desires at any particular moment of awareness—along with our 'commentary' on them and on our actions. Besides our awareness of some *content*—something we are analyzing or interpreting, including some sense of self—consciousness also includes the *state of being aware*." According to him, three kinds of consciousness have been identified by psychologists, and the activity state of consciousness divided into such three levels are anoetic (nonknowing), noetic (knowing) and autonoetic (self-knowing).

We summarize the series of spiritual activities into four levels according to physiological and psychological viewpoints.

a. The level of physiological reflex: This includes such physiological nerve activities as unconditional reflexes and conditional reflexes. This level carries out simple neural arc activities.

b. The level of perception: This governs perception about what is going on inside and outside the human body, including the analyzers in each part of the central nervous system. It accomplishes analysis of different senses.
   The two levels of nerve activities mentioned above accomplish transmission and analysis of information through nerve pathways (i.e., nerve fibers).

c. The level of consciousness: This level is established on the basis of the two levels mentioned above. It analyzes and synthesizes all kinds of input information and abstracts concepts from them. So the functions of this level include the activities of forming concepts as well as applying concepts and vocabulary, such as association, retrospection, analysis, synthesis, memorization and deduction. This level is closely related to the above-mentioned two levels and can affect their functions. On the other hand, the previous two levels are also the important factors that bring about consciousness activities.

d. The level of consciousness-in-itself (including subconsciousness): This is in the deepest "background level" of the entire nerve activities and is usually covered by the level of consciousness, flooded by consciousness activities and can manifest itself only when consciousness activities stop. It does not use concept activities and directly combines with information, so there is neither analysis nor concept in it. When concepts start to function, consciousness activity begins. The consciousness activity carried out without using concept described by Albert Einstein refers most probably to this.

iii) In Medicine

In Western medicine, consciousness is the waking conscious state of humans and also the functions of identifying and dealing with the relationship with the surrounding environment. In traditional Chinese medicine in ancient times, consciousness referred to the entirety of thinking activities, including human perception, reflection and thinking of the external world.

It is said in *Ling Shu* in *Huang Di Nei Jing* "What is used in reflecting things is called *xin* (heart). What is kept and can be recalled in *xin* is called *yi* (mind, recollection). What is stored and maintained in *yi* is called *zhi* (志 memory, record). Change in the mind that takes place on account of *zhi* is called *si* (thinking). To think further on account of *si* is called *lv* (consideration, thinking in an extended scope and much in detail). Dealing with things on account of *lv* is called *zhi* (智 wisdom)."

Therefore ancient masters of traditional Chinese medicine believed that thinking activities included the six levels of *xin* (heart), *yi* (mind), *zhi* (memory, record), *si* (thinking), *lv* (consideration) and *zhi* (wisdom), which can now all be called mental activities or consciousness activities.

iv) In Buddhism

The term *yi shi* ("consciousness" in brief) in the Chinese language originated from Buddhism. There is a treatise in Buddhism on consciousness called *Wei Shi Lun* (*Cittamatra*), which includes *Cheng Wei Shi Lun* (*Vijnaptimatratasiddhi-shastra*) and *Wei Shi Er Shi Lun* (*Vidyamatra-siddhi-vimsakakarika-shastra*). In Buddhism, *Yi* and *shi* have different meanings. *Yi* refers to that humans think about things in the mind. *Shi* means to distinguish, differentiate and understand things. Both *yi* and *shi* are the functions of "heart". They are the embodiment of the functioning of the realm of heart. Although *xin*, *yi* and *shi* differ from each other, they are in fact one entity. It is stated in *Ju She Lun* (*Abhidharma-kosha-shastra*) that "The three names of *xin*, *yi* and *shi* differ in what they mean but their body is the same." It is also said in *Da Cheng Yi Zhang* (*Mahayana Encyclopedia*), "*Shi* is another name of *shen*.

When *yi* and *shi* are combined, they become one special term called *yi shi*. In *Cittamatra*, it is believed that *yi shi* ("mental consciousness" in this context) is the sixth *shi* of "the controlling mind of the eight vijnanas" (or eight kinds of consciousness), which includes eye consciousness, ear consciousness, nose consciousness, tongue consciousness, body consciousness, mental consciousness, atma-graha or (familiar intelligence) self adherence consciousness, alaya or foundation consciousness". It is built on the basis of *yi gen* or indriya and is part of all human psychological activities.

Yi gen (the root of *yi*) or indriya is the seventh *shi*, i.e., atma-graha consciousness, and its function is to "analyze, understand and recognize Dharmadhatu or Dharmad-realm" as a heart king. In Buddhism, it is believed that different kinds of environment and terms in the universe correspond to *liu gen*—the six roots of the eyes, ears, nose, tongue, body and mind. For example, scenes to which the eyes correspond are called *se* or color, and *se* stands for existences with form and mass that are not penetrable according to familiar intelligence. Things in the universe to which indriya corresponds are called *fa* or Dharmad. The term *fa* represents the existences with names (and name is concept), e.g., the concepts of tables, chairs, cups are all called *fa*, which is formed with the union of hetu-pratyaya (primary cause and secondary cause). *Yi shi* is just to receive and use concepts. To distinguish, understand and know these concepts is the functioning of *yi shi*.

Yet it needs to be pointed out that *fa* as stated here is rather broad in meaning. What is stated in Buddhism that accommodates all existences in the universe—the six "roots" (the eyes, ears, tongue, nose, body and mind), the six shad samvrta or "dusts" (sight, sound, taste, smell, touch, concept) and the six shad vijnana or *shi*s (consciousness or perceptions produced from the eyes, ears, tongue, nose, body and mind) are all called Ashtadashadhatu or the 18 realms (which include the realms of six "roots", the realms of six "dusts", and the realms of six *shi*s). The 18 realms are all called *fa* or Dharmad, upon which *yi shi* (mental consciousness) can have effects. In addition, physical, verbal and mental behaviors also result from the functions of consciousness. Even making the wish and being determined to cultivate *Dao*, to rid oneself of confusion and to understand and prove the truth of life is also the functioning of consciousness.

v) In Daoism

Unlike Buddhism in which terms and concepts are fully expounded, Daoism emphasizes practical applications. Therefore, many terms and concepts in Daoism are not explicitly explained. The term *yi shi* is not found in Daoist classics. What is similar to *yi shi* in Daoism is *yuan shen*, *shi shen* and *zhen yi*, which will be respectively stated as follows:

a. *Yuan shen* (primal spirit): It is stated in *Mai Wang*, "What is *yuan shen*? When the mental activity from within does not occur and the mental activity from without does not enter, that which is independent and in self mastery is called *yuan shen*." It is said in *Huang Ting Wai Jing Jing Shi Zhu*, "*Yuan shen* is the mentality in 'heart' when it is neither in motion nor still and is lively and vivacious." *Yuan shen* is born with human life and all spiritual activities are established on the basis of *yuan shen*.

b. *Shi shen* (perceptive spirit): This refers to thinking and spiritual activities (including the occurance of various emotions) after birth. *Huang Yuanji* said, "Having the heart (referring to consciousness activity) is *shi shen* and having no heart (the state when no consciousness activity occurs) is *yuan shen*."(*Le Yu Tang Yu Lu*).

c. *Zhen yi* (true mentality): In *Wu Liu Xian Zong*, "*Yuan shen* does not move and is the entity; *zhen yi* gets connected and is in use. *Yuan shen* and *zhen yi* are actually one and the same thing that can either be called *yuan shen* or *zhen yi*. Therefore, *zhen yi* is the right awareness in the void state, the subtle 'sense' one is aware of that is rather fine and exquisite." We think that the two of them should not be equated.

In the relationship between *yuan shen* (prenatal primal spirit), *shi shen* (postnatal perceptive spirit) and *zhen yi* (true mentality), *shi shen* makes use of *yuan shen*'s magical paranormal perceptibility (*yi* relies on *shen* for its motion), and *yuan shen* is flooded in *shi shen*. In a sense, *shi shen* is the specific embodiment of *yuan shen* and in *shi shen* resides *yuan shen*. When *shi shen* calms down, the nature of *yuan shen* will reveal itself. The consciousness activity that can perceive "the nature of *yuan shen*" differs from common consciousness activity and is called "true mentality."

In ancient Chinese's opinion, *yuan shen* and *shi shen* were opposed to each other. They believed that *yuan shen* was the foundation for the cultivation of *Dao* and *shi shen* was the cause of life and death. Jing Cen once said, "The *Dao* learners do not know the truth. This is because they have only known the perceptive spirit (*shi shen*). *Shi shen* is the cause of life and death for endless years. However, it is regarded as their own original body by the ignorant."

ii. The Concept of Consciousness in *Zhineng Qigong* Science

According to the science of *Zhineng Qigong*, consciousness is a special form of motion of substance. It is the activity state of the special substance of *Yi Yuan Ti* and the content and process of motion in it. Consciousness in *Zhineng Qigong* includes three parts: Consciousness activity in the narrow sense—the motion of concepts in the subjective world; consciousness activity in the broad sense—the reflection of all life activities in the subjective world; and super intelligence consciousness activity. Consciousness comes from the reflection of different relevant information (social information, natural information and life information of one's own body) about the internal and external environment in the human body's life process.

In a human being's different life stages or levels, the content and expression of consciousness activities are not the same. For example, though a baby cannot speak or use concepts, its *Yi Yuan Ti* can give instructions to command its body's movement, and the baby also has the functional ability of analysis with senses. The consciousness activities during this period are essentially related to the information connected to sense and movement. The consciousness activities in young children enter the stage of reflecting images of existences in an integrate manner, namely, "thinking in terms of images" as it is usually called. Adults use concepts to carry out their thinking activities, which is logical thinking as generally called. When *qigong* practice reaches a certain degree, concept activities in the practitioners' consciousness will turn into actively observing or instructing their own life motions with *Yi Yuan Ti*. When the intermediate and the advanced stages are reached in *qigong* practice, holistic super intelligence consciousness abilities will be displayed.

II. Classifications of Consciousness Activities

Classifications of consciousness activities in different subjects have their own distinctive characteristics due to their different focus in research. For example, in Buddhism, consciousness activities are divided into *wu ju* consciousness[1], *wu tong yuan* consciousness[2], *wu hou* consciousness[3] and *du tou* consciousness[4]; in Daoism, they are divided into *yuan shen* (primal spirit), *shi shen* (perceptive spirit) and *zhen yi* (true mentality); in philosophy, we find a division into perceptual knowledge and rational knowledge; in psychology, there are intuitive motor

Note:

1. Consciousness that accompanies the five senses of sight, hearing, smell, taste and touch.
2. One of the two sub-categories of *wu ju* consciousness that occurs with the five senses and is closely related to what is sensed and perceived.
3. Consciousness not simultaneously born with the five senses of sight, hearing, smell, taste and touch, yet not separated from them and taking place successively after these senses occur.
4. Consciousness in settled meditation, in distracted state of mind and in dreams.

thinking, thinking in images, and logical thinking; and so on and so forth. *Zhineng Qigong* classifies consciousness activities as follows according to the need to give guidance to *qigong* practice and to enhance consciousness activities:

i. Classifications according to Content of Consciousness Activities

i) Classifications according to Mode Content of Consciousness Activities
a. Intuitive Motor Sensory Consciousness Activity

This is most evidently shown in babies and the characteristic is the combination of their consciousness activities with movement or senses of their bodies. For example, when babies are touching and playing with toys, their consciousness instructions and their bodily actions are directly combined. There is no conceptual activity of touching or playing with toys in their mind. When babies respond to stimulation, it is also like this. The responsive behavior of a person in an emergency state still belongs to this type. This state conforms to the requirement in *qigong* practice of integrating mind and body in oneness and is able to enhance life motion. The fact that human beings in their primitive stage could display unbelievably great strength and brisk actions is an example of this. However, in babies, it is naturally displayed instead of being consciously applied with initiative.

b. Consciousness Activity Concerning Images of Existences

Babies, when they have already acquainted themselves to a certain extent with existences around them yet still have not acquired the ability to abstract, think mostly in terms of specific images in their consciousness activities. The advantage of this method of thinking lies in its distinctiveness of image, immense amount of information and easiness of *qi* mobilization. It is direct and comprehensive and, in solving some problems, simpler and swifter compared with logical calculations. People of the artistic type think mostly in terms of images. It is believed in modern psychology that most left-handed people think in terms of images and the display of this function concentrates in the right brain. In *qigong* practice, more importance is attached to thinking in terms of images.

c. Consciousness Activity that Uses Concepts

This is the thinking activity that solves problems with abstract concepts or theoretical knowledge. For example, application of mathematics and logical deduction involve mostly symbols or concepts in the content of consciousness activities. Since such content becomes separated from concrete things, it is called abstract thinking or logical thinking in psychology. Since concepts and vocabulary have formed the second signal system, it is not easy for this kind of consciousness activity to cause *qi* change in familiar intelligence people's life activities (as they have not yet learned to use it). There is also a difference between formal logic and dialectical logic

212

here.

d. Consciousness Activity that Observes and Perceives Life Activity

This is mostly related to applying consciousness in *qigong* practice—to concentrate attention on the life activity related to *qigong* practice and to observe and perceive the motion process as well as the various kinds of change it brings. This kind of consciousness activity is equivalent to *guan* (meditative inward observation) in Buddhist shamatha-vipashyana or *hui* (enlightened wisdom) in *Yin Ding Sheng Hui* (generating enlightened wisdom in meditative stability) in Zen. It is believed in *Zhineng Qigong* that this kind of consciousness activity is a process of concentrating one's mind. It is a process of strengthening life motion with spirit. In this process, there exist no such concept activities as judgment or deduction as in general logic, but the direct "observation" and perception of different kinds of scenes of change. In traditional *qigong*, such content that is observed and perceived is called inward vision (*nei jing*), while the entire process of observation and perception is called inward sight (*nei shi*).

e. Super-Intelligence Consciousness Activity

This kind of consciousness activity is the *qigong* functional ability at a more advanced level or the consciousness activity peculiar to people with paranormal abilities. It is established on the basis of *Yi Yuan Ti*'s super extraordinary information and can directly capture the holistic information about existences. Inspirations in people with familiar intelligence also belong in this category of consciousness activity. This is a kind of highly efficient consciousness activity which can enable people to know the mysteries in the world of nature more deeply and more comprehensively. It can bring humans and nature into a state of deep harmony.

All these five modes of consciousness activities exist in adults to different extent and can be displayed in different degrees in different situations.

ii) Classifications according to Object of Consciousness Activities
a. Consciousness Activity Related to Science

This refers to the logical dialectical motion for abstract concepts that conform to rationality to use each other as medium. This kind of consciousness activity goes on in *Yi Yuan Ti*'s "Second Reflection Space". Generally speaking, consciousness activities of science, except for their consumption of *hunyuanqi* in the brain, do not have much impact on the human body's life activities.

b. Consciousness Activity Related to Arts

This refers to the holistic style of consciousness activity that uses emotions as its medium and is integrated with such psychological activities as perception, understanding, imagination and cognition. This kind of consciousness activity also goes on in "The Second Reflection

Space" in *Yi Yuan Ti*. As it is mostly associated with thinking in images and is connected with *Yi Yuan Ti*'s "First Reflection Space", consciousness activities of arts can usually bring about different changes in the human body's *hunyuanqi*.

c. Consciousness Activity Related to Morality

This refers to all kinds of consciousness activities centered on the mode of value judgment about human life. It is the consciousness activity going on in the stratum of *Yi Yuan Ti*'s Reference System and is the fundamental level of consciousness activity that can have direct impact on human emotions and sentiments.

d. Consciousness Activity Related to *Qigong*

This refers to the consciousness activity that improves and perfects one's own life activities. It is the consciousness activity that penetrates "The Second Reflection Space", "the First Reflection Space" and the Reference System. *Qigong* consciousness activity guides human beings to transcend into more advanced levels.

ii. Classifications according to Direction of Consciousness Activities
i) Outward Consciousness Activity

The outwardness feature of consciousness activity, in the first sense, refers not only to the outwardness in reflecting objective existences around, but also to reflecting changes in one's own life as far as *Yi Yuan Ti* is concerned. Outwardness, in the second sense, refers to the phenomenon that consciousness activity often expands around one central question and can make one's thought ramble far from the central point. Cognition, judgment, association, deduction, even invention and creation and so on in people with familiar intelligence all belong to the outward application of consciousness. The outwardness feature in consciousness activity leads to the outwardness feature in the mechanism of human body's *qi* and makes the *qi* unable to be stored inside and tend to be easily consumed.

ii) Inward Consciousness Activity

This is the consciousness activity in *qigong* practice (Introspective psychologists also use the method of introspection, a kind of inward consciousness activity). It is just the opposite to the outward consciousness activity in direction and also has two levels of meaning. One is to make the content of consciousness activity become focused and simple, namely, to make the myriad of wandering thoughts in one's mind simplified and concentrated on one point and gather the extending consciousness activities inward. The other is that the target of consciousness activity is directed toward the life of one's own body (including the body and *qi* channels) and is then further directed toward one's own *Yi Yuan Ti*. This kind of consciousness activity can make one's *qi* collect inward and then become cultivated and enhanced and can strengthen the commanding power of consciousness activity over life activity.

Further classification finds that there are inward consciousness and outward consciousness in both familiar intelligence and super intelligence.

Outwardness in familiar intelligence consciousness refers to the phenomenon that consciousness is directed toward external existences and that thinking extends outward, and this belongs to the outwardness of outward consciousness activity. Concentrated thinking about things or existences belongs to the inward aspect in outward consciousness activity. Concentration in the spirit or merging one's mind into one's own life activity in people's *qigong* practice belongs to the outwardness in inward consciousness activity. When one's mind is concentrated on observing and perceiving changes in the entity and phenomenon of one's mental activities, it is the inwardness of inward consciousness activities.

Outwardness in super intelligence consciousness refers to the phenomenon that the function of super intelligence is directed outward, and this is the outwardness of outward consciousness activity. The consciousness activity involving judgment that uses the familiar intelligence Reference System in *Yi Yuan Ti* belongs to the inward aspect in outward consciousness activity. The consciousness activity of people with paranormal abilities who have not experienced arduous *qigong* practice has not surpassed this sphere. In their consciousness activity which involves the functional state of target and direction, there is still the difference between that which can direct and that which is targeted. Such consciousness activity belongs to inwardness of consciousness activity but has not yet reached the truly inward realm.

It is only when the attribute of the observing and perceiving function of consciousness activity has broken away from the Old Reference System that the truly inward realm is achieved. If familiar intelligence consciousness and super intelligence consciousness can combine and the outwardness and inwardness are interactively used and become merged together, the wonderful magical realm is reached. In the magical-realm consciousness activity in super intelligence, there is already no difference between outwardness and inwardness. Inward is outward and outward inward. It has reached the realm in which the outwardness and the inwardness become the same.

iii. Classifications according to Level of Correlation between Consciousness Activity and Life Activity

Consciousness being described here refers to consciousness activity in the broad sense, that is, the reflection of all life activity in *Yi Yuan Ti*, which includes:

i) Biological Level Consciousness

This mainly refers to the information regarding metabolism at the level of cells (including cranial nerve cells) reflected in *Yi Yuan Ti*. It is reflected into *Yi Yuan Ti* the moment *Yi Yuan Ti* is formed. With the development of individual human life, the information newly displayed is continuously reflected and makes the information at the biological level in *Yi Yuan Ti* gradually improve. It should be pointed out that different information in the objective world can also be reflected in *Yi Yuan Ti* at the same time. It is simply that the amount of the information is less and thus more difficult to be extracted. In the two cases mentioned above, whether it is the life information regarding the individual human being or the natural information about the outside world, both are reflected in *Yi Yuan Ti* according to the accommodating feature of *hunyuanqi*. Unlike the information reflected in *Yi Yuan Ti* later after birth that has become intensified with the functioning of different tissues and organs of the human body, these reflections in *Yi Yuan Ti* are still in an inactive and unperceivable state that cannot be extracted. It should be noted that, though these reflections are not perceivable in the familiar intelligence state, they affect *Yi Yuan Ti*'s vitality.

ii) Life Level Consciousness

This refers mainly to the reflection in *Yi Yuan Ti* of the functional activity of the various tissues and organs of the human body and the information regarding different kinds of substances (including hormones) produced in the course of these activities. As such information about the functional activity of the tissues and organs and about the different substances accordingly produced differs from the information regarding the metabolism of cells, it can be specifically reflected in *Yi Yuan Ti* and can become cognizable and extractible information once it has experienced proper amount of repeated intensification (Motor sensory consciousness belongs to this).

iii) Consciousness Related to the Activity of Emotions

This is the consciousness activity established on the basis of biological level and life level consciousness. It is the internal feeling displayed in *Yi Tuan Ti* about the internal state of whether or not the motion of human body *hunyuanqi* is going on smoothly and it is shown in certain life activities. This is a complicated life process. Consciousness of morality is just established on the basis of this level.

iv) Consciousness Related to the Stratum of Thinking

This mainly refers to such thinking activities as thinking in images and logical thinking.

iv. Classifications according to the Nature of Subject or Object of Consciousness Activities
i) Subject Consciousness (or Consciousness of Self)

This refers to the concentrated display in *Yi Yuan Ti* of the function of maintaining the normal activity of one's own life. It can be both evident and latent consciousness activities. It is the reflection of the structure and function of the *hunyuan* whole entity of a person in this person's

Yi Yuan Ti. Subject consciousness is gradually improved and consummated with the continuous evolution of human life activity. Human subject consciousness has evolved from lower organisms' function of maintaining their own *hunyuan* whole entities. Human beings' subject (self) consciousness penetrates all the "stratums" of biological level consciousness, life level consciousness, emotional level consciousness and thinking level consciousness, making consciousness at different levels subordinate to one's own need. Whether a person is in the state of sleep or unconsciousness, so long as this person is still alive, the subject consciousness of this person will function to a certain extent.

ii) Object Consciousness

Other consciousness activities aside from subject consciousness are all object consciousness activities.

III. Formation of Consciousness Activity

Human consciousness activities gradually come into being with changes in *Yi Yuan Ti*—with the formation of Self *Yi Yuan Ti* and Partial Identity *Yi Yuan Ti*. They are gradually established in human beings' practice and continuous contact with the objective world and in the process in which the sensory organs perceive the objective world with the inculcation of language from adults. Language is closely related to the establishment of consciousness. Consciousness and language have experienced a development process from simplicity to complexity. This process, no matter it is in the evolutionary history of the human race or in the evolutionary history of human individuals, experiences a rather long process before it is gradually accomplished. The formation process of consciousness is a process in which the *hunyuanqi* of existences in the outside world merges and combines with *Yi Yuan Ti* and causes changes in it.

i. The Evolutionary History of Consciousness in the Human Race

If we say that all existences in the universe are the content of the objective world naturally formed through continuous division, evolution and interaction of primordial *hunyuanqi*, then human consciousness is the content of human subjective world formed with the continuous *hunhua* of Pure Original *Yi Yuan Ti* in the process of social practice, such as labor and language communication. Although the history of human consciousness is shorter than that of all existences in the universe, it has also gone through a very long period of millions of years. It is true that we cannot examine quite detailed formation process of human beings' consciousness, yet it is still possible that we can make necessary analysis based on the historical data available about ancient society and ancient human beings.

Although human civilization has a history of merely thousands of years so far, according to investigation, human beings had already separated themselves from the world of animals for millions of years. For a very long time, human beings had been in an uncivilized age. They had only simple language and crude tools and knew almost nothing about nature and themselves. The state of human spirit in this stage was equivalent to the period of Self *Yi Yuan Ti*. Human beings at this period were characteristic of communal living. They worked together and lived together. A life pattern like this, when reflected in *Yi Yuan Ti*, formed the consciousness of being "for living" and "for the public", namely, the maintenance of one's own subsistence and the submission to the public needs of the group. The individual and the group were an integrated entity. Even matrimony was communal marriage among members of the entire group. People's thinking during this time was very simple and had not become civilized yet. However, their simple consciousness actively commanded their life activities, and the two of them were closely combined. Everybody was a member of the group and obeyed its "disciplines", thus the human nature of "being free and conscientious" were displayed.

With the development of tools, language communication was also gradually enriched and great evolutionary progress was shown in the vocal organs. Various information in *Yi Yuan Ti* also increased day by day, which reinforced the cranial nerve cells' further differentiation and improvement and enhanced *Yi Yuan Ti*'s functions. Associative thinking using simple vocabulary developed gradually into thinking in images carried out with the use of systematic language. When thinking in images was formed, human beings were able to carry out rudimentary independent thinking. Consciousness activity and life activity became gradually separated. Primitive totem worship was just born on the basis of such level of consciousness. With the need of human beings' development such as the improvement of the force of labor and the reinforcement of group life, not only did the pattern of marriage become gradually reasonable, but the notions of morality and ethics also gradually formed. Furthermore, appreciation of beauty in Nature developed to the stage of creation of beauty with labor and this was the beginning of human beings' art in its early stage. All this further reinforced the improvement of *Yi Yuan Ti*'s functions. Conceptual thinking gradually substituted thinking in images and written language was accordingly born in this world.

With the emergence of private property, the notions of "for living" and "for the public" in the realm of consciousness began to vanish and were replaced by the mentality of private ownership. "For the private" was an extension of "for living" in *Yi Yuan Ti*. It accelerated human beings' development on the one hand but, on the other hand, guided human consciousness to self-centeredness. The desire to possess things outside themselves became the dominant force in humankind. We know that when humans take possession of things outside themselves, the things in turn take possession of humans. Thus the innate human species nature of "being free and self-awakened" began to be lost and humans were gradually led astray into a state of being driven by external things.

In addition to this, in the process of pursuing and possessing all kinds of material things, correspondent familiar intelligence in humans became sufficiently developed, which made humans and things acquire mutual acknowledgement and confirmation, and the relationship between humans and things was fixed at the level of familiar intelligence. Human beings up to now still obstinately believe that the familiar intelligence relationship between humans and things is absolutely and invariably correct, and this is the display of what we call Partial Identity *Yi Yuan Ti*. It hinders human beings from knowing laws at a deeper level of the objective world. In fact, consciousness activities of various logical calculations advocated in science today have not revealed and cannot reveal the nature of things in a direct manner. They are but the demonstration and deduction according to the apparent relationships between different parts of things. This is an indirect method of knowing the nature of things from without to within.

In sum, human beings' consciousness was gradually formed in the development process from apes to humans. Readers might ask why it took such a long time for the formation of consciousness in the human species. This is because the evolution from apes to humans was a natural process, a process in which brain *hunyuanqi* carried out *hunhua* with the outside world and new content was gradually created in human brain *hunyuanqi*. This is a process concerning the continuous evolution of various human functions and the continuous improvement and evolution of human internal tissues and structures. There is a complicated changing process here in which changes in *hunyuanqi* became deposited in concrete substances. Once these changes have become deposited in concrete substances and are embodied in genetic substances, new individuals will be able to replay the formation process of consciousness in a relatively short period.

ii. The Evolutionary History of Consciousness in Human Individuals

The formation process of an individual human being's consciousness is very similar to the process of the formation of consciousness from apes to humans. However, the two formation processes are also essentially different. This is because, though a newly born infant still does not have specific consciousness activities, in its genetic information there is already the condition for the generation of consciousness activities—*Yi Yuan Ti*. Moreover, the language and consciousness of people around the baby are the newborn individual's environment which can be reflected in the baby's *Yi Yuan Ti* and directly form the content of the Reference System. None of this, however, existed in the environment of the apes. Therefore, under the influence of human consciousness, the newborn individual rapidly forms its *Yi Yuan Ti*'s Reference System similar to the Reference Systems of the humans in the baby's environment. Moreover, the newborn baby at the very beginning not only receives natural information, but also proves and cognizes objective existences on the premise of receiving the information about consciousness and language. It is rather essential to understand this point. Otherwise it is apt to confuse human beings with animals. Details about this will be elaborated as follows:

i) The Process of the Formation of Consciousness in a Newborn Individual

After a baby is born into this world, dramatic changes in the baby's internal and external environment different from that in its mother's womb take place. The density increase of CO_2 in the baby's body not only raises the degree of excitation of the brain cells, but also brings about the motion of breathing. The internal organs begin to exercise their respective functions. All this destroys the even state of *qi* innate to and inside the human fetus. On the other hand, the objective world which the infant contacts also differs from the even and balanced state in the mother's womb and shows a greatly diverse feature.

With the continuance of the baby's independent life activity, each sensory organ begins to receive different kinds of stimulation from changes in the human body's internal and external environment. The same characteristic of a certain substance would bring about the same effect in *Yi Yuan Ti* under the same condition, and this is sense. With repetition of the effect in *Yi Yuan Ti*, not only is the connection between *Yi Yuan Ti* and the sensory organs established, but the information obtained with the sensory organs also becomes intensified in *Yi Yuan Ti* and leaves in it a certain "imprint". Synthesis of different kinds of "imprint", e.g., the imprint of space, concrete objects, people in the environment as well as the baby itself, etc., then becomes unconsciously the background or the Reference System for the baby to deal with the relationship between subject and object. On this basis, the sensory organs send different information about substances to the brain, and *Yi Yuan Ti* can reflect the features, similarities and differences of substances. Hence is born the perception of things. This is the process in which different kinds of material features of the objective world are internalized into *Yi Yuan Ti*, the subjective world. Although sense and perception directly reflect specific images of objects and belong to the preliminary stage in the cognitive process, a simple mode is already established in *Yi Yuan Ti*, and this is the foundation for consciousness activities afterward to evaluate objective things.

Besides, when effects of sense and perception occur in *Yi Yuan Ti*, responsive reactions are generally required in order to give answers to the objective stimulation. As babies have not formed consciousness activities that use concepts, all the information that *Yi Yuan Ti* sends directly combines with intuitive movement, which is motor intuitive thinking as referred to in psychology. In *Zhineng Qigong*, the above-mentioned content and process of the activity in *Yi Yuan Ti* is called sensori-motor consciousness. During this period, since the baby's *Yi Yuan Ti* still maintains to some extent the magical void and even state of Pure Original *Yi Yuan Ti*, different kinds of changes that take place both inside and outside the baby's body will all be reflected in the baby's *Yi Yuan Ti*. Such changes reflected in *Yi Yuan Ti* then react upon the brain cells and make material substantial changes occur in them. Hence the Reference System internalized into *Yi Yuan Ti* with the content of the objective world becomes settled and fixed.

With the baby's contact with the outside world continuously developing in depth and scope, changes in brain cells facilitated by *Yi Yuan Ti* continue and functions of the brain cells correspondingly improve. When the different information about one existence can be connected in *Yi Yuan Ti* and forms a relatively integrate reflection of the objective existence, i.e., an image, synthesis function of *Yi Yuan Ti* begins. This is called in psychology thinking in terms of images. It is believed in qigongology that thinking in images is a kind of holistic consciousness and is an effective method to know the holistic features of things. Since images have a certain relative independence, that is, they can be shown in *Yi Yuan Ti* independent of concrete things, image-based memory begins. When *Yi Yuan Ti* has the ability to abstract and can abstract from phenomena concepts that represent the nature of existences, human beings' consciousness activities enter the stage of logical calculations, or logical thinking as it is called in psychology. The modes of thinking mentioned above exist in and are applied by adults to different extents and people may show talents in particular modes according to their characters and professions.

ii) The Function of Adults' Speech in the Formation of Consciousness

Adults' speech plays a crucial role in the process in which babies' consciousness is formed because speech is not only an objective stimulation of sound, but also the sound signal that represents specific existences. Babies receive stimulation of objective existences entirely in accordance with their genetic physiological features and there is no restriction to the information that their *Yi Yuan Ti* receives. Why can the vast quantities of information in *Yi Yuan Ti* be intensified or become settled and stored? Continuously receiving stimulation is a necessity, but what is more important is the input of adults' speech. Teaching received from adults can improve babies' attention to objective existences. The combination of speech with specific things increases the information amount of stimulation and also simplifies the cognitive process.

According to psychology and dialectical materialism, the cognitive process of unknown things all goes through such processes of sense, perception, image and concept. Babies know none of the objective existences and we do not know how long it will take them to understand objective existences one after another if they follow the above-mentioned cognitive process of sensing, perceiving, forming images and abstracting concepts. As babies live in environment with other people, their cognition of the objective world takes a shortcut by using language. Therefore, babies' consciousness activity and their study of language complement each other.

Babies in the early stage can only make simple vocal sounds. In their contact with the external world, with the formation of their *Yi Yuan Ti*'s Reference System and the establishment of their stress response to external stimulation, they gradually learn to speak and slowly combine language from the adults with concrete existences. For example, when the adult says "kitten", the baby will notice and look at the cat. After the baby has learned to speak, even though "kitten" in the baby's mind is still not a conceptual word but a symbol of a specific something, this is already the direct association between language and consciousness and it is helpful for image-based memory. The influence of this upon *Yi Yuan Ti* is great, and the change that it causes in the brain cells is enormous.

Meanwhile, changes in the brain cells and *Yi Yuan Ti* facilitate babies' development into normal human beings. This is the critical factor for a newborn baby to develop into a normal human being. According to modern science, the brain of the newborn baby has not developed into maturity yet and it will continue to develop from the first to the third year after the baby's birth. During this development process, the environment is of utmost importance. If a newborn baby leaves human society and enters the world of animals, this baby will not be able to obtain the imprint of human information (consciousness and language, etc.) but the imprint of the information about animals. In such an environment, though the baby has in its brain all the genes for developing into a human being, because what is reflected in the baby's *Yi Yuan Ti* is not the reflection related to human life but the reflection related to animals' life activity, material changes correspondent to this in the baby's brain cells will inevitably take place, which fixes the information about animals in all the baby's life and makes the baby develop and change according to the characteristics of animals. Reports on the wolf child, the tiger child, the monkey child, the lion child and the elephant child, etc., published abroad years ago are convincing proof of this.

Readers may wonder: Many functions in humans are innate, yet why can the environment change them? Modern psychologists' experiments prove that the development of innate abilities in humans and animals can be reinforced only when they have received certain stimulation from the environment and have acquired individual life experiences. Some experiments will be quoted here to illustrate this point.

A. H. Risen's experiment: Put a newborn monkey into a dark box and feed it there. As it is deprived of the stimulation from the outside world, its eyesight is not as good three months later as the eyesight of monkeys living in normal conditions. The little monkey's eyesight is restored after it lives in the normal condition for two or three months. If it lives in the box for eight or nine months, its eyesight cannot be regained.

M. K. Harlow's experiment: Put a newborn monkey with two "toy" monkeys. One of them is made of steel wires with a bottle of milk in its hand; the other is wrapped with cotton and soft materials, such as fur and leather, and looks very similar to a real monkey. As a result, the little monkey likes to stay with the monkey wrapped with fur and leather. This shows that the newborn monkey makes the choice to stay with the big monkey probably not based on food, but to receive a certain amount of information, as feeling is also information stimulation. And such young monkeys, when they grow up, are not able to take care of their own baby monkeys. This is because they have not received care from their mothers when they were young and the mode of taking care of their offspring has not been established in the monkeys' mind.

K. Lorenz's experiment: A normal newborn gosling will walk after the mother goose after it is born. Lorenz's experiments show that the newborn gosling also walks after a wooden goose. If the gosling is shut in a black box immediately after it is born and is given no chance to contact other geese, it will follow neither its mother goose nor the wooden goose 24 hours later. This shows that the gosling's behavior of following its mother goose is not inborn, but an experience or imprint of an early stage, and each kind of animal has a certain imprint period. If this period expires, the animal's innate functional ability will not be exhibited as it has not received the necessary imprint stimulation from the environment of the outside world. A newborn chick's imprint period is 10 to 16 hours, a newborn gosling's imprint period is 24 hours and a newborn puppy's imprint period is three to seven weeks.

The experiments stated above decisively show that the innate behavior of animals (and also humans) needs the initiation of early experiences. If the newborn animals are deprived of stimulation from the environment, correlated functional disorders will be caused in them.

IV. Structural and Functional Stratums of Consciousness
i. Functional Stratums of Consciousness Activity

Consciousness activities are the display of *Yi Yuan Ti*'s functions. They are the content and process of *Yi Yuan Ti*'s activities, whereas the motion of consciousness activities is carried out against the background of *Yi Yuan Ti*'s Reference System. As previously mentioned, the mode and content of consciousness activities are multiple-leveled, these different consciousness activities are, therefore, carried out at different levels in *Yi Yuan Ti*'s Reference System. This is what we call the structure and stratum of consciousness activities. According to the content of consciousness activities as well as their levels in *Yi Tuan Ti*, we divide the structure of consciousness activities into three stratums (which are different from the four stratums in physio-psychology previously stated and they shouldn't be confused).

i) The Perception-and-Reflection Stratum

This stratum belongs to *Yi Yuan Ti*'s "First Reflection Space", which is equivalent to the total of the first sensory area and the second sensory area in psychology but also significantly different from it. It is directly affected by changes in the *hunyuanqi* of the cranial nerve cells. When the nerve cells receive various information and changes in them take place, the *hunyuanqi* related to the nerve cells will change correspondingly so that the information will be reflected in *Yi Yuan Ti*. This stratum can integrate scattered information to form holistic information. The processes of human sense perception, perception and thinking in terms of images are carried out in this reflection space.

Due to the mutual transmutation relationship between *Yi Yuan Ti* and cranial nerve cells, the influence of *Yi Yuan Ti* upon a person's own overall life activity is also accomplished through changes in *Yi Yuan Ti* which have effects on nerve cells. The process of sensory motor thinking is realized through this process. The process of memorization and association related to thinking in images are also accomplished here. This stratum can be directly connected with life activity through the function of nerve cells, so emotions are also the activity content in this stratum.

All the related information can be stored and even enter the Reference System and can also be extracted and sent outward as instructions. This stratum exists both in humans and in higher animals. It is the fundamental content of *hunhua* of *Yi Yuan Ti* with external information and is accomplished with the direct participation of the Reference System.

ii) The Concept-and-Abstraction Stratum

This stratum belongs to *Yi Yuan Ti*'s "Second Reflection Space." It is peculiar to human beings and is formed on the basis of the first stratum. When the multi-level compound feature of reflections of existences in *Nao Yuan Ti* (Brain Oneness Entity) of animals reaches a certain extent, *Nao Yuan Ti* begins to be able to carry out self reflection about changes that take place in itself. To be specific, a second "reflection" in another stratum of *Nao Yuan Ti* is carried out after the reflection of a concrete existence in the outside world is formed in *Nao Yuan Ti*. The reflection this time differs from the reflection of the first time. The reflection of the first time is a realistic reflection of a concrete existence; the reflection of the second time is the processing of the first reflection which replaces it with a simple symbol. This is concept (vocabulary) formed through abstraction. When *Nao Yuan Ti* has a function like this, the animals in which this change takes place have evolved into humans and we call it *Yi Yuan Ti* (Mind Oneness Entity) instead. Concepts formed through abstraction can be stored and can also be sent outward as instructions. Logical thinking as well as invention and creation in humans all take place in this stratum.

When this stratum is established, concepts can be abstracted from image information in the first stratum and can also be directly obtained from the learning process of spoken and written language. Since the content in this stratum is a transformed re-reflection of the content in the first stratum, it contains less specific information about existences and therefore has less impact on the *hunyuanqi* of brain cells. This is the reason why it is usually not easy for logical thinking to have effect on life activity. Nevertheless, the formation process of concepts is after all the result of abstracting specific reflections. Therefore, in the process of an integrate concept activity, holistic information about this something reflected is also contained. It is simply that the energy that the information carries is drastically reduced after abstraction. The First Reflection Space and the Second Reflection Space are both involved in the formation of the familiar intelligence Reference System and help it exercise its functions.

iii) The Wisdom-Illuminating Stratum

Thinking at this level goes on in the stratum of the super-intelligence Reference System. Thinking activities at this level are beyond-logic thinking or called "substantial-image thinking". This is the thinking different from both thinking in images and observant inward-directed thinking. In beyond-logic thinking or "substantial-image thinking", both the information received and the information sent are holistic substantial real existences. When normal-intelligence Reference System and super-intelligence Reference System immerse into each other and combine into one, human beings' thoroughly integrated Reference System is formed. Thinking in the wisdom-illuminating stratum must be carried out on the basis of super intelligence. With super intelligence alone yet without super intelligence Reference System, thinking in the wisdom-illuminating stratum cannot be carried out. For further illustration of the relationship between the perception-and-reflection stratum, the concept-and-abstraction stratum and the wisdom-illuminating stratum in consciousness activities, it can be schematically shown in a diagram (See Fig. 9). (See page 226)

ii. Time-and-Space Structural Stratums of Consciousness

Consciousness, as the content and process of activities in *Yi Yuan Ti*, transcends time and space as far as its essential attribute is concerned and there shouldn't be a structure of time and space in it. The time-and-space structure we talk about here is a term adopted for convenience in explanation. It is just like the three-colored quarks in the microcosmic realm which do not mean the difference in color but are used as a special term for the purpose of illustration.

According to the theory of *hunyuan* consciousness, human consciousness is a kind of special transcending-time-and-space structure established in *Yi Yuan Ti* by human beings in their entire life process. It is centered on the consciousness of self and the Reference System and forms a self system that integrates the levels of human biological metabolism, life, sentiment, thinking and morality into an organic consciousness whole entity, with each level growing on the basis of the former and in a relationship of mutual restraint. The consciousness whole entity is closely connected with the entire process of life activity, in which the biological metabolic level is in the basic stratum and morality is in the highest stratum, as is shown in Fig.10. (See page 227)

V. Generation Process of Consciousness Activity in People with Familiar Intelligence

Consciousness activity includes a rather broad range of aspects, such as cognition, memory, thinking, emotion and formation of instruction. In modern psychology, these consciousness activities are studied and illustrated mostly from the perspective of their modes as well as the different changes that they bring about. Psychology has not touched upon the nature of consciousness activity itself. The theory of *hunyuan* consciousness can give substantive

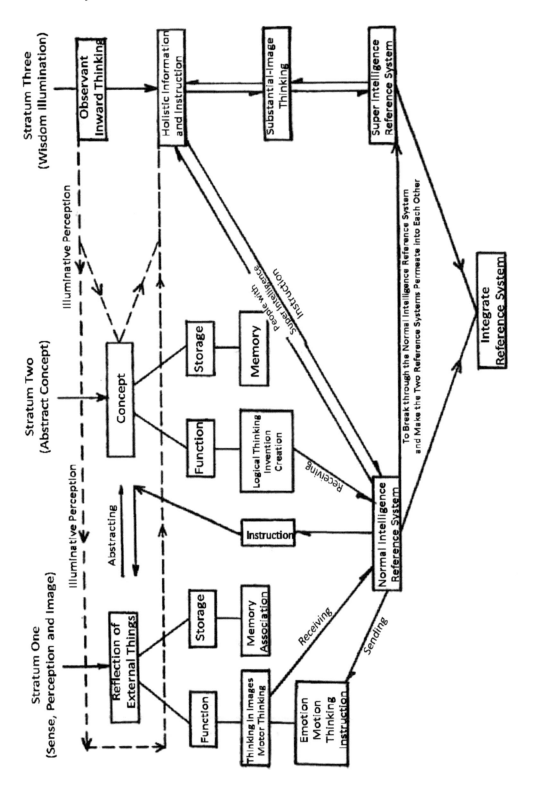

Fig. 9 Schematic Diagram of the Three Stratums of Consciousness Activities

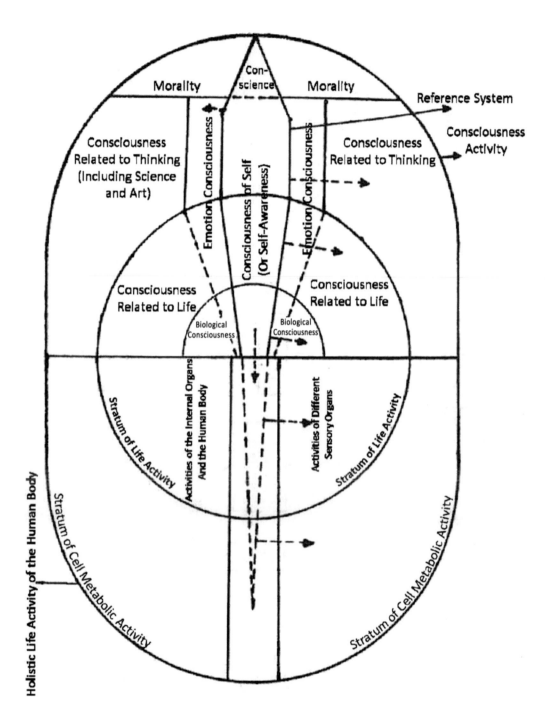

Fig. 10 Schematic Diagram of the Time-and-Space Structure of
Consciousness and Its Relationship with Life Activities

explanation of all these consciousness activities at the level of *Yi Yuan Ti*. In consideration of the length of this book, only the consciousness activities of instruction and emotion directly related to *qigong* science are introduced here.

i. The Generation Process of Consciousness Activity

Why different kinds of mental activities occur in the human brain and how these mental activities change step by step and function in the realm of consciousness are questions that *qigong* science researchers have to answer. According to the theory of *Yi Yuan Ti*, an integrated mental activity includes the following content: initial start, directional gathering, commencement activity, formation of the content of mental activity, acting on life activity, giving feedback, adjusting the point of commencement, correcting the content of consciousness. They will be explained respectively as follows:

i) Initial Start

Initial start refers to the initial cause of consciousness activity. In the initial cause that forms consciousness activity, whether it is the mental activity inexplicably occurring to the mind, the mental activity occurring on account of direct impact of external stimulation or that occurring due to direct guidance of subjective consciousness, there exists a process of orientating toward what is being thought about in the mind. Here we emphatically analyze the mental activities that occur unconsciously. Where does the power of initial start come from? We know that *Yi Yuan Ti* is a multi-level and "multi-faceted" information receiver. All kinds of information both inside and outside the human body can be reflected in *Yi Yuan Ti*, while the degree of how much each piece of information can be reflected in *Yi Yuan Ti* is affected not only by the dynamics of nerve cells but also by the amount of external information. Meanwhile, it is also determined by how important this piece of information is for human holistic life activities. Interaction and integration of a lot of information are going on in *Yi Yuan Ti* at any time. Some of the information is flooded and buried; some is intensified and displayed. The information displayed functions as frequency lock on what is around it and forms a temporary information center. This is the initial start of mental activity.

ii) Directional Gathering

When the temporary information center has formed in *Yi Yuan Ti*, the even state in *Yi Yuan Ti* is destroyed, which makes *Yi Yuan Ti* directionally gather toward the information center. This not only further intensifies the information center, but also makes the information related to it in *Yi Yuan Ti* connect together and forms a holistic information structure sequence. This is an integration activity before a mental instruction is formed and is mainly shown in change in information. Yet according to deduction based on the theory of *hunyuan* whole entity, there should be slight change in energy in the entire brain during this process. Once mental tendency is formed, commencement activity begins.

iii) Commencement Activity

When the holistic information structure sequence is settled, the mental instruction will be formed in *Yi Yuan Ti* and act upon related cranial nerve cells, which brings about series of physical and chemical changes in the brain neurons. This can be shown not only in bioelectricity in the brain, but also in relevant bio-chemical change. The signal of change in neuron, on the one hand, is transmitted to the effector organs (e.g., the internal organs and muscles, etc.) and, on the other hand, feeds back into *Yi Yuan Ti* and gives responsive chain stimulation to the original mental activity, thus bringing about series of consciousness activities.

iv) Effect Activity

After signals of the nerve cells are transmitted to effectors, activities in the effectors begin according to neural command. Various changes brought about by the activities of the effectors can be transmitted to the cranial neurons through the afferent nerves and make the brain cells change accordingly, thus bringing about change in their *hunyuanqi*. The series of changes can be synthesized in *Yi Yuan Ti* and bring about further change in the original mental tendency, which makes the time-and-space position of the original point of commencement change, form new instruction in consciousness activity and enter new cyclical effects.

The explanation given above is the formation and exercise processes of instructive consciousness activity. If it is not instructive but other general thinking activity, the consciousness activity can change of its own accord in accordance with the integrate content of the mental tendency based on the nature of the content of the mental tendency after the mental tendency is formed.

We know that any mental activity in a person with familiar intelligence is carried out by means of vocabulary, and vocabulary has certain meanings. That is to say, concepts expressed with vocabulary have their specific content; they have certain time-and-space structures. Therefore, once a mental activity is formed in *Yi Yuan Ti* (Mind Oneness Entity), its holistic time-and-space structure will be inevitably shown in the Mind Oneness Entity. When this time-and-space structure is expanding outward and around, content related to it in *Yi Yuan Ti* will be connected. This can be both related content directionally chosen and a rambling association process naturally born. Generally speaking, when attention is concentrated on a system with a certain order in it, directionally chosen association is easy to appear due to the content in the system which is an organic integrated entity. If the thinking is related to isolated events or aimless thinking or mental tendency, disorderly or unsystematic distraction will easily appear.

ii. The Process of How Emotions Occur

Emotions are a comprehensive display of relatively complicated life activities shown essentially in psychological feelings and are also one of the major areas studied in psychology. However, they have not been satisfyingly defined so far. We tend to define emotions as follows: Emotions are the unbalanced psychological state caused by certain stimulation (either external stimulation or internal physical and mental condition) felt by an individual from inside. They are shown in facial expressions and usually take happiness, anger, fear and sadness as their fundamental patterns of expression. In an emotional state, there is not only the feeling of subjective psychological change, but also the accompaniment of physical change in the individual.

i) How Emotions Occur

From the perspective of the evolutionary history of the human race, emotions result from the evolution of animals and humans. We know that, when animals adapt themselves to different stimulations, changes in the life motions of individual animals take place. These stress changes are retained and passed on to their descendants and facial expressions gradually form: Facial expressions are not only different expressions of whether different existences can satisfy the needs in their life activities, but also the integrate expression of the modes of specific life motions.

Although traces of the same fundamental emotions as those in animals are still retained in human beings, human emotions are not only more complicated in their modes of expression than those in animals, but their demand has also developed from the simple need for materials to the need for culture and morality based on the need for materials. Even their need for materials to satisfy and maintain life activities also differs in nature from that in animals.

From the perspective of the evolutionary history of human individuals, human emotions come, on the one hand, from heredity, e.g., facial expressions showing fundamental emotions begin from infanthood and facial expressions in babies all over the world are basically the same; on the other hand, they come from postnatal acquisition, e.g., sentiments shown in subtle facial expressions are not completely the same in different races.

As to how a specific emotion is generated, according to the theory of *hunyuan* whole entity, it is the special display of the internal feeling that occurs when drastic change takes place in the *hunhua* between human body *hunyuanqi* and *hunyuanqi* in the outside world.

ii) The Mechanism of How Emotions Occur

According to the theory of *hunyuan* whole entity, there are two fundamental patterns in the mechanism of generation of emotions.

One is that changes in human body *hunyuanqi* cause changes in *Yi Yuan Ti*'s Reference System and bring about emotions, and this was called by the ancient Chinese "*qi* unified to change mind". We know that the human body is a whole entity with form (*jing*), *qi* and spirit (*shen*) unified in a trinity. The three of them have their own mode of balance. When human body *hunyuanqi* is affected by the *hunyuanqi* outside the human body and loses its own original balanced state, a special distorted motion is displayed. This new motional state, which either stimulates or inhibits the continuance of life motion, will be reflected in *Yi Yuan Ti*'s Reference System and bring about different internal feelings. Generally speaking, drastic upward and outward motion of *qi* can generate anger in consciousness; drastic inward and downward motion of *qi* can generate fear in consciousness; *qi* that gathers toward one point can bring about deep thinking; *qi* that stampedes and dissipates can cause shock and fright; *qi* that flows smoothly can bring about gladness; *qi* that flows unsmoothly can cause sadness. The reason why *qigong* masters can adjust patients' emotions by means of changing the motional state of their *qi* lies in this. Since facial expressions are the holistic display of the mode of the human body's life activities, every *qigong* practitioner can use the motion of different kinds of facial expressions to change the motion of *qi* in their bodies so as to adjust their emotions.

The other pattern is that changes in *Yi Yuan Ti*'s Reference System cause changes in human body's *hunyuanqi* and bring about emotions, and this was called by the ancient Chinese "mind unified to change *qi*". It is developed on the basis of "*qi* unified to change mind". This mechanism results from the evolution of human beings' emotions and is a symbol of the fact that human beings have separated themselves from the emotions of animals.

The so-called "mind unified to change *qi*" refers to the fact that concentration of consciousness causes change in *qi*. This can not only be brought about by material factors, but also by spiritual factors. These objective factors first act on *Yi Yuan Ti*'s Reference System and, through the Reference System's identification and according to the judgment response based on the impact of the event on one's own physical and mental condition, bring about internal feelings of emotions which then cause change in the human body's *qi* and form. The old Chinese saying of "joy in the heart and smile on the face" can be regarded as a description of this mechanism. Of course, a person accomplished in self-cultivation can keep all kinds of emotions and feelings in the heart and shows none of them from the outside. "Showing neither joy nor anger in the countenance" is a description of this state. Yet for average people, when an emotion has occurred in their *Yi Yuan Ti*, change in *qi* in their body usually follows. The following is what is said in traditional Chinese medicine about this: "Anger makes *qi* go upward", "gladness makes *qi* flow smoothly slow", "thinking (too much) makes *qi* gather toward one point", "sadness makes *qi* exhausted", "fear makes *qi* go downward", and "fright makes *qi* in disorder".

When changes occur in qi in the human body, a certain change in facial expressions will also occur. The change and motion in facial expression further facilitate the flow of qi and blood and the change in *ju san* (gathering and expanding, including condensing and dispersing) of qi, which either strengthens or lightens the internal feelings of emotions. From this we can see that, in the sentiment mode of "mind unified to change qi", the functional link of "qi unified to change mind" is included. Yet it is shown not as the initial cause of an emotion but as the function of feedback. Here we can also see from analysis that, in the sentiment mode of "qi unified to change mind", when the internal feeling in *Yi Yuan Ti* has formed, it will further influence the change in an individual's body and qi so as to affect the continuance of the emotion.

VI. Consciousness Activity Is a Special Mode of Motion
i. Consciousness Activity Is the Most Advanced Mode of Motion of Substance

Friedrich Engels pointed out in *Dialectics of Nature*, "The motion of matter is not only rough mechanic motion, pure displacement, but also heat and light, electric pressure and magnetic pressure, chemical composition and decomposition, and life and consciousness."(*The Complete Works of Marx and Engels*, Vol. 20, p. 370) In this paragraph, Engels drew a brief conclusion about the motions of substance in the universe and summarized the complicated and varied motions of substance into four categories: Physical motion, chemical motion, the motion of life and the motion of consciousness. This is also the sequence in which the motions of substance in the universe develop from low level to high level.

i) Physical Motion

This is the most fundamental mode of motion in the material world. It is universally existent whether it is in the macrocosmic realm or the microcosmic realm, e.g., mechanical motion, displacement motion and heat motion in classical dynamics related to the macrocosmic realm; quantum motion (e.g., photons, electrons, etc.), change in field (such as electro-magnetic wave) as well as transmutation between mass and energy related to the microcosmic realm. Physical motion is the basic content in the motions of all existences in the universe.

ii) Chemical Motion

This refers to synthesis and decomposition in inorganic and organic matter. This motion goes on at the levels of atoms and molecules. Chemical motion is inevitably accompanied by a certain physical motion. Yet physical motion cannot substitute for the special content of chemical motion.

iii) The Motion of Life

The motion of life is a special phenomenon peculiar to organisms that are constituted of macromolecular nucleic acid, protein, etc.. It has the abilities of metabolism, reproduction and adaptability to environment. Physical motion and chemical motion are included in life motion. But the physical and chemical motions included in life motion are already different from independent physical and chemical motions because the living body is "a highly advanced unity in which dynamics, physics and chemistry are united into an entirety, and this trinity cannot be further separated."(*Dialectics of Nature* in *The Complete Works of Marx and Engels*, Vol. 20, p. 594)

iv) The Motion of Consciousness

This refers to the mode and content of *Yi Yuan Ti*'s activity formed on the basis of the special material composition of the cerebral cortex. It is the reflection of objective existence and is the display of life having evolved into an advanced stage. A healthy nervous system is the foundation upon which the motion of consciousness is established. Engels pointed out, "*Vertebrates.* Their essential character: *the grouping of the whole body about the nervous system.* Thereby the development of self-consciousness, etc., becomes possible."(*Dialectics of Nature* in *The Complete Works of Marx and Engels*, Vol. 20, p. 653) It should be pointed out that consciousness activities are established on the basis of physical and chemical changes in the nervous system. Yet these changes and consciousness activities are not the same. It is just as what Engels said, "One day we shall certainly 'reduce' thinking experimentally to molecular and chemical motions in the brain; but does that include all the nature of thinking?"(*Dialectics of Nature* in *The Complete Works of Marx and Engels*, Vol. 20, p. 591) Although Engels didn't point out the specific attribute of consciousness activity, he already pointed out that the nature of consciousness activity cannot be equated to the physical and chemical motions in the brain cells. Elaboration of consciousness in the theory of *Yi Yuan Ti* in *Zhineng Qigong* further extends on this basis.

What is the specific attribute of consciousness activity then?

ii. Features of the Entity and Attribute of Consciousness Activity

As it has been previously stated, consciousness activity is the content and the process of the activity in *Yi Yuan Ti*. It is not only the content of activity in the subjective world of humans, but also a kind of objective existence. It can affect the objective world. However, consciousness activity is a special mode of motion different from the modes of motion stated in modern science.

i) Special Features of the Entity and Attribute of Consciousness Activity

Consciousness activity is a kind of special motion of time-and-space structure (information). Consciousness motion is not conveyed with concrete substances that have form and mass nor with formless energy that has no form or mass (like light, electricity, magnetism, heat, etc.). It exercises its functions by means of *Yi Yuan Ti* or *hunyuanqi*. Consciousness activity can be the thinking activity going on in one's own subjective world and can also be applied to act upon existences in the outside world.

We know that consciousness activity in the narrow sense refers to the activity using concepts. The formation process of concept is a process in which substances in the objective world are reflected, abstracted and processed in *Yi Yuan Ti*. In this process, features of mass and energy in concrete substances are extracted, with only the content of information left. Consciousness motion is just this kind of activity of information which has only the feature of time-and-space structure. This kind of time-and-space structure has the holistic feature that can exert functions on the holistic time-and-space structures (information) of other things or existences. It can also gather *hunyuanqi* to act upon the *hunyuanqi* of other existences. What's more, it can act on other concrete substances through substances that have form. The latter belongs to familiar intelligence and the first two belong to super intelligence.

ii) Special Features of the Motion of Consciousness Activity
a. *Quan Xi* Feature
Consciousness activity is the motion in *Yi Yuan Ti*, whereas information in *Yi Yuan Ti* is infinite. As previously mentioned, the human brain has approximately 14 billion neurons and each neuron contains 10^{10} base-pairs, every three of which form one codon. Therefore, the amount of information included in *Yi Yuan Ti* is innumerable. It is on account of this that *Yi Yuan Ti* can not only reflect the information about different kinds of things in the objective world, but can also construct various information on initiative. The *quan xi* feature mentioned here, instead of being inborn in the human brain, is formed after birth. It is formed in the *hunhua* process that *Yi Yuan Ti* interacts with existences in the outside world.

b. Voluntariness
a) Voluntariness in Speed
As the speed of the motion of *Yi Yuan Ti* is faster than light, consciousness activity has a feature of being faster than the speed of light. However, when a specific mental activity has formed, the mental activity can also be controllable at will. Generally speaking, free thinking that occurs in consciousness activity is mostly faster-than-light motion. For example, when a person thinks of something remote (the sun or a certain star, for instance), the person mentally reaches it as soon as such thinking occurs in his or her mind no matter how far away this something is. A sensitive person can feel that a special kind of substance is injected onto the target being thought of. This kind of motion usually does not carry energy. Controllable mental activity, however, is usually the mode of motion when a specific mental activity has combined with the *hunyuanqi* of a certain thing or certain things and is not pure mental activity. In other words, it is the motion in which mental activity involves and mobilizes *hunyuanqi*.

b) Voluntariness in Making Choices

When consciousness captures a target, it is free from the influence of such factors as direction, distance and location. So long as the target is correctly chosen in the mind, consciousness can have effect on it. For example, in remote therapy or in experiments involving samples with the experimenters off-site, so long as the healers or experimenters correctly choose the target in the mind, their consciousness will directly act on the target no matter how far the distance is and regardless of the influence of such factors as change in direction. Meanwhile, other things around the target are not affected.

There is a slight difference between conducting experiments with samples and remote therapy. In doing experiments with samples, the experimenters must know something about the place where the samples are kept. Otherwise, it is not easy for transmitting *qi* to produce an effect successfully in the target in the distance. In remote therapy, the healers can transmit information and *qi* to exert effects on the patient by using the information in the patient's relative who comes to ask for help (without the need of knowing the position of the patient). This is using the information in the consciousness of the patient's relative to transmit information and *qi* to the patient.

c) Voluntariness in Making Changes

Due to the holistic information feature of consciousness activity, consciousness activity can construct time-and-space structures of different things and different time-and-space structures can make *hunyuanqi* display different changes. The accomplishment of this changing process is, on the one hand, a condensing and dispersing (including gathering and expanding) change in *hunyuanqi* and, on the other hand, a settling-in-form change in structure. The different changes shown in "meeting *qi* with *yi* (mental intention) and "meeting form with *qi*" stated in traditional *qigong* refer just to this process.

Furthermore, the motion of consciousness can also transform the nature of *hunyuan* (merged-in-oneness) energy of *hunyuanqi* into single-form energy, such as electricity, magnetism, heat, light and mechanical force, so as to do work of different kinds. Consciousness activity can have effect on many kinds of samples or instruments and the reason lies in this. The "strength of consciousness" that people commonly talk about refers mostly to this.

c. The Feature of Penetration

Since *Yi Yuan Ti* is a kind of *hunyuanqi* similar in feature to primordial *hunyuanqi* and can mutually accommodate all levels of *hunyuanqi* (including concrete substances), consciousness activity that is established upon the basis of *Yi Yuan Ti* can also mutually accommodate all levels of *hunyuanqi*. When such a feature of consciousness activity is combined with its feature of making choices (mainly choice of direction), the feature of penetration is shown. It should be pointed out here that the feature of penetration is the innate feature of consciousness activity. The moment you think of something, you have already penetrated it with your consciousness. It is not that you first think of something and then it is penetrated by your consciousness. If you think of something first and then you think that you will penetrate it with your consciousness, you cannot usually do it as your consciousness has made this something fixed and enclosed.

d. The Feature of Independence

Although the formation process of consciousness activity is influenced by the human body's internal and external environment and is under the restriction of the activity of neurons, consciousness activity has a certain independence. That is to say, consciousness activity and stimulation in the environment as well as activity in the brain cells do not show a linear relationship of cause and effect. This feature is more evidently shown in people who are more advanced in their accomplishment in self-cultivation.

For example, people who are well-trained in their *qigong* practice can "look but not see and listen but not hear" in a chaotic environment. They can go in and out of a large fire without a scorching feeling or getting burnt. They can carry on their consciousness activity according to the requirements of *qigong* practice. Even people who are not *qigong* practitioners yet who have strong self-control over themselves can also carry out their systematic thinking activity according to what they prescribe in their own mind, free of the disturbance of chaos in the outside world. All this shows that consciousness activity has a certain independent feature of its own.

Section IV Functions of Consciousness

Consciousness activity is the most advanced and complicated mode of motion. It can exert effects not only on human life activity but also on natural substance in the objective world and causes them to change.

I. Guiding Function of Consciousness Activity on the Life of the Human Body

As to the nervous system guiding the life activity of animal and the human body, it had long since been pointed out by Engels, "... the nervous system, when developed to a certain extent (because of posterior elongation of the head ganglion of the worm) takes possession of the whole body and organises the body according to its needs." (*Dialectics of Nature*) This has already been proven by physiologists and is gaining more and more attention in medical science. The guiding function of consciousness on the human body's life activity has not been recognized, however. As it is well-known, the human body receives information from the outside world through the sensory organs and sends corresponding information accordingly to adjust the balance between the human body's life activity and the surrounding environment through the activity of relevant tissues and organs. This has become a fixed conclusion in modern science. This process in humans, however, is all carried out under the command of the higher nervous center, or to be more precise, this is a process in which *Yi Yuan Ti* receives and sends information. Some of the information in this process becomes overt consciousness activity while some becomes latent consciousness activity.

In order to have a deeper understanding about this, let me try to explain it by taking the example of the learning of skills. No skill can be learned merely from books. To take swimming as an example, swimmers must establish their consciousness activity about the entire code of all the coordinated swimming actions of their bodies in their real swimming practice. It is only by means of this that the instruction of guiding the whole body to move harmoniously can form and that the skill of swimming can be displayed.

Yet consciousness commanding life activity of the human body is not inborn, but comes about with the formation of consciousness activity. The newborn baby guides its life activity by means of the function of life reflexes (unconditional reflexes) inherited in its body. With the information about sense entering *Yi Yuan Ti*, association between different kinds of information is gradually established, thus conditional reflexes are formed. Meanwhile, corresponding changes also take place in the structure of the brain cells. The guiding function of the nervous system on life activity at this moment, though it is direct reflection of the information about objective concrete existences, already belongs to the domain of consciousness in the broad sense. When language is formed and the responsive mode of the second signal system in *Yi Yuan Ti* is established, a lot of information concerning life is contained in concepts or vocabulary.

Since people believe that concepts (vocabulary) are high-level abstraction of concrete existences and have already separated from specific realistic substances, the guiding function on life using concepts (vocabulary) is neglected. In fact, any word or concept concerning life activity has gathered in real life activity abundant information about life. Therefore, when such content of consciousness activity occurs in the mind, it directly influences the continuance of one's own life activity. As for the channels through which consciousness commands life activity, it has been explained in "Formation of Human *Hunyuanqi*" in the previous chapter on *Human Hunyuanqi* (Chapter IV). The influence of emotions on the human body's life activity will be stated in The Theory of Morality (Chapter VI). Here the influence of consciousness on life activity is described from the level of *Yi Yuan Ti*.

When consciousness is formed in *Yi Yuan Ti*, be it the consciousness activity of emotion or thinking, a synchronous "resonance" phenomenon occurs in *Yi Yuan Ti* in a certain scope and to a certain extent. This change will influence the change of the brain neurons' *hunyuanqi*. The cranial nerves transmit the information into the *hunyuanqi* in the entire human body through the spinal nerves that permeate into the tissues of each part of the human body, which makes the *hunyuanqi* of the whole human body change according to the content of information contained in consciousness activity and therefore shows the guiding function of consciousness on life. This is shown as follows:

i. In the Realm of Overall Human Life Activity

Overall human life activities refer mainly to social activities concerning clothing, eating, housing, transportation as well as labor processes. All these activities are the exhibition of the overall activities of the life of the human body, yet all these life activities are carried out and accomplished under the control of consciousness activities. Human consciousness activities are the forerunner of life activities, and the life activities of the flesh are the means to realize the demand in consciousness activities. Not only do human labor and social communication purposefully go on according to plan and consciousness instruction, but even each action in the overall life activities of humans also follows the law of mind guiding *qi* and *qi* guiding the body.

When specific information activity occurs in *Yi Yuan Ti* (the instruction of body movement, for example), it can be transmitted to specific tissues along specific routes and brings about the effect of information regarding consciousness instruction—relevant body movement. There is a complicated *qi* gathering process in the generation of this movement because any movement of the human body needs the nourishment of *qi* and the source of *qi* is partially the *qi* stored in a particular part of the human body and mostly the *qi* mobilized from *dantian*. When the instruction of movement is given in consciousness, it is carried out simultaneously through many channels. *Qi* should be instantly gathered to the position of movement and changed into force (*li*), and this is what is usually called force *qi* or *li qi* in the Chinese language. The mechanism of change in this motional process is called in qigongology mind guiding *qi* and *qi* guiding form (the body).

There exists a complicated coordination process and an integrated mode in any body movement. For example, in an athlete's action of throwing a javelin, it is not only the movement of the hands, but also very good cooperation of the action in the arms, body and legs. The *qi* needed to accomplish all the action can be distributed under the instruction of consciousness after the set pattern of the action of throwing a javelin has formed in *Yi Yuan Ti*. The *qi* is then transmuted into *li* to accomplish different involved actions. As these compound actions are accomplished under a unified and integrated instruction, a cohesive force can be formed which displays an integrated or holistic power.

ii. In the Realm of Human Physiological Activity
i) Consciousness Can Change the Strength of Human Force

In daily life, forces are needed in such activities as walking, dressing, eating and working.

Generation of these forces results from the function of the human body's locomotor system. However, human consciousness activity can change the strength of human force in certain circumstances, namely, it can strengthen or weaken human force. Generally speaking, the strength of a human being will increase when the person's spirit is highly concentrated or when the person becomes excited because of strong desire for achievement; whereas the strength of a human being will be reduced when the person's spirit is distracted or when he or she is depressed or feels fearful.

There are many such examples in daily life. For example, things that are hard to achieve in normal situations get successfully done when people who want to do them become very earnest in urgent situations. Objects that cannot be ordinarily moved are now moved. There is also such example in China during wartime: The guerrillas fighting at night and chased by their enemies all jumped unexpectedly over a small river as wide as two *zhang* (about 22 feet). They were called by people at that time soldiers of Bodhisattvas. The soldiers were said to be protected by the Bodhisattvas and couldn't fall into the river. The actual fact was that the guerrillas, being chased by those who closely followed them, had no other mental activity in their brain, not even the conceptual activity of "I must jump over it". As soon as the intention of "jump" occurred to their mind, they jumped. This is the unity of body and spirit. In this mental activity that was so highly concentrated, dramatic changes took place inside their bodies and a great, unimaginable force commonly called "urgent force" appeared. If their emotions began to surge when they saw that there was a river in front of them, thinking that they must jump to the bank beyond, bring glory to their country and destroy the enemy, etc., their force would have been distracted and scattered and they would not be able to jump over the river. I once expressed this opinion to some of these old soldiers who had taken part in the liberation war of China. They agreed with me, saying, "That's right. We did not have the time to think that much and jumped over the river without knowing how we did it."

Sportsmen usually cannot break the world record in their daily practice. When it is time for the match, if their competitive state is good and their state of mind is great, they can break the record and achieve the best result. Of course, if their state of mind and competitive state are not good, they might not even be able to accomplish what they can normally do in their daily practice.

These are the display that consciousness increases strength in physiological state.

In the pathological state, spiritual activity can also produce enormous force. Some weak and thin female patients, when they are in a fit of hysteria, cannot be controlled even by several strong young men. This is a "crazy force" which is called in medical science generalization of focus of excitation in the cerebral cortex. That is, when the sick excitation center occurs, induced inhibition around it is not formed; instead, the excitation center floods outward and around and shows a synchronous excitation in the brain, which makes a united action occur in the whole body and produces an overwhelming force beyond the patient's control.

Quite a few people have experienced similar phenomena mentioned above in their daily life. It is simply that they have never carefully thought about it or studied it.

As for *qigong* masters and acrobats who have undergone assiduous training, they can "hold up two tons of weight" after they become concentrated. Two precast concrete panels can be put on their bodies, with over ten people standing on them, which altogether amounts to several thousand kilos in weight. Some *qigong* masters can break stone tablets with their heads. This requires not only the hardness of the heads, but also the necessary strength. Some *qigong* masters can even tow a car of *Shanghai* brand from behind when it already has its engine started and stop it from moving forward after they begin transmitting *qi*. This tremendous force comes not from the outside world but from inside their own bodies after they begin to use their consciousness.

Spiritual factors can also reduce strength, e.g., a person gets tired and inefficient easily at work when he or she is low in spirit or is depressed and disheartened; intense fright can make strength disappear to nowhere. There are some people who are actually robust in body, yet, when seeing gangsters taking out their knives, they are so frightened that they cannot stand and even urinate in their pants. This is what is called in traditional Chinese medicine "fright makes *qi* in disorder" and "fear makes *qi* go downward", that is, because *shen* (spirit) has lost its mastery and *qi* has lost its commander, the force accordingly disappears.

ii) Spirit Can Make Change Happen in Substances in the Human Body

The meaning of spirit changing into substance in *qigong* cannot be equated to the meaning of spirit turning into substance in philosophy. Spirit turning into substance in philosophy refers to the social effect brought about by a right thought after it is grasped by the mass of people and put into social practice. It is realized through the indirect practice in society. Spirit changing substance in *qigong* is directly realized in the human body itself. One's own spirit directly changes into one's own substance. This is because consciousness activity can change and affect the law of metabolism in the substances in the human body that have form. For example, consciousness activity can change synthesis and decomposition processes in the substances in the human body. This kind of change exists in both physiological phenomena and pathological phenomena and is even more evident in *qigong* practice.

In the physiological phenomenon in average people, talking about tasty food can make people's mouths water. For example, the classic story of Quenching Thirst by Thinking of Plums: Thirst attacked Cao Cao's soldiers when he was leading them on their way to a battle. Cao Cao said, "A grove of plum trees are in front of us not far away. You will not be thirsty any more when you taste those plums." On hearing this, his soldiers' mouths all watered and the thirst was stopped.

Most children in China have tasted sugar-coated haws (and other fruits) on sticks (*tang hu lu'er*), red haw skin (*guo dan pi*) and haw jelly cakes (*shan zha gao*). When parents say that they are going to buy them these tasty things, saliva can instantly run in the children's mouths. In medical science, this changing process is called conditional reflex that is caused by the second signal system. We say that this is a process in which spirit changes into substance and we are referring to the cause and nature of how this process is realized. Of course, there is a complicated process of physiological change and bio-chemical change in this process.

Another example is that in the case of couples who are newly wed or who have not been together for a long time, the wives can become pregnant in the safe contraception period. This is because the surge in emotion and the highly excited spirit make the ova become mature ahead of time. This is a more evident example that spirit causes the substance in the human body to change.

Besides, there are also a lot of examples about spirit changing the substance in the body to show a pathological state. There is a legend in China called Wu Zixu Passed the Guarded Passage of Shao. In the story, Wu Zixu was tormented with such acute worries and sorrow that all his hair turned white overnight, and this is a material change caused by the spiritual factor. In daily life, though it is not as typical or evident as this, some people who are worried about something and are sleepless the whole night look gaunt the next day and show no spirit or vigor in their countenance and their weight can even decrease. However, if they meet their old friend whom they have not seen for years and talk all through the night, it is not easy for them to lose weight. In some people, whenever they get angry, their blood pressure increases. Those who get terribly angry even expectorate blood. In the story of Zhu Geliang Irritated Zhou Yu Three Times, Zhou Yu was so enraged that he expectorated blood and died. This is because in an anger like this, the sympathetic nerves became excited and both the liver and the blood vessels contracted. This affected the returning flow of the blood to the liver and made the veins become drastically dilated. The blood pressure in the veins increased and the blood flowed out. During the defensive period of the Second World War in the ex-Soviet Union, many people who lived in Stalingrad suffered from ulcers. One of the key reasons for this is the nervousness in their spirit that stimulated the excessive secretion of sympathin and caused the imbalance in the secretion in their bodies so that changes occurred in their relevant body parts and showed a morbid state.

In the *qigong* state, material changes that take place in the human body are miraculous. When the hard-form *qigong* masters mobilize *qi*, they can make a big lump appear on their arms that can roll along their arms under the skin. The lump is formed by condensing the *qi* in the *qigong* masters' bodies with their mental intention and its hardness can make it resistant to knives and bullets. Some hard-form *qigong* masters can give performances called "silvery spear stabs the throat" and "the point of a hayfork moves the mill" (The *qigong* masters put the point of the steel fork against their navels, lay their bodies horizontally on it and rotate). Statistics shows that the navels of the *qigong* masters who perform "the point of a hayfork moves the mill" have to sustain more than 20,000 kilograms of pressure per centimeter. This can neither be explained in dynamics nor in physics or in medical science. The fact that the masters' navels become hardened results also from their applying consciousness which causes material change in their bodies.

Another example is what happened in a *qigong* class in *Jiamusi* in 1988. In the *qigong* class, there was a lady who thought that her breasts were small. At that time, I had already given lectures on applying consciousness, transmitting *qi* to cure diseases, organizing *qi*-field effects and so on and encouraged the learners to apply what they had learned in their own practice. This lady, according to the theory that consciousness can make *hunyuanqi* "condense into form", used her consciousness to grow her breasts bigger. However, she did not use her consciousness appropriately and, instead of growing her breasts bigger, she grew on her breast a hard swollen lump as big as a walnut. She wrote a note, telling me about her trouble. I said to her, "You could use your consciousness to grow it from nowhere, now use your consciousness again to dispel it and it will be gone." This lady was very good at receiving information and immediately thought, "That's right. The tumor is gone. It's gone… " Then she touched her body and found it already disappeared.

iii) Consciousness Can Change the Sensitivity of Senses in the Human Body

Different kinds of information in the objective world can act on the sensory organs of the human body at any time. Whether a certain amount of stimulation can cause sensory effect in a person or not as well as how much it can be felt by the person depends on the person's state of consciousness activity. That is to say, the intensity of how much stimulation one can sense may not be in proportion to the amount of objective stimulation. Whether a person pays attention to the external stimulation or not and the person's internal psychological state when paying attention to it will increase or decrease the intensity of senses caused by the stimulation.

a. Reduce the Intensity of Senses
a) Distraction Can Decrease the Intensity of Sense

The old Chinese saying goes, "Listen yet do not hear, look yet do not see, eat yet do not taste as the heart is not there." What is said here is that if one's attention is not with the ears, one will have no auditory response to the sound from the objective world; if one's attention is not with the eyes, one will have no visual response to the light in the objective world; if one's attention is not with the tongue, one will show no reaction of taste to the tasty food one is eating. This is what people can often experience in daily life. Other examples are: When absorbed in reading, one may not hear other people's calling from nearby; when dedicated to deep thinking, one may respond to nothing in the surrounding environment. At this moment, even tasty food is "as tasteless as wax" to the person lost in thinking according to some writers' description. The sense of pain is slight if one is injured without having noticed it. It sometimes does not even cause any feeling of pain if one is unaware of the injury. In fierce battles, soldiers who sometimes get shot without noticing it can still act normally and can even continue their struggle with the enemy. In daily life, though it is not as apparent as it is in war, the pain that people sense when they get injured while not having paid attention to it, e.g., to get part of the nail cut off when chopping vegetable, is rather different from the pain felt when one is aware that he or she is being wounded.

b) Resistance in Consciousness Can Reduce the Intensity of Sense

If we say that decrease in intensity of sense due to unawareness is a natural physiological process, then being consciously resistant to reduce intensity of one's sense is an active psychological suggestion process that takes place on one's own initiative. To face objective stimulation indomitably, courageously and confidently with a resistant will and sentiment can decrease and weaken the feeling caused by the stimulation. The story of Scraping the Bone to Eradicate Poison in Chinese history is about Guan Yu in the period of the Three Warring Kingdoms who was shot by a poisonous arrow and needed to be treated with an operation. Due to the "heroic" psychological temperament nourished in him, Guan Yu looked down upon any "enemy" and adversity and did not weigh in his mind at all about the suffering of having to be cut open and scraped to the bone with a knife. Therefore, he experienced his operation while playing chess in perfect composure.

Let me give you one more example. During the Cultural Revolution, a military surgeon in a certain PLA unit, in order to condemn anesthesia by means of acupuncture as "fake science", encouraged the young soldiers to carry forward "the spirit of revolutionary hard bone" and to criticize acupuncture anesthesia with their actual action of "great dauntlessness". As a result, twelve soldiers in perfect health had their appendices cut off without taking any anesthetic measure. We think that this course of action is undoubtedly undesirable, but from this example we can see the enormous effect that consciousness has in reducing the intensity of senses caused by stimulation.

b. Increase the Intensity of Senses
a) Attention Can Increase the Intensity of Sense

In both the physiological sense and the psychological sense, attention can increase the degree of excitation of nerve function, so the nervous system, when receiving stimulation, can reflect it more accurately. This is in fact the reinforcement of the process of passive reception. From the perspective of the theory of qigongology, actively paying attention to objective existences can enhance the interaction between the subjective world and objective world. It increases the sensitivity of subjective feelings and also the amount of stimulation of objective information, which therefore displays the selective reinforcement function of consciousness on receiving various surrounding information.

For example, when talking with people around you, you suddenly notice that a distant crowd of people are talking and this arouses your concern and attention. Although the voices of the people in the distance are weaker than the voices of the people around, when you strain your ears and attentively listen, you can vaguely hear the relatively low voices in the distance instead of clearly hearing the louder voices of the people nearby. Moreover, if a person's attention has been concentrated on closely observing things for a long time, it can greatly improve the sensitivity of this person's perception and makes very subtle things clearly heard or seen. In the story of Ji Chang Learning to Shoot Arrows recorded in Chinese historical books, Ji Chang began his practice with "refining eyesight". He hung a flea up and gazed at it without blinking. After three years of practice like this, the flea Ji Chang saw "became" as big as a wheel. Ji Chang could see very clearly each tiny little part of the flea. People might consider this as an exaggerated story intended to amuse people. In fact, it is not. A lot of things in *qigong* are in compliance with this law of concentration.

b) Psychological Suggestion Can Increase the Sensitivity of Sense

As to subtle stimulation from the outside world, if one senses it with initiative and suggests to oneself that "I can definitely perceive it", the sensitivity of sense in this person will be increased. This process is even more liable to be displayed in a quiet state. For example, people know the painful feeling of being cut with a knife on the skin. If the person to be operated on (especially a timid one) is told that he or she will be directly operated on with a knife without using anesthesia and is shown the sharp knife for the operation, severe pain can be caused in this person after he or she is blindfolded even if only a small wooden stick is used to prick his or her skin. In the process of *qigong* practice, if psychological suggestions are given to practitioners, particularly by *qigong* masters, various *qigong* feelings are liable to occur and various *qigong* functional abilities can easily form.

iii. The Influence of Consciousness Activity on the Health of the Human Body

In the above-mentioned three aspects of the functions of consciousness in human physiological activities, the influence of consciousness on the health of the human body has been more or less touched upon. Here a detailed introduction will be given to the functions of consciousness on disease.

Human mental activities have a commanding and adjusting effect on the life activities of the entire human body, and this is what "spirit is the commander" means. A joyous spiritual state can make people healthy whereas a chaotic spiritual activity can make people ill.

A disordered spirit can make a healthy person fall ill. For example, if someone has quarreled with someone else at work and has become very angry, after going back home, this person feels suffocated in the chest and bloated in the stomach and is unable to eat. He or she feels oppressed in the heart and dizzy in the head and even experiences an increase in blood pressure. The reason for these symptoms is but "anger". It is caused by disorder in the spirit after this person has become enraged.

Let me give you another example. I have a friend who works as a technician. He once ate gourmet powder and then mistakenly believed that he had eaten caustic alkali. His colleagues gave him an enema at once with 10 percent dilute hydrochloric acid solution. He was only halfway home riding his bike when he felt pain in his throat and stomach. When he managed with all his effort to get home, he was already soaked with sweat and his face was as yellow as old wrapping paper. He was immediately rushed to the hospital to have an injection and take some medicine. Three hours had passed before his symptoms were little relieved. He got a certificate for sick leave for three days and went back home. When he brought the sick leave certificate to the factory the next day, his colleagues told him that what he ate was not caustic alkali but monosodium glutamate. Why could eating gourmet powder bring about such reactions? This was because he knew the strong corrosive effect of caustic alkali, and, what made the matter worse, the scene of so many people giving him an enema with dilute hydrochloric acid disordered the function of his brain. The information about discomfort in the stomach was sent to the cerebral cortex and reflected as pain that instructed the stomach to contract. The contraction of the stomach made the brain feel more severe pain, which caused a vicious circle of negative feelings. This is nerve function that can cause disease in a person who has no disease at all.

A disordered spirit can worsen a minor disease. In 1984, a gentleman in the Construction Bureau in Beijing had a severe pain in his stomach, so he went to the hospital to have a health check. After having an X-ray taken, he accidentally knew that it was cancer, already incurable. He got extremely concerned. After he went back home, his pain became more acute and the suffering could not be alleviated even by taking injections of Sauteralgyl. He lost about 10 kilograms of his body weight in over 20 days. Then he went to the tumor hospital for a group consultation. Doctors diagnosed with gastroscopy and biopsy that he did not have cancer, but an ulcer complicated by calculus in the stomach. With the burden on his mind relieved, he gradually recovered in two months.

Moreover, a person's mental activity can even make someone without any disease die a sudden death. Before liberation in 1949 in China, a hospital once conducted an experiment on two death penalty criminals. The criminals were asked to expose their arms with their eyes screened and blood was drawn from one criminal. The blood from him dropped with a pitter-pattering sound into a basin. The other criminal was told that his blood was also being drawn. There was also the same sound in the basin. The first criminal, owing to severe loss of blood, fell dead to the ground. After a while, the second criminal also fell to the ground dead. In fact, they did not draw blood from the second criminal. His death was totally because of mental effect.

Beijing Evening News once reprinted a piece of news having been published in *Thames*. A British sanitation worker often worked under a high-tension wire and was always worried that the high-tension wire might break one day and shock him to death. One day, unfortunately, a wire fell down and this worker immediately fell dead. His autopsy showed that his liver and heart were both damaged, as the case in those who are electrocuted to death. However, what fell down was not that live high-tension wire. The cause of his death was also completely due to mental effect.

There are also examples that a calm unruffled spirit alleviates serious disease and enables dying people to survive. In the ex-Soviet Union in the 1950s, it was reported that doctors in a hospital once operated on a patient with brain cancer. They only cut off the central focus and sewed the cut up. The authority of the hospital gave strict orders that nothing about the state of the illness could be let out. They merely told the patient that the operation of the brain tumor was very successful. Happy with the news, the patient left the hospital and then traveled far and wide, jovial and cheerful in his mind. A year and a half later, he returned to the hospital for a reexamination. The doctors were amazed that an undoubtedly dying patient unexpectedly survived and enjoyed very good health. This was also totally a mental effect. He did not know the condition of his illness, so he had no burden in his mind. He was calm in his mood and what his brain received was all good information. The magazine of *Science, Technology and Life* also published a case: A patient with terminal breast cancer who could not afford to be operated on just traveled around. Over one year later, the disease did not get worse and she survived as well.

A good spiritual state can enhance human life force to the utmost extent. According to a newspaper report, four miners were trapped in a mine for fifteen days in the earthquake in *Tang Shan* in 1976 and they survived. The reason for their survival, if analyzed from the aspect of spiritual factors, is that, firstly, they lost the notion of time. The pitch dark made them unable to see their watches and they did not know how long had passed, so they did not have any dread of the passing of time. Secondly, an old worker frequently gave them positive encouragement which calmed them down and filled their hearts full of hope of coming out alive. What's more, the special surroundings consumed extremely low energy in their bodies. The metabolic processes slowed down and they eventually managed to survive.

Another piece of news was published in *Liberation Daily*. In Peru, a deaf-mute young man was buried in a cemetery by his family after he took an injection for his high fever and went into suspended animation, a state that his family did not understand. The next year, workers who repaired the tomb found him sitting in the tomb alive. This news caused a sensation throughout

South America. According to analysis later, the reason for his survival is that, firstly, a deaf-mute person such as him who has little contact with the outside world is habitually quiet and peaceful in the mind; secondly, he had in his mind that he should be calm, peaceful and not anxious or worried, and in the dark tomb he did not have the notion of time; thirdly, there was not a bit of stimulation from the outside world in the tomb and it was the most advanced tranquil environment; fourthly, the metabolism in his body was extremely low. He was quiet, relaxed and not worried in his mind, and this was a state equivalent to practicing *qigong*. Therefore, it is understandable that he could survive.

II. Function of Consciousness on Substances in the External World

Functions of consciousness on the life of the human body stated above can be considered mostly the same as consciousness or psychological effects in psychology. However, functions of consciousness talked about in qigongology extend far beyond the domain of psychology, as the effect of consciousness on substances in the external world stated here.

In the School of Engineering and Applied Science at Princeton University, some experiments were conducted and two of the results of the experiments were published. In one experiment, two small mirrors stood face to face and an electronic monitor was used to closely observe the change in the distance between them. A random selection of people were then encouraged to think that the distance between the mirrors was shortened. They were told to think hard that the mirrors were becoming "closer, closer, even closer." This experiment was done with thousands of participants. A conclusion was drawn with statistical analysis that human mental intention can cause changes in the position of the little mirrors. Although they were only moved by much less than one ten-thousandth of a centimeter, there was after all some movement. In another experiment, a random selection of people were asked to raise the temperature of an electronic thermograph with their mind. This was an experiment that also involved thousands of participants. A positive conclusion was also reached through statistical analysis. The temperature was increased by much less than one thousandth of a degree. These are demonstrating examples that familiar intelligence people can change the condition of objective substances in the external world with their consciousness.

As for people who practice *qigong*, particularly those who have acquired super abilities in their practice or those who are endowed with paranormal abilities from birth, their performances can more clearly illustrate the function of consciousness on external substances. For example, Mr. Zhang Baosheng once did an experiment as a demonstration for a government official. This leader held a health ball in his hand. Zhang Baosheng covered the official's hand with his hand. After Mr. Zhang transmitted *qi*, the ball in the official's hand disappeared. A while later, a bodyguard in the front yard came to return the health ball, saying that he found it all of a sudden in his pocket. A gentleman on the spot could not believe what had happened. Mr. Zhang then teleported the gentleman's watch into a thermos. After that, they poured all the boiled water out of the thermos and the watch was poured out as expected. Other *qigong* masters have also done a lot of experiments telekinetically, such as moving matchsticks, miniature radiotelegraphic receivers and transmitters, etc., cutting cakes and oranges open, cutting a ball of wire entirely from the middle to make it become many short wires, and changing the structure of molecules—all purely with their mind. All these experiments indisputably show with solid facts that consciousness has enormous energy. As to how great an effect it can exert upon both the subjective and objective world, this still remains a significant project demanding immediate exploration.

The consciousness function which is the most difficult to understand is "consciousness configuration". When someone is thinking in mind about the image of a certain structure or form and making it seem to be a real image in the space, people with paranormal abilities can perceive it. For example, if some people are asked to simultaneously imagine a triangular form in the air, people with paranormal abilities can notice and perceive this image. Sometimes the different kinds of images in a person's mind can also be perceived by people with paranormal abilities. In the Third Session of the Academic Symposium held by Beijing *Qigong* Research Association in 1986, a *qigong* master successfully perceived all the images being thought about in the consciousness of Mr. Liu Jianhua, the honorary chairman of Beijing *Qigong* Research Association. What's more, Ma Weiqing, the especially gifted child can write simply with his mind instead of a pen or a writing brush and the many performances he has given are all acclaimed with success.

Besides, consciousness configuration can also be shown in one's own physiological changes. Walter B. Kolesnik wrote in his article *Learning Approaches and Their Application in Education*: "More often than not people find that there exist certain behavioral patterns in a family. Children, not only in their appearance, but also in the aspect of their social attribute as well as in such aspects as their sentiment and intellect, usually bear resemblance to their parents."(translated from Chinese into English) This might be said by some people to be "the result of heredity." We think that heredity does play a role in it, but not all resemblances in the family result from heredity.

An item of news was reprinted from *Reader's Digest* (September, 1988) in *China Youth*: "The research conducted by Robert Zajonc, doctor of psychology at the University of Michigan in America, and three of his postgraduates shows that couples, though they looked quite different when they got married, become alike in their looks after having lived together for 25 years, and the happier the marriage, the more similar their looks."(translated from Chinese into English)

In addition to this, the author of this book has observed some children adopted by infertile couples. These children who were taken by their adopted mothers soon after birth resembled their adopted mothers in their facial features when they reached the age of four or five, and the deeper the parents' love to the children, the more similar they are in appearance. These living examples show that, when the image (or other information) of a person is firmly reflected in the realm of consciousness in another person, it can subconsciously guide the other person's life activity to develop toward this mode.

In fact, that the nervous system guides directed development exists not only in humans, but also in animals. Engels said, "*Vertebrates. Their essential character: the grouping of the whole body about the nervous system. Thereby the development of self-consciousness, etc., becomes possible. In all other animals the nervous system is a secondary affair, here it is the basis of the whole organization; the nervous system, when developed to a certain extent (because of posterior elongation of the head ganglion of the worm) takes possession of the whole body and organizes the body according to its needs." (Dialectics of Nature)*

In the human body, more advanced nerve activity—consciousness appears. Consciousness activity, therefore, also occupies the whole organic body to a certain extent and organizes and changes the state of the organic body according to consciousness information. In Buddhism, there is the saying of "mind generating the body." The great many remarkable effects in *Zhineng Qigong* are accomplished with this function of consciousness. For example, an accomplished *qigong* master concentrates on thinking of a person's image and meanwhile internalizes this image into his own image (He should totally forget about his own image in this process). At this moment, other people around him, if they narrow their eyes and look at him (the light should be dim), will see that the *qigong* master has changed into the person that he is picturing in his mind. If you often think of a particular image of a person, you can make directed change happen in your own image. This can simply be called *qigong* cosmetology.

The examples given above show the process of change that familiar intelligence people's consciousness activity brings about to the existential condition of substance in the subjective and objective world. People who practice *qigong* can make this change even greater. People with paranormal abilities can make still greater change.

III. How Consciousness Exerts Functions on Substances

The function of consciousness activity upon human body's own life activity is realized essentially through *Yi Yuan Ti*'s *ju san* (condensing and dispersing, including gathering and expanding) motions which change human body *hunyuanqi*'s mode of motion, and this has been illustrated in Chapter IV Human *Hunyuanqi*. Here the focus is on the illustration of the function of consciousness upon external things or existences.

i. The Mechanism that Consciousness Exerts Effects upon External Existences

i) Consciousness activity transmutes *hunyuanqi* into specified forms of energy and acts upon external existences. As it was stated in "Consciousness Activity Is a Special Mode of Motion" in Section III in this chapter, consciousness activity can turn human body *hunyuanqi* or the *hunyuanqi* in the natural world into simple energy forms, such as sound, light, electricity, magnetism, heat and mechanical force, so as to act on existences in the objective world.

ii) Consciousness activity directly produces an effect in the holistic time-and-space structure of objective existences and makes them change. We know that consciousness activity has the holistic information feature and in each conceptual activity all the information about this concept is contained. Therefore, when a conceptual instruction is sent out, it can directly act upon the holistic time-and-space structure of the target and can also gather the *hunyuanqi* in Nature to act upon the target according to the information regarding instruction. In this process, both the making-choice feature and the driving-force feature of consciousness are having effect.

ii. Factors that Influence the Effect of Consciousness Activity upon External Things
i) The Degree of Concentration in Mental Activity

We know that in a familiar intelligence person's *Yi Yuan Ti* there are many excitation centers and the activity content is multi-faceted. They check each other and make it hard for *Yi Yuan Ti* to form a concentrated single-minded activity that displays the power of consciousness. If strength can be increased at the commencement point of mental activity till the mental activity penetrates all the stratums and forms a concentrated single-minded mentality, a unified all-dimensional holistic mental activity can form and the effect of super intelligence information can be exercised. The ardent wish in the subjective world and attaching great importance to the change in the target are the critical point for increasing the strength of commencement of mental activity and for forming a concentrated mental activity.

ii) The Degree of Accuracy in Choosing the Target

Knowing more profoundly and accurately about the target and knowing more details about it are helpful not only for making a precise choice about the target but also for close integration of consciousness activity with it so that effective *hunhua* of consciousness with the target can be reinforced.

iii) The Degree of Initiative and Brightness of *Yi Yuan Ti*

The degree of how exquisite *Yi Yuan Ti*'s own entity and attribute are has an enormous impact on consciousness activity that forms the extensively correlated unified whole entity in *Yi Yuan Ti*. Generally speaking, calmness in spirit can avoid or reduce the interference of body *hunyuanqi* in *Yi Yuan Ti* and can increase *Yi Yuan Ti*'s initiative and brightness degree.

iv) Wish and Expectation in the Subjective World

This is the commencement drive of consciousness activity. The ardent wish not only is related to the concentration of mental activity but can also improve the degree of accuracy in making choices about the target.

Section V Features of the Theory of *Hunyuan* Consciousness

The theory of *hunyuan* consciousness includes the theory of *Yi Yuan Ti* and the theory of consciousness activity. It is creatively established on the absorption of the essence of the theories of consciousness in Buddhism and Daoism, on the assimilation of the achievements in modern science, medicine and psychology and on the author's assiduous *qigong* practice (including the *qigong* practice of many other *Zhineng Qigong* practitioners). It endows consciousness with the standpoint of materialism, changes the "stand-upon-the-head" concepts of consciousness in Buddhism and Daoism, and enables consciousness to enter into the realm of science. It illustrates, on the one hand, the active functions of consciousness upon people's life activities and points out, on the other hand, the deficiency in human beings' consciousness activities today. All this not only lays a foundation and creates a condition for the application of consciousness promoted in *Zhineng Qigong*, but also allows *Zhineng Qigong* to present itself with a brand new scientific look to the world. In order to further our knowledge and understanding of the theory of consciousness, its features are summarized as follows:

I. The Theory of *Yi Yuan Ti* Is a Further Development of the Theory of *Hunyuanqi*
i. *Yi Yuan Ti* Is a Spiral Upward in Development Compared with Primordial *Hunyuanqi*

It is believed in traditional *qigong* theory that human spirit belongs to human primordial ancestral *qi* and human primordial ancestral *qi* is the display of the primordial ancestral *qi* of heaven and earth (*Dao*), that human primordial ancestral *qi* and the primordial ancestral *qi* of heaven and earth are the same.

The theories of *hunyuanqi* and *Yi Yuan Ti* in *Zhineng Qigong* are the theories having developed forward. It is not to return to the starting point of the primordial ancestral *qi* of heaven and earth, nor is it to develop into an enclosed circle, but to spiral upward. This can be indicated as follows: formless and imageless primordial *hunyuanqi* →(condensed into) concrete substances and the *hunyuanqi* that fills the substances→ independent *Yi Yuan Ti*. The features of *Yi Yuan Ti* seem to have repeated some features of primordial *hunyuanqi*, e.g., both primordial *hunyuanqi* and *Yi Yuan Ti* are extremely fine and even *hunyuanqi* and can generate complex existences. However, what comes from the mergence and change of primordial *hunyuanqi* is all existences in the objective world while what comes from the mergence and change of *Yi Yuan Ti* is the content of consciousness in the subjective world. In addition, all existences that generate and develop from primordial *hunyuanqi* are all natural beings while consciousness activities that result from the mergence and change in *Yi Yuan Ti* have initiative. Therefore, *Yi Yuan Ti* and primordial *hunyuanqi* are not exactly the same. *Yi Yuan Ti* has become a spiral ascent after it has gone through a negation of negation in the process of development. When primordial *hunyuanqi* has developed into *Yi Yuan Ti*, it is not the summit of development, nor is it that *Yi Yuan Ti* returns to the original starting point, becomes an enclosed circle and does not take any step forward. On the contrary, after *Yi Yuan Ti* has formed, it continues to develop forward and displays a series of changes: From Pure Original *Yi Yuan Ti* to Self *Yi Yuan Ti* and to Partial Identity *Yi Yuan Ti*. The practice of *Zhineng Qigong* is not to return to the original point but to continuously develop forward in order to unfold all *Yi Yuan Ti*'s functions: To develop from Partial Identity *Yi Yuan Ti* to Complete *Yi Yuan Ti* and to Developed Integrate *Yi Yuan Ti*.

ii. Practicing *Qigong* Is Not to Make Consciousness Return to the Primordial State but to Develop It Forward

In the advanced level *qigong* in the past, whether it was in Buddhist or Daoist *qigong*, practitioners claimed to a certain extent to "return" or called "to return to the simple and truthful state". It is believed in Daoism that from the primordial ancestral *qi* (or called primordial *Dao qi*) evolve heaven, earth, human beings and all other existences, that human spirit (*shen*) evolves from *yuan shen* (primal spirit) to *shi shen* (perceptive spirit) and that the human body and its *qi* also evolve from their prenatal state to their postnatal state. In their *qigong* practice, practitioners should not only "return from their state of being old to their state as a child", "return to their prenatal state", but should also "return" to the chaotic state without consciousness and perception, and at last, return to the level of primordial ancestral *qi*.

In Buddhism, it is believed that Bhutatathata (genuine nature) in which there is no life or death in dharmadhatu or dharma-realm is bound by karmas of illusions arising from primal ignorance so creatures and existences enter into transmigration of life and death and have lost their original true nature. To cultivate oneself is to rid oneself of delusions and return to the true state, to break through one's benighted state and restore the original Buddhist nature, and finally, to enter the "forever tranquil enlightened land".

Although what is advocated in Buddhism and Daoism are not the same, their ultimate propositions both concern the idea of returning to the original starting point. It is under the guidance of the thought like this that practitioners in the past, when having achieved a certain level in their cultivation practice, gave up their lives willingly in the illusion of returning to the beginning void realm. Hence it became a great tragedy for *qigong* practitioners.

Let us think about some of the accomplished *qigong* practitioners in the past. They indeed had a lot of magical powers during their life time, yet none of them reappeared after their death. These people, before their death, were all full of mercy and fraternity and taught people earnestly and untiringly. They all took the promotion of the cause of *Dao* as their own responsibilities. Why don't they show themselves again and teach their students as well as the great multitudes after their death (or after their having achieved *Dao* according to the way they put it)? This merely shows that they did not actually achieve *Dao*, but were really dead.

Even in the tranquil meditative state in *qigong* practice, it is still impossible to return to the void primordial ancestral *qi* state because the tranquil entity and attribute of *Yi Yuan Ti* being perceived in meditation is still a form of dynamic existence with the act of perception in it even without any other mental activity. Therefore, the *Yi Yuan Ti* perceived is the state of *Yi Yuan Ti* with perception in it, and perception is also a kind of consciousness activity. The realm attained with perception in which the perceiver and the perceived are combined is still a dynamic balance instead of the void state of tranquil ancestral *qi*. If without the existence of perceptive consciousness, the practice then becomes "stubborn emptiness", which is opposed in both Buddhism and Daoism. From this it can be seen that the so-called returning is impossible whether it is in the fundamentals or in the state of *qigong* practice.

According to the theory of consciousness in *Zhineng Qigong*, consciousness of human beings today is established on the basis of Partial Identity *Yi Yuan Ti*. To practice *qigong* and get rid of the bias and partiality in *Yi Yuan Ti* lies in the tapping of super intelligence, in changing the familiar intelligence Reference System, and in adding to it the content of super intelligence information so that the partial and biased state will gradually give way to a state of prodigious perfection. This is the further development of *Yi Yuan Ti* and the enhancement of humans into a more advanced stage.

II. A Reasonable Attitude toward Familiar Intelligence Consciousness Activity and Super Intelligence Consciousness Activity

Consciousness activity of familiar intelligence is called illusioned differentiation in Buddhism. It is partiality and stubbornness, and adherence to it is the key obstacle that blocks the attainment of truth. It is called perceptive spirit (*shi shen*) in Daoism and is regarded as a thief on the way of the cultivation of *Dao*. Both Buddhists and Daoists try their best to advocate the abdication of familiar intelligence, and the method to break away from it is to do away with knowledge, abdicate intelligence and return to the state without consciousness and perception. They think that it is only when they thoroughly break away from familiar intelligence consciousness activity that consciousness activity of super intelligence can manifest, and that only the world in the consciousness activity state of super intelligence is real and the world corresponding to the consciousness activity state of familiar intelligence is unreal. In a word, consciousness activities in familiar intelligence and in super intelligence are considered to be antagonistic.

The relationship between familiar intelligence consciousness activity and super intelligence consciousness activity is considered in the theory of *hunyuan* consciousness as follows:

i. Both Kinds of Consciousness Activities are the Display of *Yi Yuan Ti*'s Functions with Only Little Difference between Them

Familiar intelligence consciousness is the consciousness activity formed in *Yi Yuan Ti* by synthesizing the restricted amount of information about existences that is obtained with normal sensory organs. Super intelligence consciousness is the consciousness activity formed with the aid of holistic super extraordinary information. Although familiar intelligence consciousness and super intelligence consciousness differ in their means of obtaining information, their outcomes in *Yi Yuan Ti* are both holistic. Therefore, it is believed in *Zhineng Qigong* that both the world perceived in the state of familiar intelligence consciousness and the world perceived in the state of super intelligence consciousness are real.

ii. They Differ in Their Process and Means of Sending Information

The information that super intelligence consciousness activity sends out is holistic information that can directly act on oneself (subject) or on object outside oneself so as to establish super extraordinary connections with objective concrete existences and enables object to display super extraordinary features. The information regarding familiar intelligence consciousness, though holistic in *Yi Yuan Ti*, is divided into several partial pieces of information when being sent out to act on the target by means of different organs. Since the information being sent out restores the original partial attributes as it is being received by the sensory organs, the connections that familiar intelligence consciousness establishes with objective concrete existences are familiar intelligence connections, and what object displays is familiar intelligence features. Nevertheless, the two different kinds of manifestation (familiar intelligence and super intelligence substantial states) in objective concrete existences are both their features, and the nature of the objective concrete existences does not change with the different states they manifest.

iii. There Is No Impassable Chasm between the Two Consciousness States

Although familiar intelligence consciousness activity and super intelligence consciousness activity are shown in different ways, their information in *Yi Yuan Ti* is both holistic, only with difference in their degree of completeness. For example, images that belong in the consciousness activity of familiar intelligence, though they have synthesized features of different aspects of the objective existences they reflect, are not all the information about the existences because even the image of the same existence differs in different people's reflections because of the differences in their experiences, methods of observation, educational and cultural backgrounds and so on. This shows that in the process when an image is synthesized into being in *Yi Yuan Ti*, some information is intensified and some is dropped. It is because image of familiar intelligence consciousness activity is not able to reflect all the information about an existence as super intelligence consciousness does that it cannot display the super extraordinary feature of holisticness. Nevertheless, an image after all reflects to a certain extent all the characteristics of an existence, thus even familiar intelligence consciousness activity also has some slight effect on objective existences. Experiments conducted in the School of Engineering and Applied Science at Princeton University in America provide forceful evidence for this by showing that familiar intelligence people's consciousness activity can increase the readings in electronic thermograph.

With this point clearly in mind, we can reflect existences correctly through comprehensive observation and sober-minded thinking without any restriction of stereotype to form integrated images of existences in *Yi Yuan Ti*, and this can also display the features of super intelligence consciousness activity. As to this, it is said in *Guan Zi*, "Concentrated in the mind, attentive in the heart, upright in the ears and eyes—this leads to the perception of what is far beyond." It is also said "Now that the whole body is upright and the *qi* and blood flow peacefully well, the entire mind is concentrated and there is no indulgence in the use of the ears and eyes. Therefore, what is far is just as what is near and knowledge is borne of this right way of perception."

We also discover in the practice of teaching *qigong* that paranormal abilities are more easily tapped and liable to appear in those who are good at thinking in images. All this fully shows that familiar intelligence consciousness activity can be transformed into super intelligence consciousness activity under certain circumstances.

iv. To Actively Tap the Functional Ability of Super Intelligence Consciousness Activity (or Super Extraordinary Consciousness Activity)

Functional abilities shown in super extraordinary consciousness activity used to be called "*shentong*" (miraculous connecting capabilities), which is claimed in both Buddhism and Daoism as the abilities that should be "capably acquired yet wisely unused", for fear that using *shentong* should affect the improvement in one's *gongfu* practice. They also believe that the functional abilities of super extraordinary consciousness activity come as the result of eradication of the disturbance of familiar intelligence consciousness activity and that familiar intelligence consciousness is the illusioned postnatal perceptive spirit that should not be used in their *gongfu* practice. Therefore, the two kinds of consciousness activities are absolutely antagonistic.

According to the theory of *Yi Yuan Ti*, the two kinds of consciousness activities are both commanders of the form and the *qi* of the human body. Although they differ in their modes of commandment, they can change into each other under certain circumstances. In light of this, *Zhineng Qigong* advocates tapping super extraordinary consciousness activity by means of actively enhancing familiar intelligence consciousness activity. For example, in *Zhineng Qigong*, the mechanism of how the super extraordinary consciousness state is generated and the specific method to practice it are taught and learners accordingly practice—this is in fact carried out in the state of familiar intelligence consciousness activity. Yet, as a result, paranormal abilities of sending and receiving holistic information are acquired. When these functional abilities are used to enhance the functions of each part of the practitioners' bodies, the progression of *qigong* practice is greatly facilitated. These functions can be applied not only to oneself, but also to other people in treating disease and helping improve *gongfu*. What is even more important is that the tapping and attainment of super intelligence is crucial in order for *Yi Yuan Ti* to enter the complete level.

Whether or not familiar intelligence consciousness will become an obstacle to tapping super intelligence consciousness lies critically in whether or not familiar intelligence consciousness is fossilized. If the diverse knowledge acquired with familiar intelligence consciousness activity is fossilized and is regarded as the standard to evaluate all existences, the harm of this apriorism in philosophy will form in *Yi Yuan Ti* an advantageous excitation center of the acquired knowledge which depresses and repels the reflection of new information, thus hindering the generation of super extraordinary consciousness. If consciousness activity based on acquired knowledge does not fossilize, *Yi Yuan Ti* will manifest its innate attribute of "revealing things when they come" and "becoming void when they have left". There will be no "hindering the reflection of the new information with the retained information". Therefore, *Yi Yuan Ti* will enter a crystal clear realm of brightness and super extraordinary consciousness activity will be successfully tapped and developed.

III. Consciousness Activity Needs Further Improvement
i. The Problem of the Divided State of Familiar Intelligence People's Consciousness Activity (or State of Mind) Remains to Be Solved

It is pointed out in the theory of *hunyuan* consciousness that familiar intelligence people still do not know the inward application of their consciousness. They have not formed the regular motions of their life activities and consciousness activities yet. Their *Yi Yuan Ti* is not only constantly disturbed by the information sent to it from the different sensory organs of the eyes, ears, nose, tongue and skin, thus showing many excitation centers, but the internal information at different "stratums" inside *Yi Yuan Ti* is also vying with each other. The disturbance of emotions to *Yi Yuan Ti* is particularly evident. All these severely affect the concentration and unity of consciousness activities and reduce the state of mind to a divided state. This is the morbid state of human beings today. It makes humans unable to bring their intellectual power into full play; it affects human health; it makes *Yi Yuan Ti*'s familiar intelligence unable to be fully applied; it makes super intelligence unable to be tapped... In sum, it makes humans unable to form a genuine integrated whole entity. This is the fundamental reason why *Zhineng Qigong* emphasizes consciousness cultivation.

ii. The Level of Human Beings' Consciousness Activity Still Awaits Improvement

Human consciousness activities develop gradually from low to high levels and this is a law having been proven in the evolutionary history of both the human race and human individuals. The theory of *hunyuan* consciousness reveals in the broad sense all the levels of consciousness activities, which include emotion at the level of impulse, sense, perception and thinking in images that belong in perceptual consciousness, formal logical thinking and dialectical logical thinking that belong in rational consciousness, and observant inward-directed thinking and super intelligence thinking that belong in intelligence consciousness. However, human beings' consciousness level up to now has only reached the level of rational consciousness, and this is the achievement that results from the evolution of we do not know how many tens of thousands of years. It is also due to such achievement that we can not only know the laws of Nature but also have a foundation laid and a condition created to know the content of all the levels of human consciousness.

Just as Engels pointed out, "Only after these different branches of the knowledge of the forms of motion governing non-living nature had attained a high degree of development could the explanation of the processes of motion represented by the life process be successfully tackled."(*Dialects of Nature* in *The Complete Works of Marx and Engels*, Vol. 20, p. 408)

It is pointed out in the theory of *hunyuan* consciousness that rational consciousness is the method and content of thinking based on familiar intelligence whereas intelligence consciousness is the mode and content of thinking in the domain of super intelligence (with observant inward-directed thinking as the transition from rational consciousness to intelligence consciousness). It is only when human beings have learned to apply intelligence consciousness that they can transcend into the more advanced stage.

iii. Partiality and Limitation in Familiar Intelligence People's Consciousness Activity

It has been pointed out in changes of *Yi Yuan Ti* and in the formation process of consciousness activity that consciousness activity comes into being on the basis of Self *Yi Yuan Ti* in its process of developing into the partial identity subjective world. Therefore, consciousness activity has in itself certain partial and narrow features.

i) Mistaken Identity of One's Internal "Master"

As it has been previously stated, the Reference System which is formed with all kinds of information internalized into *Yi Yuan Ti* with the help of senses and perceptions and is used to deal with the relationship between subject and object has become the background and the evaluation mode for people to know and deal with objective existences. It has become the foundation for people to guide their conduct and behavior. People usually consider this as their own internal "master". For example, the judgment people make about good or bad, right or wrong of somebody or something is generally considered as the judgment made by their *self* (internal master). In fact, this evaluation and judgment result from the working of their mode of evaluation. Therefore, these modes and the Reference Systems are neither their internal masters nor belong to their consciousness activities, but merely the reason in their *Yi Yuan Ti* for the generation of consciousness activities. They are called indriya (or root of consciousness) in *Cittamatra*. Consciousness activities are the content and process of the judgment and evaluation that take place in *Yi Yuan Ti*. These consciousness activities are the activity established upon indriya, so they are not the master of a person's own being, either. The internal master of a person's own being is the part of initiative function in *Yi Yuan Ti* that can evaluate, identify as well as send instructions. Consciousness activities are the changes that take place after this function starts (Of course, the occurrence of such changes is also the function of *Yi Yuan Ti*), and indriya is the background against which the established familiar intelligence mode of evaluation judges and evaluates existences, while the mode of evaluation and identification is the grounds for the activity of consciousness to identify existences. *Qigong* practice is intended to recognize one's own true internal master, to overcome bias and partiality, to break through the confinement of these stereotypes and modes in one's consciousness and to attain freedom in the realm of consciousness.

ii) The Consciousness Activity in People of Familiar Intelligence Is Based on the Functioning of Normal Sensory Organs

We know that *Yi Yuan Ti* has the functions of receiving and sending information, including familiar intelligence information and super intelligence information. In babies' development process, all kinds of information can enter their *Yi Yuan Ti*. It is because the information sent out by the language and behavior of people around them is the information based on familiar intelligence function that only familiar intelligence information in their *Yi Yuan Ti* is enhanced. As super extraordinary information related to super intelligence has not been reinforced to become internalized and settled in their *Yi Yuan Ti*, only familiar intelligence information is condensed in their *Yi Yuan Ti*'s Reference System. This kind of change in *Yi Yuan Ti* has an active guiding effect on babies' development. So functions at the level of familiar intelligence information in the human body as well as in the related tissues and structures correspondingly grow whereas functions at the level of paranormal abilities as well as in the related tissues and structures are inhibited and gradually disappear. Thus the relationship between humans and the natural world is restricted to the level of normal sensory information and humans' freedom in the world of nature is severely confined. This is the result of the limitation and restriction of consciousness in people with familiar intelligence.

Effective ways to solve this problem are a) to tap and develop paranormal abilities through *qigong* practice so as to restore *Yi Yuan Ti*'s original state of receiving and sending super extraordinary information and b) to change *Yi Yuan Ti*'s Reference System. It is only through this that the old cognitive and perceptive mode can be genuinely broken through and the existent norm of using familiar intelligence alone can be eradicated. For example, according to physics at the familiar intelligence information level, every kind of concrete substance has its feature of taking a position, a certain threshold value of impenetrability and its manifestation of form. Take a table for instance, its features of occupying a position and being impenetrable are shown in that articles on the table cannot penetrate into or through its surface. However, at the level of super extraordinary functional information, objects on the table can penetrate into or through the surface of the table without obstruction. Quite a few people now with paranormal abilities can make coins on the table fall through the table onto the ground as soon as the coins are put on the table. This results from the different kinds of information in *Yi Yuan Ti* that interact with the objective substances in the natural world: Interact with them with super extraordinary information and one brings about a result at the super intelligence level; interact with them with familiar intelligence information and one produces a result at the level of familiar intelligence senses (the physical level). Human beings' consciousness activity must be further improved on the contemporary basis and enter into the information level of super intelligence. This is the very content of cultivation of consciousness.

IV. Formation of Consciousness Activity Is a *Hunhua* Process according to the Theory of *Hunyuan* Consciousness

i. Formation of Consciousness Activity Is the Mergence and Change of the Prenatal with the Postnatal

The prenatal here refers to *Yi Yuan Ti* that is peculiar to human beings. It is a fundamental sign that differentiates humans from animals and it comes with humans when they separated themselves from the world of animals. Although the degree of consummation of *Yi Yuan Ti* increasingly grows with the development of human beings, the fundamental functions of *Yi Yuan Ti* (e.g., the functions of reflecting object and oneself) have already become a genetic factor to pass on from generation to generation. That is to say, for a new individual human being to be born, *Yi Yuan Ti* will form in it as long as it develops to a certain stage (about the seventh month of the human fetus). The formation of *Yi Yuan Ti* is directly related to the growth of cranial nerve cells. In the later stage of human fetus, *Yi Yuan Ti* already has the functions of receiving, storing and processing, and sending information, and this is the prenatal condition for the formation of consciousness activity. It should be pointed out here that this prenatal condition refers to *Yi Yuan Ti*'s functions of being able to receive information and reflect existences. It is not the prenatal framework of consciousness or the prenatal framework of knowledge as some scholars claim.

It is pointed out in modern medical science that the cranial nerve cells continuously develop in the first one or two years after the newborn baby comes into this world, and this development process is positively correlated to the establishment of consciousness activities. According to the theory of *hunyuan* consciousness, the development period of the brain neurons after the fetus is born as a baby into this world is the process in which *Yi Yuan Ti*'s Reference System is formed. The development of the cranial nerve cells, particularly the development of dendrites, comes as the result that Pure Original *Yi Yuan Ti* carries out *hunhua* with the *hunyuanqi* reflected in it and condenses it into substance having form. The information reflected at the very beginning that has merged and combined with *Yi Yuan Ti* and has become fixed by the nerve cells then becomes the foundation for *Yi Yuan Ti* to reflect objective existences—the Reference System. It becomes the identification mode for things or existences to be reflected afterwards. All human beings' consciousness activities take place and continue on this basis.

The theory of *hunyuan* consciousness points out that the newborn baby's Pure Original *Yi Yuan Ti* has only the prenatal condition for the formation of consciousness activity, that consciousness activity and consciousness structure are established after birth, which results from the mergence and change of Pure Original *Yi Yuan Ti* with the consciousness activity of adults and particularly with their language. Without the *hunhua* with human consciousness, formation of the newborn individual's consciousness activity would be impossible. This has already been proven in the case of the wolf child. Our point of view as this seems to be identical with the theory of Tabula Rasa in empiricism in modern psychology yet differs from it in nature. Higher animals also have their nervous systems and their animal psychology. Why can't they form human consciousness on the basis of their "blank slate" with the aid of human education and instruction? This is because of the prenatal difference between animals' *Nao Yuan Ti* (Brain Oneness Entity) and human beings' *Yi Yuan Ti* (Mind Oneness Entity). It is true that if animals (such as chimpanzees) are successively raised and educated by human beings for some generations, the animals' *Nao Yuan Ti* is to evolve into *Yi Yuan Ti* in not a very long time (compared with millions of years of evolution from anthropoids into humans) and their consciousness activity in the genuine sense will form.

According to the theory of *hunyuan* whole entity, the essential difference in any existence is formed in the process of *hunhua* or mergence and change. That is to say, there are differences in nature between all existences in the universe, and the evolutionary process of the universe is a continuous *hunhua* process between the prenatal and the postnatal. The postnatal of species characteristics that have already formed is the prenatal of an individual of the species, and the postnatal activity of an individual of the species is the realistic existence of species characteristics in mergence and change. The prenatal and the postnatal are not only inseparable, but they transform into each other and are thus reinforced. This is the concept of evolution in the theory of *hunhua*. Consciousness activity also develops in *hunhua* or mergence and change.

In addition, consciousness of an individual and consciousness of society also relate to each other in dialectical mergence and change since individual consciousness forms in the background of social consciousness. The influence of consciousness and language of adults on Pure Original *Yi Yuan Ti* stated above is in fact the specific embodiment of the influence of social consciousness upon *Yi Yuan Ti*. Social consciousness needs always to be expressed through people or things. All that in society will be internalized into the individual's *Yi Yuan Ti*, with some of the social consciousness becoming the content of *Yi Yuan Ti*'s Reference System, and some becoming the material for the individual's consciousness activity, which may generate new thoughts different from existing social consciousness after it experiences *hunhua* in the individual's *Yi Yuan Ti*. When the new thought becomes effective to a certain extent, it will become part of the content of social consciousness and can be fixed and maintained with written language and art. Individual consciousness and social consciousness dialectically *hunhua* and display a relationship of chain of causes and effects that mutually reinforce.

ii. The Process of Consciousness Activity Is a Holistic *Hunhua* Process of the Brain with *Yi Yuan Ti*

According to the theory of *hunyuan* consciousness, though *Yi Yuan Ti* has a certain degree of relative independence, it is a whole entity closely related to the human brain. This, therefore, determines the merging or *hunhua* change between *Yi Yuan Ti* and the brain in the process of consciousness activity. For example, when external objective things are reflected in *Yi Yuan Ti*, though this is accomplished through change in the brain neurons that are characteristic of functional localization, the *hunyuanqi* of the brain neurons in which changes have taken place will become part of *Yi Yuan Ti* once it enters *Yi Yuan Ti*. That is to say, information change that comes from part of the brain can be displayed as change in the entirety of *Yi Yuan Ti*, and the consciousness activity thus formed also spreads all through *Yi Yuan Ti* that exists in a mutually accommodating manner with the whole brain. This is, perhaps, the basis for the theory of equipotentiality in modern psychology. When changes in *Yi Yuan Ti* reach a certain degree, concrete substances will gather into form. This process of "generating substance that has form from that which has not form" is accomplished in the central area of *Yi Yuan Ti*. This is the critical point for what is reflected and impressed in *Yi Yuan Ti* to become fixed and stored, without which it is hard to form lasting long-term memory. This kind of material change can form protein particles in the nerve cells and can also form directed development in the dendrites of nerve cells. These concrete substances with the characteristic of memory then influence *Yi Yuan Ti* by means of *hunyuanqi*. Achievements in theories of biomolecular memory and synaptic memory in modern psychology can be regarded as an evidence of the point of view above.

V. Consciousness Activity Is the Manifestation of Human Nature

Both humans and animals have material level activities, such as metabolism and species multiplication, and also simple thinking activities at the spiritual level, such as neural reflexes, senses and emotions. It is because of this that some Western psychologists consider humans and animals as the same. The theory of *hunyuan* consciousness tells us that this is a misunderstanding in history which will be dissolved with the disclosing of the mystery of *Yi Yuan Ti* and its laws of functions and activities.

Although humans also have animal-level life activities such as looking for food, taking food, excretion, seeking a mate and copulation, these life activities are implanted with advanced-level spiritual content such as morality, rationality and artistry by human beings. That is to say, these life activities are endowed by human beings with the feature of consciousness. Human senses and sentiments have also become the senses and sentiments that are felt, under the control of (human) spirit at a higher level. The life that human beings have is not only to satisfy the needs at the level of life activity, but also to satisfy the needs at the level of human spirit, and human beings are continuously pursuing and creating their own spiritual needs. In this process, there is always the aspiration in spirit first followed by practical actions, the actions of the human body in the category of life activity. Here life activity not only goes on under the control of consciousness activity but becomes the means to satisfy spiritual activity. From this it can be seen that the commencement point of human life activity, i.e., the motivation of action, differs from the passive instinctive reflexes of animals; it is a kind of behavior command that is generated consciously, freely and with initiative in the realm of consciousness. This special feature of consciousness activity is established on the basis of life activity and this activity of consciousness is peculiar to human beings. Therefore, it is the essential point that tells human beings and animals apart.

We have previously illustrated that the process and content of human consciousness activity is restricted by *Yi Yuan Ti*'s Reference System, or that consciousness activity is carried out against the background of the Reference System. Meanwhile, we have also pointed out in the partial and restricted feature of consciousness activity that mistaken identity of one's "internal master" is in fact mistaking the activity in the Reference System for one's inner master. That is to say, one cannot take the consciousness activity in one's Reference System as ones' internal master. But here we say that consciousness activity in the Reference System is the nature of human life activity. Isn't this contradictory? The answer is no. When we say that consciousness activity is the nature of human life activity, we are comparing it with the nature of the life activity of animals. When we said previously that the Reference System is mistaken as one's internal master, we were comparing it with the nature of *Yi Yuan Ti*'s activity—the free, self-awakened, self-knowing nature with initiative and self-mastery.

The Reference System has, to a certain extent, "locked" *Yi Yuan Ti* into a certain mode and makes its free, self-awakened, self-knowing nature with total initiative full of self-mastery follow the mode of the Reference System. Or rather, the commencement activity in consciousness activity is restricted by the consciousness mode of science, arts and morality existing in the Reference System, which makes *Yi Yuan Ti*'s innate dynamic void nature with initiative marked with the consciousness mode of science, arts and morality that human beings have acquired. It is true that any consciousness activity has a certain content and a consciousness activity without any content in it does not exist. The content of *Yi Yuan Ti*'s Reference System continuously changes and improves with the advancement of human beings, and human consciousness activity also continuously goes toward a more accurate and truthful realm.

The partiality and limitation in consciousness previously talked about results from lack of the content of super intelligence information in the Reference System, which leads to partiality and limitation in consciousness of science and arts. The partiality and limitation in consciousness also results from the fact that the mode of value judgment in the domain of morality established in the Reference System is based on private possession, which consequently forms the mode of moral consciousness with all being centered on oneself. This severely confines human thoughts and cages the nature of human life in the sphere of being desirous to possess.

People might ask why consciousness of morality imposes such a great force of constraint on human life nature. This is because morality belongs to a relatively fundamental stratum in the Reference System. It concerns the motivational commencement of human behavior and conduct and also people's attitudes of love, hatred, acceptance and rejection of things. All this not only becomes the commander of human life activity, but in a certain sense surpasses the life motion of the human body that has form. For example, some people devote their lives to the cause of righteousness and their names go down in history; some turn traitors to protect their own lives and their names pass from generation to generation as a byword of infamy. As to the influence of consciousness on life activity, it has already been elaborated in Functions of Consciousness (Section IV). The influence of morality on the nature of human life activity will be comprehensively explained in the next chapter of The Theory of Morality.

Chapter VI The Theory of Morality (*Dao De*)

The theory of morality in the theory of *hunyuan* whole entity develops upon the foundation of the concept of morality in traditional Chinese culture. It describes the kernel content of the entity, usage and change of *hunyuanqi*. It differs not only from the concept of morality in modern ethics but also from the concept of morality in traditional *qigong*. If we say that the theory of consciousness is an extension of the theory of *hunyuanqi*, the theory of morality is a further extension on the basis of the theory of consciousness.

Section I Synopsis of Morality (*Dao De*)

I. Origin and Meaning of the Term Morality (*Dao De*)
i. The Meaning of Morality in Ethics

Ethics as an independent branch of learning did not form until modern times and it takes morality as its target of research. The term morality in ethics comes from "moris" in Latin, which originally refers to custom and convention and also to norm of conduct as well as assessment of good and evil of behavior in its extended sense. Morality in modern ethics includes morality motivation, morality consciousness, morality conduct and morality evaluation. In Marxist ethics, morality is the special ideology established on the economic basis. It is the sum of adjusting people's norm of conduct with public opinion in society, with customs and conventions and with personal conscience while using good and evil as the standard of evaluation. Furthermore, morality is the product of a certain socio-economic relationship. It changes with the change in society and in the economic basis. Although morality can adjust people's conduct, it differs from the compulsory nature in laws and politics. It exerts functions in a person's inner world through public opinion and conscience. Therefore, it is only part of but not all of the content of people's conduct. Yet it affects people more profoundly and extensively than laws and politics.

ii. The Meaning of Morality *(Dao De)* in Chinese Culture

In ancient Chinese civilization, the meanings of the two Chinese characters of *Dao* (道) and *De* (德) were not the same. For example, in *Xin Shu* (Part I) in *Guan Zi*, it is said, "That which is void and formless is *Dao*; that which generates and grows all existences is *De*." The two Chinese characters of *Dao* and *De* (道德) were sometimes also used together as one term. For example, it is said in *Nei Ye* in *Guan Zi*, "So long as it is *Dao*, it must be thoroughly ubiquitous and fills all; it must be commodious and accommodating; it must be sturdy and sound. Guard kindness and never let it go; drive away carnal desire and nourish the meager. Now that the ultimate is known, one returns back to *Dao De*." In *Wu Du* in *Han Fei Zi*, it is said "In ancient times, people vied with each other in their pursuit of *Dao De*; in medieval times, people chased after resourcefulness". Although in *Zhuang Zi* we can find "I am ashamed of my lack of *Dao De*. Therefore, aiming at the sublime, I dare not say I have conducted anything benevolent or upright; dreading what is below, I dare not conduct any excessive or eccentric deed." However, this *Dao De* is still not morality in ethics, but is what is said in *Zhuang Zi*, "Being void in tranquility, peaceful in positive apathy, silent in inaction—this is the root of heaven and earth and the utmost of *Dao De*."

In ancient Chinese civilization, ethical morality (*Dao De*) belongs to *li* (ritual) claimed in Confucianism. For example, in *Li Yun* of *Li Ji*, "How can all these be rightly done without *li* when the sages govern their seven emotions[1] and cultivate their ten righteousnesses[2], when they adhere to integrity and cultivate harmony, when they politely decline what they have the right to possess and avoid entering into rivalry with others?" In *Yue Ji* of *Li Ji*, it is said, "Uprightness and innocence are the essence of *li*; being solemn, respectful, reverential and compliant is the formulation of *li*."

Li in ancient Chinese civilization, however, not only is the norm of conduct for different kinds of interpersonal relationships, but also has an adjustment function in one's own life activities to make one's own life activities healthier. Therefore, it is the important content in self-cultivation. For example, it is said in *Li Yun* of *Li Ji*, "Ritual and righteousness are the most important aspects of people's being with which they adhere to honesty and cultivate harmony, consolidate the integrity of muscles and skin and the unity of tendons and bones. It is the most important in the cultivation of life, sending away the dead and offering sacrifice to gods and ghosts. It is the grand entrance through which people achieve heavenly *Dao* and comply with successful human relationships."

Moreover, the fundamental difference between *li* (ritual) in ancient Chinese civilization and modern ethics lies in that *li* in ancient Chinese civilization not only is the ultimate of human beings, namely, "*Li*, the ultimate of *Dao* of humanity"(*Li Shu* in *Shi Ji*), but also connects with *Dao* and *De* of heaven and earth, as is said in *Xun Zi*, "Study ends at the accomplishment of *li* and this is called the ultimate of *Dao De*".

It is true that some people might doubt whether the *Dao De* stated by Xun Zi is the *Dao De* of Nature and the universe. Then let us have a look at the description of *li* in *Li Yun,* "Therefore,

Note:

1. Seven emotions: Gladness, anger, sadness, fear, likes, dislikes and desire.
2. Ten righteousnesses: The father is merciful, the son is pious; the elder brother is friendly, the young brother respectful; the husband responsible, the wife faithful; the elderly benevolent, the younger obedient; the emperor kind, the ministers and subjects loyal.

li must originate from *tai yi* (utmost oneness). It divides and becomes heaven and earth. It moves and turns into *yin* and *yang*. It changes and exhibits four seasons. ... " It is also said that "*Li* is the truth that is unchangeable." It is stated in the prelude of *Li Ji Zheng Yi*, "*Li* administrates heaven and earth and manages human ethical relationships. This is because it began from before the division of heaven and earth. Therefore, it is said in *Li Yun* that '*Li* must originate from *tai yi*', so before the division of heaven and earth *li* exists." This is to say, *li* is related in a certain sense to *Dao De* in ancient Chinese civilization. In order to make this point clearly understood, a detailed analysis about the connotation of *Dao De* in ancient Chinese civilization must be made in order to grasp the general idea about it.

The term *Dao De* in ancient Chinese civilization is the combination of the two Chinese characters of *Dao* (道) and *De* (德). *Dao* refers to the root and origin of the cosmos and Nature and is a kind of special existence. *De* is the functional manifestation of *Dao*, the root and origin. Rather comprehensive elaboration about *Dao De* had been given by Daoist System in the pre-*Qin* days.

Let us first begin with *Dao*. It is said in *Dao De Jing*, "There exists something in a merging state born before heaven and earth. Silent and infinitely vast, independent and unaltered, it is motional all around without an end. It can be regarded as the mother of heaven and earth. Its name I do not know, so formally I call it *Dao* (literally way, truth) and somehow manage to name it as *Da* (greatness)." In *Zhuang Zi*, it is said, "*Dao* has great love and integrity in it, yet it is in inaction and has no form. It can be passed on yet cannot be forcibly taught. It can be attained yet cannot be seen with the naked eye. It is the origin and root itself and has existed since the remotest antiquity before the existence of heaven and earth. It is as magical as gods and ghosts and generates heaven and earth. Riding above *taiji*—this is still not its highest point. Reaching below *liuji*—this is still not its deepest place. It was born before heaven and earth and this should not be considered as being long. It is older than the remotest ages and this should not be considered as being aged."

These statements show that *Dao* is a special kind of independent existence relying not on any specific things. It is boundless and infinite in space and has neither beginning nor end in time. It is the root and origin for the generation of all existences.

What is *De*, then? To put it in simple words, *De* is the functional attribute of *Dao*. It is *Dao*'s external manifestation. It is said in *Dao De Jing*, "*Dao* generates it. *De* grows it. Substance shapes it. Tendency accomplishes it. Therefore, all existences respect *Dao* and value *De*. The respectability of *Dao* and the preciousness of *De*—this is not prescribed yet naturally so."

This passage means that all existences in the universe and Nature are generated by *Dao* and raised by *De*. Various kinds of specific things, after they have gone through a certain development, will accomplish their ending at last. Since all kinds of things are generated and formed with *Dao* and *De*, all existences respect *Dao* and cherish *De*. The reason why *Dao* and *De* are respected, precious and noble is not ordered or entitled since they are naturally so.

Thereupon, it is further pointed out in *Dao De Jing*, "Therefore, *Dao* generates it. *De* raises it. To grow and rear it, to develop and mature it, to nurture, uphold it and cover and protect it. To generate it yet not possess it, to do what is required yet not rely on what is accomplished, to grow it yet not control it—this is the most profound *De*."

This passage means that, though the function of *Dao* is to be able to generate all existences, to raise them and to develop them into maturity and to uphold their weight and shelter and protect them, *Dao* generates all existences yet does not possess them, has many a good deed yet does not rely on the deeds as its own merit, grow and nourish all existences yet does not control them. For *Dao* to grow and nourish is something that is natural. This attribute is the greatest *De*.

De is more clearly stated in *Zhuang Zi*, "When oneness manifests and form has not come into shape yet, that on which things rely to generate and grow is called *De*." (*Tian Di* in *Zhuang Zi*)

From this it can be seen that *Dao* and *De* in ancient Chinese civilization are the most fundamental substance in the universe and its most fundamental function, i.e., its entity and use. Its entity is "silent and infinitely vast, independent and unaltered, motional all around without an end"; its use is natural—with all attributed to nature and taking being natural as its use. *Dao* has its own function that can make all kinds of changes happen in all existences and this function is *De*.

Although some differences exist between *Dao* and *De*, it is hard to separate them because they are the two aspects of one same thing. Therefore, it is pointed out in *Xin Shu* (Part I) in *Guan Zi*, "*De* is the shelter of *Dao* and things rely on it to generate and live" and "To take inaction as *Dao* and sheltering *Dao* as *De*. Hence there is no discrepancy between *Dao* and *De*."

From this we can observe that *Dao* and *De* as stated by the ancient Chinese people are one and the same thing that not only generates all existences but can also be embodied in all existences. It is explicitly pointed out in *Zhuang Zi*, "What is *Dao*? There is the *Dao* of heaven and the *Dao* of humanity," and "Is there any place where there is no *Dao*?" When *Dao* is embodied in human beings, it becomes the *Dao* of humans. It is precisely because of this that Confucianists in the later generations directly called *li* in the domain of ethics *Dao De*. Kong Yingda in the Tang Dynasty stated in *Li Ji Zheng Yi*, "*Dao De* we talk about today embraces all things in the broad sense and includes talents, attainments and good deeds in the narrow sense. Be it in the broad or narrow sense, *Dao De* should be applied in accordance with truth and reason. The reason why a person's talents, attainments and good deeds can be considered as *Dao De* is because with talents and attainments it is easy for one to do things smoothly and successfully, and with good deeds carried out with one's body one attains truth and reason. Hence they are called *Dao De*."

So the integrate concept of *Dao De* in Chinese culture is formed, namely, humans and nature as well as society are one whole entity, and this is the important content in the classical concept of whole entity with heaven and humanity in one entity.

iii. The Meaning of Morality (*Dao De*) in *Zhineng Qigong* Science

The concept of morality in *Zhineng Qigong* has its starting point in the theory of *hunyuan* whole entity. It bears a lot of similarity with the concept of *Dao De* in ancient Chinese civilization. According to the concept of morality in *Zhineng Qigong*, morality (*Dao De*) is the natural feature shown in the entity and function of *hunyuanqi*. It is pointed out in the theory of *hunyuan* whole entity that everything is determined by its own *hunyuanqi* and the entity and function of *hunyuanqi* constitute the different attribute in everything. Why then do we call it (the natural feature shown in the entity and function of *hunyuanqi*) *Dao De* (morality)? On the one hand, this is because of the influence of traditional Chinese culture and, on the other hand, it is set forth from the relationship between fundamental human attribute and morality (which is to be stated in particular part later). As it has been previously stated in the concept of whole entity, though there exist multitudinous things in the universe and nature that differ from each other in numerous ways, they are a whole entity closely connected with human beings. In the science of *qigong*, it is emphasized that humans strengthen human relationship with Nature through cultivation of morality so as to form a harmonious whole entity between humans and heaven (nature). Since human morality belongs to consciousness activity while *Yi Yuan Ti* is the special form of expression of *hunyuanqi*, in order to highlight the identical feature between the morality activity of human beings and the activity of *hunyuanqi* of all existences, in order to make it easy to give guidance and to explain the mechanism of *qigong* practice in *Zhineng Qigong* science, we not only inherit the domain of *Dao De* prescribed in traditional Chinese *qigong*, but also further develop it to make it a more integrated concept of *Dao De* (morality).

In the traditional theory of *qi*, it is believed that formless and ubiquitous *Dao* and *De* can be perceived and practiced through *qigong* exercise and that *De* itself is cultivation and refinement. For example, it is said in *Zhuang Zi*, "*De* is the cultivation of achieving harmony," namely, the cultivation that makes a person's behavior and conduct attain harmony is *De* (According to *Zhong Yong* which is translated as *Doctrine of the Golden Mean*, "When gladness, anger, sadness and joy do not occur, it is the state called Golden Mean; if gladness, anger, sadness and joy occur yet they are moderate under control, it is called harmony"). In the ancient Chinese character of *De*, there are such three parts of "human, direct and heart". This means according to *Shuo Wen* "to get it externally from people and internally from oneself." In *Xin Shu* (Part II) of *Guan Zi*, "Do not let things disturb your sensory organs, nor the sensory organs disturb your heart, and this is called internal *De*."

In addition to this, ancient Chinese people believed that through self-cultivation they could also attain *Dao*. In *Nei Ye* of *Guan Zi*, "*Dao* has no residence. It settles where the heart is kind, peaceful and affectionate, where the mind is calm and *qi* is well adjusted." It is also said, "*Dao*'s sentiment is like this—it loathes voice and sound. Those who cultivate tranquility in their heart can achieve it." This shows that if people can cultivate their heart and calm their mind according to the requirements of *qigong* practice, and if they love others with a kind heart, their *qi* will be adjusted, with which they can further attain *Dao* and perceive *Dao*'s entity and attribute.

It is said in *Dao De Jing*, "The grand *De* is like this: It follows nothing but *Dao*. *Dao* as matter is 'faint' and indistinct. Indistinct and 'faint', there is image in it. 'Faint' and indistinct, there is substance in it. Profound and far-reaching, there is essence in it. The essence is rather true and integrity is in it." This passage means that both the body and heart of people who have attained immense *De* in their *qigong* practice follow and comply with the laws of *Dao*. And they can perceive the state of *Dao* on account of their *qigong* practice. That is, *Dao* as a substance is in a material state that is "obscure". It is not that there is nothing in it. There seems to be image in it yet you cannot clearly see it. The special entity and image of this material state are, as it were, the image of nothingness and the form of no shape. This material state is very calm, profound and far-reaching, and there is no difference between motion and stillness in it. It is vast, silent, and very indistinct. In this material state, there is essence, yet essence not in the sense of essence in *jing* (essence), *qi* and *shen*, but rather the essence of *qi*. It is the quintessence of *qi*. "The essence is rather true and integrity is in it"—in this essence there is also integrity and the motion in integrity (Integrity here is equivalent to information, substance that is more exquisite).

These statements are about ancient Chinese *qigong* practitioners' experience and perception of *Dao* after they began with their *qigong* practice.

In addition to this, ancient Chinese people believed that they could embody the characteristic of *De* of being able "to generate it yet not possess it, to do what is required yet not rely on what is accomplished, to grow it yet not control it" when they reached a certain level in their *qigong* practice.

It is said in *Dao De Jing*, "Holding oneness with your *ying po* (spirit), can you not mentally depart it and be very concentrated on it? 'Rolling' the ball of *qi* in your palms, can you be as tender as an infant? To inspect your most profound realm of mind, can no flaw be found in it? Loving your subject (referring to *qi*) and governing your kingdom (referring to body)—can this be accomplished without your consciously knowing it? To open and close your gate to heaven (*tian men*, an 'acupoint' on the head), can you be without anything feminine (*yin*, that which has image or concept)? Knowing all, holding insightful views and being able to accomplish whatever

269

you want to accomplish—can this be done in inaction? Generate it (which refers to all existences) and grow it. To generate it yet not possess it, to do what is required yet not rely on what is accomplished, to grow it yet not control it—this is the most profound *De*."

To put it in modern language (the paragraph above was written in ancient Chinese language in the Chinese original), this passage is talking about whether we can reach the realm in our *qigong* practice of making our spirit always hold oneness, guarding the golden mean and not departing from the golden mean state. Can we reach this level? The most important thing about *qi* lies in its softness instead of hardness. Can we be as soft as infants? *Xuan lan* or utmost view refers to the realm of the deepest and most profound mental state. In the most profound realm of mind, can we have no mental activity about any image or concept and also no flaw in the spiritual world? Our consciousness is clear and tranquil, bright and pure. Can we be like this? "Subject" and "kingdom" are respectively the metaphor of *qi* and the human body in ancient China. For *qi* and the body, it is not that we must do such and such, but that we do not intentionally obey the rules and laws in *qi* and *qi* change and yet we never break these rules and laws. It is the state that we do not intentionally obey the rules and laws yet they are naturally followed in our actions, the state of doing nothing intentionally and yet nothing (that needs to be done) remains undone. Can we be like this? Can we be without any *yin* in our mind but full of pure *yang* when doing things? To be very clear about many things yet with no intention to differentiate—can we accomplish this? If we can accomplish these points, we will be able "to generate it yet not possess it, to do what is required yet not rely on what is accomplished, to grow it yet not control it" and reach the realm of the most profound *De*.

The theory of *hunyuan* whole entity in *Zhineng Qigong* Science is founded upon the theories of "*Dao*", "*yuan qi*" and "*hunyuanqi*" of the ancient times. The *hunyuan* particle and primordial *hunyuanqi* we talk about are similar to the *Dao* and *yuan qi* talked about by the ancient Chinese. With *Zhineng Qigong* practice, the *Dao* of primordial *hunyuanqi*, which is "faint" and indistinct, can be perceived. Furthermore, the *hunyuanqi* in the human body can also be perceived with *qigong* practice. Even the transmutation relationship between *hunyuanqi* and concrete substances in the human body, the association between *hunyuanqi* and ideology and so on, and also the domain of morality in the human body can also be perceived. What's more, after *qigong* practice, the interrelationship between human body *hunyuanqi* and different levels of *hunyuanqi* in the world of nature can also be perceived. The purpose of our *qigong* practice lies in solving the fundamental problem of our own life motion, that is, to perceive the state of motion and the laws of motion of this fundamental substance of *hunyuanqi*. This is the fundamental point in *qigong* practice.

The theory of *hunyuan* whole entity points out that human consciousness activity is the motion of human brain *hunyuanqi* and is the true manifestation of human nature, and that morality consciousness takes up a decisive position in consciousness activity. The critical point of *Zhineng Qigong* practice is the improvement of the level of human body *hunyuanqi* through the cultivation of morality so as to influence the nature of *hunyuanqi* in the natural world. Although morality is the manifestation of the fundamental attribute of human *hunyuanqi* and has in it the characteristics of morality in ethics and somewhat differs from the manifestation of the entity and attribute of *hunyuanqi* of natural things, we still accept the traditional term in ancient Chinese civilization and call some of the characteristics (mainly the laws of growth and change) in the entity and attribute of the *hunyuan* entities of humans and natural things *Dao De* (morality). This is helpful both for applying consciousness in *qigong* practice and for the explanation of theories.

II. Classifications of Morality (*Dao De*)
i. Natural *Dao De*

Dao De stated here is *Dao De* in the broad sense which refers to the characteristics of the entities and attributes of the *hunyuanqi* of all existences themselves as well as the display of their *hunyuanqi*'s laws and functions in Nature. All existences carry out their motions and changes according to their own special prescriptive features, and this is the display of the entities and functions of the *hunyuanqi* of existences when mergence and change in their *hunyuanqi* takes place. This is also the display of *Dao De* stated in traditional *qigong*.

It is said in *Nei Ye* of *Guan Zi*, "*Dao* has no root or stem, nor leaf or foliage. Yet all existences rely on it for their generation into being and for their accomplishment of growth, hence it is given the name of *Dao*." Since the realization process of the entities and functions of existences is carried out in Nature and meanwhile their entities and functions are realized according to their own natural laws, namely, these laws are not imposed upon them by the outside world, this is called natural *Dao De*. None of the existences in the natural world lack this state of *Dao De*. For example, the performance of inorganic substances as well as their various kinds of motion and change, the attributes and functions of the growth and development of organisms as well as their laws and changes all belong in natural *Dao De*. The development stage in the human fetal period, the different kinds of physiological and bio-chemical structures and the instinctive physiological functions after a human being is born also belong in the domain of natural *Dao De*.

According to the theory of *hunyuan* whole entity, the change of the *hunyuanqi* of things among all existences is not isolated but is going on with the *hunyuanqi* of other existences outside themselves against the immense background of primordial *hunyuanqi*. That is to say, in the realization process of natural *Dao De*, there must be the participation of primordial *hunyuanqi* to stimulate the change and vitality of natural *Dao De*. As for this, traditional *qigong* theory also has something to say, "That which is called *hunyuan* is in fact *qi*. The thing that is distributed all

271

over heaven (the universe) is called *hunyuan*. The *qi* of *hunyuan* is actually the fundamental *wind* (not the wind of the flowing of air as we say today). The force of the *wind* (*qi*) is the greatest. It can uphold the sky, the earth, *san cai* and *wu xing*. The sky, the earth, *san cai* and *wu xing* cannot be greater than this wind. The *qi* and the wind are innately one body. To make flowers bloom and willow branches grow, to form seeds and bear fruits—all of this is due to the four *qi*s of spring, summer, autumn, winter and the eight *wind*s."(*Yuan Qi Lun* in *Yun Ji Qi Qian*)

ii. Social *Dao De*

Social *Dao De* instructs people to deal with appropriately the relationship in their social life among themselves and the relationship between individuals and collectivity (country). In social *Dao De*, there exists a series of norms and standards about people's conduct. Social *Dao De* makes use of many concepts, such as good and evil, beautiful and ugly, and honest and hypocritical, to restrict, standardize and evaluate people's conduct. Social *Dao De* is determined by the integrate feature of society. In their own development process and in the process of lasting struggle with Nature, human beings have displayed the value of humankind's existence and have protected the existence of the species of humanity. In this process, they combine together in certain patterns. In dealing with the relationship among people, certain norms and standards of conduct, i.e., social *Dao De*, must be followed.

To view social *Dao De* from the perspective of *qigong*, all this is the entity and the application of humans in the whole entity of society. Although social *Dao De* varies in accordance with society, class, race, historical conditions and so on, there is one common point, that is, the function of *Dao De* lies in maintaining the stability of human beings' society so as to benefit the existence of humans as a species. People in primitive society did not have very strong hands and teeth as those in some animals and their bones and tendons were not as strong, either. The reason why they could manage to survive was because, aside from their wisdom and hands, they relied on the power of collectivity—communal living at that time. And in order to live together, each member's behavior and desire had to be kept within bounds, as was the embryonic form of law and morality.

If we disregard the specific content of social morality so different from each other and consider from the perspective that morality helps to maintain the existence of the species of humanity, social *Dao De* and natural *Dao De* are in this sense unified. Or rather, social *Dao De* is the reflection of the nature of *hunyuanqi* at the level of society. It is the entity and application of the mode of value judgment in *Yi Yuan Ti*'s Reference System. It is the important content of the nature of the human species and human social attribute. Therefore, humans as social beings must nurture their own *hunyuanqi* starting from social *Dao De*.

iii. Natural Social *Dao De*

Natural social *Dao De* (or social natural *Dao De*) refers to the special *Dao De* state peculiar to humans in a particular development stage. A newborn baby as merely a natural human being who has not established human consciousness or has not entered society is only a common objective existence in the universe and Nature, and all the functional abilities that a human being has at this moment belong in the domain of natural *Dao De*. After consciousness is established when the human being begins to deal with social relationship and gradually learns to deal with the relationship with other people until he or she applies social *Dao De* into his or her own life activities and takes it as his or her standard and norm of conduct, this person becomes a social human being and enters into social *Dao De*. Natural social *Dao De*, however, refers to that, when natural *Dao De* is turning into social *Dao De*, when natural *Dao De* is in a period of transition to social *Dao De* and social *Dao De* has not completely formed yet, when the baby lives naturally and is involved in social *Dao De*, a special *Dao De* state is formed which differs from both natural *Dao De* and social *Dao De*, and we call this stage social natural *Dao De*. Social natural *Dao De* generally refers to the period of babyhood and early childhood of human beings when the concepts of good and bad, kind and evil have not formed and the dividing line between the individual and other people has not been drawn. This is mainly shown as follows:

i) There Is No Differentiation between Public and Private or Yours and Mine.

In babies' mind, concepts of possessing material and owning material have not formed yet and there are no such concepts of yours or mine. For example, as far as food and toys are concerned, babies pick up the food and eat it so long as they want to eat; they pick up the toys and play with them so long as they want to play, without caring about who the food and toys belong to. There is pretty much a flavor of "each taking what he needs" in it. When they do not need the food and toys or do not want them any more, they put them down and leave. If any other person wants them at this time, they don't care.

ii) Truthful, Honest, Candid, Pure and Innocent.

It is pure materialistic reflection with babies. They simply reflect what the actual situations are and do not tell lies, nor can they exchange banter. Whatever they have done, whether it is right or wrong, they dare to tell the truth. It is genuinely above-board. They truthfully express as much as they can about what they know of the objective things around them. For example, when parents tell them to tell a lie—to say that they do not have such and such a thing that they actually have in their home, the baby would say, "Mom said there is not such and such a thing in my home." If you ask further, "Your mom said there was not such and such a thing in your home. Can you please tell me 'Does your family have it?'" The baby would say, "Yes, we have." Sometimes the baby would talk like this, "There is such and such a thing in my home. Mom said 'Don't tell anyone.'" This is because mother has said something and that is an objective fact, therefore, the baby will naturally reflect for you; meanwhile, there is something in the baby's home and this is also an objective reality that the baby will also naturally reflect.

iii) Babies Have No Worry, Misgiving or Fear, No Sentiment of Gratitude or Grievance; They Are Always Happy and Glad without Unhealthy Thoughts or Ideas in Their Mind.

In babies' hearts, all kinds of objective things are novel, fascinating and will catch their interest. Even if someone in their family dies, they do not feel sad or worried. Instead, they are happy with the "hustle and bustle" of the funeral arrangements. In the time before 1949, when people fled away from their hometowns in times of calamity, their hearts were weighty with sufferings and their faces shrouded with sorrow and pain; yet babies shouted with glee, "Now we are taking a big train!" and "Now we are taking a big ship!" They are interested in everything without associating it with happiness or sadness. Besides, babies do not recall with nostalgia things that have happened in the past, whether they were good or bad. Even with things that involve adults' gratitude and grievance, babies leave them behind. The Chinese folk adage goes that children bear no grudge; after a quarrel or fight, in the twinkling of an eye, they become good friends again. This is very close to the cultivation realm required in Buddhism of "having no heart of the past, present and future."

iv) Babies Have a Compassionate and Equal Heart.

A baby, when seeing other little friends crying over mishaps, will be influenced and cry with them and will also be happy and even hop and skip with happiness on seeing or hearing that other little kids are happy about something. This is quite similar to what is said in *Lao Zi* (*Dao De Jing*) "Taking the heart of people as the heart of one's own." Besides, in babies there is no such thinking as being special. They are all the same, with no difference in remoteness or intimacy between them. Even young babies who have never met each other before will play happily together. Among these little friends is equality and they treat each other fairly. There is no differentiation between high and low or noble and humble. Equality and friendliness are shown among them.

v) Babies Love Being with Groups of Children.

Babies do not like staying alone. If a baby is put in a room alone, it will soon fall asleep. Babies like group activities and they are free and easy in group activities. They usually enjoy playing together around one central event. What has been expressed here and above is what people often call "the heart of an innocent child".

From what has been enumerated above, natural social *Dao De* in the stage of babyhood and early childhood seems to be very sublime at first sight, and some *qigong* masters in ancient times even took this babies' state as the target of pursuit in their *qigong* practice. With a careful analysis, however, we will find that this state is an embodiment of babies' physiological condition having not been fully developed. In babies' mind, a lot of information about objective existences has not yet been impressed, or at least not deeply impressed if there is any impression. The differentiation development of the brain functions has not been finished. Many physiological associations in the brain have not been established and *Yi Yuan Ti* is basically at a primordial chaotic level. Therefore, notions of association and differentiation between things are also in a relatively obscure state. Consciousness activity and the *Dao De* state during this time are merely in the embryonic stage, which is, in a certain sense, similar to those in primitive humans. This is the *Dao De* state that all human beings have to experience. It is only in the later process of development that social *Dao De* in a different stage is gradually formed.

From what has been stated above, it can be seen that social natural *Dao De* results from the interaction of the entire life entity of a young baby, particularly at a time when its nervous system has not developed into maturity, with the whole entity of the surrounding environment (both natural and social). This is the concept of whole entity of the inchoate human body life activity displayed in the realm of society, and it is the pure foundation upon which a human being constructs the mansion of human values as a human being. What is pitiable is that in thousands of years of history of human civilization, humans' true and thorough value has not been discovered by themselves, and what is noticed is but the value of material to human life, so human beings are led into the wrong track of knowing only gratifying material desires and have established corresponding morality concepts, while the pure and innocent foundation is thus stained.

iv. Social Free *Dao De*

What is social free *Dao De*? Social free *Dao De* is that, through learning and practicing *qigong*, people have rid themselves of selfish desires, have eradicated disturbance of emotions and differentiations between oneself and others, have tapped and developed their super intelligence and have rooted out the defects of partiality and adherence to partial reflections and opinions in *Yi Yuan Ti*; it is that, in dealing with the relationships between humans and Nature and between human beings, people can conscientiously and objectively comply with the holistic principle of harmony between humans and heaven and have reached the free realm of not adhering to the customary rules while never going beyond them, the realm of "following the dictates of my own heart while what I desire never oversteps from the boundaries of right." This is the true display of human beings' nature. This state bears certain similarities in some aspects to natural social *Dao De* state in babyhood and early childhood, yet it already differs from it in essence because social free *Dao De* has reached the elevated realm of being for the public and benefiting life, being benevolent to people and giving help to things, regarding others and oneself as being equal and viewing things and oneself as being the same. It is realized when the functions of the cerebrum are highly developed, when each part of the brain is extremely smoothly and thoroughly connected, when not only familiar ntelligence has been brought into play at a higher level, but potential intelligence has also been considerably tapped and developed, and when human life activity begins to truly step into the conscientious self-awakened holistic level.

It is precisely because of this that we think the wording of "returning from old age to babyhood" claimed by ancient *qigong* masters in their *qigong* pursuit is not desirable, as it is in fact impossible to return oneself into the chaotic state without postnatal perception of an infant or a baby through *qigong* practice. When the advanced level of *qigong* practice is achieved, it is by no means like that. With a little effort to analyze it, we can know its falsehood. Let us bear in mind the fact that a newborn baby is indeed ignorant, having neither knowledge nor perception in it, yet it has not even the barest ability to survive by itself. If it were not for the care from the adult, it would be impossible for the baby to remain alive. Will *qigong* practice having reached the advanced level be like this? Now that the first step of "returning from old age to babyhood" is unrealistic, "going back to simplicity and truthfulness of the void of *Dao*" will be even more impossible, like castles in the air. This is perhaps, I'm afraid, one of the important reasons why it was hard for the ancient *qigong* practitioners to become completely successful in achieving *Dao*. It is true that in society of private-ownership with opposing classes, the selfless *Dao De* of equality between oneself and others advocated in *qigong* is hard to realize. It is only when society develops into the state in which "all is for the public" that social free *Dao De* can possibly come true.

From the classifications of *Dao De* we have given above it can be seen that though the latter two kinds of *Dao De* (social natural *Dao De* and social free *Dao De*) have the content of social *Dao De* in it, this content of social *Dao De* is different from the social morality we usually talk about. From the third kind of *Dao De* (social natural *Dao De*), we can see that it is something that all human beings have in common. The babyhood and early childhood of all races throughout the world are basically like this. The children of capitalists are not born with the ability to exploit. Such a feature of *Dao De* in young children is the characteristic during the time from when a human being begins to grow to when this person is to enter social *Dao De*. Class morality in class society also evolves on the basis of this characteristic.

The realization of social free *Dao De* results in human beings' self-improvement and self-perfection after they have realized their own true values and the mysteries of nature and their own lives. The most effective way is to learn and practice *qigong*, to purify consciousness, to get rid of all kinds of selfish desires and to become enlightened on the idea of all being for the public. This will be the route that all the human beings will inevitably follow.

III. Content of Morality (*Dao De*) in Human Life Process
i. Natural *Dao De* in Humans

We know that, no matter how advanced or special the human species is, it is still part of Nature. That is to say, humans are humans of nature, for the human species also has to follow laws of natural *Dao De*. Just as it is said in *Si Qi Tiao Shen Da Lun* in *Su Qi*, "The four seasons together with *yin* and *yang* are the root and foundation of all existences. Therefore, sages nourish *yang* in spring and summer and *yin* in autumn and winter in order to adhere to the root and to flow with all existences at the gate of growth. Doing adverse things against what the root and foundation demand cuts the root and harms the truth. Therefore, *yin yang* and the four seasons accompany all existences from beginning to end and are the cause of life and death. To be adverse to this brings misfortune and calamity while following it brings no disease. And this is called having attained *Dao*."

In addition to this, the life motion and change concerning metabolism in the human body belongs in the domain of natural *Dao De* as well.

ii. Social *Dao De* in Humans

As the most advanced creatures on the earth who have subjective consciousness activities, human beings not only have the ability to actively reflect and peceive the objective natural world, but also the ability to adapt actively to the natural world and cause changes in it. Yet it should be pointed out that all kinds of human deeds and values are realized with the cooperation of consciousness and language in the communication process with people in society. If a person does not enter human society after birth, the value of this person as a human being (not only as a natural human being) can by no means manifest. It is because of this that we say humans are social humans, that the life entity of a human being embodies both the natural features and the social features. The norm that a person should follow in the communication with people is social *Dao De*. In class society, everybody's concept of social *Dao De* is restricted by their economic and political positions in society and is in a process of constant change.

iii. Physiological *Dao De*

The so-called physiological *Dao De* refers to all kinds of behavior caused by physiological changes in one's own body to satisfy the needs in one's own life. Natural social *Dao De* in babyhood and early childhood should actually be categorized in the domain of physiological *Dao De*. In babyhood and early childhood, the natural self and the social self have not formed in babies' consciousness yet; though babies and young children deal with the relationship with people around them essentially according to their physiological needs, we merely call it natural social *Dao De*. In adults, though they also retain a certain natural social *Dao De* in them, e.g., "the heart of innocent children", this natural social *Dao De*, instead of being based on their physiological needs, is the outcome of a certain content in their Reference System.

The physiological *Dao De* we talk about here refers to that, after the opposition between humans and nature and between humans and society has formed, self has already clearly established and the physiological needs of an individual are endowed with the selfness feature. In other words, an individual's behavior comes as the result of one's self directly satisfying one's own physiological needs. It includes the behavior of directly satisfying various material needs in metabolism and the consciousness activity based on various behavior caused by the effect of hormones. This part of consciousness activity seems to be correlated in a certain degree to the physiological activity in the domain of subconsciousness stated by Freud. Although it can be shown in the realm of society, it is not based upon the norm of social *Dao De*.

iv. Ideal *Dao De*

Ideal *Dao De* is a kind of moral aspiration formed through education. It inspires people to strive industriously for the realization of this beautiful prospect. Ideal *Dao De* differs with each different person. The ideal *Dao De* of *Zhineng Qigong* practitioners is social free *Dao De*.

The four kinds of *Dao De* stated above simultaneously coexist to different extent in each person's moral consciousness. Every *Dao De* behavior is a unified outcome of the struggle between the four of them.

IV. Formation of Morality (*Dao De*)

The study of the formation process of human *Dao De* needs to be traced back to the formation process of humans. In the formation of humans, there is the evolutionary history of the human race as well as the evolutionary history of human individuals. The evolutionary history of human individuals is the miniature of the evolutionary history of the human race and can be taken as the starting point for the study of human *Dao De*. The formation of *Dao De* in an individual human being experiences such stages as natural *Dao De*, physiological *Dao De*, social *Dao De* and ideal *Dao De*. From the evolutionary history of human individuals it can be seen that human life begins with the zygote formed with the combination of the sperm and the ovum from parents. The moment when the *hunyuanqi* of the zygote is formed is the very moment when the features of the new human being are determined and the features of the entity and function of the zygote's *hunyuanqi* come into existence with it. This is the beginning of human natural *Dao De*. According to *Zhineng Qigong*, human *Dao De* as well as its function and attribute is not endowed by God; it is the *hunyuanqi* formed with the combination of the sperm and the ovum that determines *Dao* and *De* of a human being. The zygote differentiates and evolves henceforth according to the laws in itself into a human embryo. When the embryo develops to a certain extent, the nervous system and the cardiovascular system begin to come into form in the human body. When the nervous system has taken its rudimentary form (at about the fourth to the sixth week in the embryonic stage), the features of an independent natural human being begin to form, which shows the features divorced from lower animals, and the embryo is now called a fetus. The fetus further develops according to human features and it gradually forms the five *zang* internal organs and six *fu* internal organs inside and the four limbs and bones outside, with the five sense organs and seven apertures (i.e., the eyes, ears, nostrils, mouth) intact. This process is also carried out according to the *Dao De* feature of the zygote. This *Dao De* is a natural human being's natural *Dao De*. The life activity at the metabolic level in the whole life of a human being continues according to this law. The function of each tissue and organ of a human being is also determined by it.

After the fetus is born as a baby into this world, though there are billions of brain cells in its cerebral cortex and all functions of the nervous system synthesize in the brain to combine into a whole entity, i.e., *Yi Yuan Ti*, they are still "empty" at this moment and do not have the ability to distinguish existences in the environment yet. A baby does not open its eyes nor is it afraid of sound and voice. It is now still a natural human being that does not have community information from society. It is just a natural specific objective existence. In the baby, there exists the metabolic function and simple unconditioned reflexes like sucking, defecating and urinating and these reflexes are instinctive, for it will also suck whether you put a nipple of a feeding bottle or one of your fingers into its mouth. These functions of the baby go on completely according to the features brought about by the life structure inherited in the human body as the baby still does not have consciousness activity and these functions go on passively according to the laws of the baby's own life activities. The baby now is just a natural human being, a human being of instinct. Such instinctive activities in the baby still belong to human natural *Dao De*. From this we can know that human natural *Dao De* is completely the outcome of species inheritance.

How does natural *Dao De* in babies evolve into social *Dao De* then? We know that *Dao De* is part of consciousness activity and it is gradually established in the formation process of human consciousness activities. It is firstly the development from natural *Dao De* into physiological *Dao De*.

Modern medicine tells us that in a newborn baby there are only unconditioned reflexes that maintain the baby's basic life activities, e.g., the sucking reflex as mentioned above, the blink reflex as well as different kinds of functional activities of its internal organs. With the unfolding of the newborn baby's individual life activities, various conditional reflexes are swiftly formed and can be shown in its facial expressions. For example, feeding the baby can get rid of its sense of hunger and guarantees the provision of nutrition for its body's development. After times of repetition, the caregiver becomes a sign of the positive stimulation mentioned above and the baby shows its happiness whenever seeing its mother (or someone else who often feeds it). A Chinese saying goes that, for a baby who does not know anything yet, "Anybody who feeds it is its mom". When the happy facial expression can be shown by the baby to the feeder, this is in fact already a rudimentary form of dealing with the relationship with people. With the input of language, the relationship between the baby and the adult is gradually standardized according to social norms.

During this period, in the contact with people and when dealing with relationships, a baby's standards of being happy or unhappy with daddy, mommy and other people are not related to the standards of being good or bad, kind or evil as generally accepted in society, but are related to whether the attention given is helpful for the continuance of its own life activities. Since there is still not much distinguishing ability in the baby's cerebrum, these tints of social features impressed in the baby's mind in the earliest stage still belong in physiological natural response, which is a level higher than the most fundamental instinctive reflexes, such as absorption (assimilation), excretion, rest, and body movements in early babyhood that are indispensable for maintaining life activities. Babies start from their physiological needs to deal with interpersonal relationships around them. There are no private considerations or called distractions, no differentiations between themselves and others or worry and anxiety. This is the transition from natural *Dao De* to social *Dao De*, and we call it social natural *Dao De*.

Features of *Dao De* in this period are basically as what has been previously stated and will not be repeated here. It should be pointed out that this *Dao De* concept is also determined by objective existence. It results from the internalization into the baby's *Yi Yuan Ti* of various information about the baby's own life activities as well as various stimulation from people around. For example, all kinds of life conditions that babies need are all prepared by their parents, and babies can have their food delivered or have themselves dressed whenever this is needed, so they needn't worry about these things. When this reality is reflected in babies' mind, it is shown in the fact that babies do not understand the right to possess or the right of ownership nor do they know mine or yours. Under the command of thinking as this, babies naturally "take what they need" and they do not keep things to themselves after they have used them. Another example is that parents always actively take care of everything about babies and directly and smoothly satisfy their needs. When such kinds of reality are reflected in their brains, babies are pure, frank, not crooked or capricious. Since none of the various differentiations have been established in babies' mind, particularly since judgment of good or bad about people around can by no means be established, babies rely only on the pure psychological state in dealing with interpersonal relationships, and this is to a great extent influenced by their own physiological state.

In light of this, a baby's physiology, psychology, emotions and life activities are closely combined. Babies deal with things according to their own mental state, while their spiritual realm is the reflection of their own vigorous life activities as well as the care, love, protection, help and consideration they have received from people around them. In babies in this period (babyhood and early childhood), it is essentially sensori-motor thinking and then thinking in terms of images. Therefore, babies are highly imitative. With this clearly in mind, we will not feel surprised at what is shown in babies' social natural *Dao De*. From this it can be seen that "Humans are born with a good nature" claimed by Confucius and Mencius and "Humans are born with an evil nature" claimed by Xun Zi, both of which adhere to the idea that morality is prenatal or that morality is something one is born with, are incorrect. A baby, after it is born, will not have human consciousness if it does not have contact with humans, if it does not receive the information from human society, or if it is not connected with human relationship. The wolf child, tiger child, lion child, pig child, etc., that have lost the *Dao De* feature of humans under the influence of particular environment can hardly be called kind in nature. As to the saying that humans are evil in nature, it is in fact that babies' instinct to strive for living is exaggerated and distorted as human private desire.

Although private desire is somewhat correlated with instinctive striving for living, they can by no means be equated as the same. In babies' process of growth and development, such tendencies as eating better and living better will also be shown and this kind of wish is also reflected in the cerebral cortex. However, this is not selfishness, but the absorption of subsistence and the display of instinct. Private desire is that the instinct for survival changes and develops into one direction of wantonly plundering material things into one's private possession, and this is what is called selfishness. The aforementioned notions of equality and being for the public are also the reflection of the instinct of striving for survival. They are derived from the goal of making one's life activity go on more smoothly and one's inward spirit more comfortable. As to into which direction the instinct of striving for survival will divide, it depends upon the surrounding social environment.

With their development, babies not only gradually learn in practical life how to live, but also establish in themselves thinking in images. All this facilitates the development of their cerebral cortex to grow gradually into maturity. When their childhood comes, the development of babies' brain cells is already nearing that in adults. The features and functions of *Yi Yuan Ti* are also reaching its developed state and consciousness activities gradually enter the stage of logical calculation. Children now begin to step out of the small circles of their families and come into the contact with society. We know that society is complicated and different people in society treat children with different attitudes. Even parents no longer satisfy children's needs unconditionally as they were satisfied in babyhood and early childhood. Such external environment, when having been internalized into children's *Yi Yuan Ti*, will inevitably change their pure and simple psychological state, and adults' moral conduct has even greater impact on them.

We know that in the living environment of private ownership in which materials are not abundant, the language and behavior of grown-ups are mostly selfish, which will influence the consciousness of babies and young children and produce the feature of partiality in them. Babies and young children change from taking actions according to the needs of their life activities to taking into possession more things in order to live better and enjoy a more convenient existence, thus selfish behavior and notion are shown in them. Having changed from being for living to being for private demands and showing privateness and selfishness, one's self and privateness are combined. The joyous feeling shown in benefiting one's life activity in the past is slowly taken up by various tendencies, with which come various desires. In the condition like this, human sentiments change and center on self and private needs. Therefore, in dealing with things and handling interpersonal relationships with people, the starting point is no longer the need of being beneficial to one's life activity, but making decisions about acceptance or rejection according to the notion of private desires.

With the formation of logical thinking as well as the influence on life activity from class status and economic state and so on in society which are comprehensively reflected in *Yi Yuan Ti*, together with the input of adults' culture and thoughts and also the influence of laws, the concepts of good and evil, beautiful and ugly, good and bad, love and hate are established and firmly imprinted in *Yi Yuan Ti*'s Reference System. They become the standard to make judgment about right or wrong, the foundation for applying consciousness and the motive power to drive the motion of consciousness. Therefore, children completely enter social *Dao De*. Human *Dao De* generates, develops and changes like this.

It should be pointed out that, though an individual human being's morality (*Dao De*) begins to form in childhood and early teens and becomes basically settled in youth, the influence in babyhood and early childhood can never be neglected. Although in this period morality consciousness cannot form yet, the morality conduct of adults as well as the rewards and punishments they give to babies' and children's behavior (the behavior that has moral significance) will leave a deep brand in their *Yi Yuan Ti*'s Reference System, and this has an immeasurable impact on the later formation of morality in them. Why some children are apt to accept the morality education of kindness and honesty while some children are apt to accept selfish and hypocritical influence—this is related in a certain extent to the impression left unconsciously in their *Yi Yuan Ti* in their babyhood and early childhood.

As to the formation of ideal *Dao De*, it results from further development to an extreme on the basis of social *Dao De*.

Section II *Dao De* and Its Impact upon the Human Body's Life Activity

Quite a lot of content is included in morality or *Dao De*. From the aspect of ethics alone, there are morality motivation, morality consciousness, morality conduct and moral judgment. Morality conduct can also be further divided into ethical morality, social public morality, profession morality and so on. Here we merely emphatically introduce some aspects of morality in ethics that are related to *qigong* science and also the relationship between morality and the human body's life activity.

I. Fundamental Points of View in the Domain of *Dao De*
i. The Concepts of Good and Evil

Good and evil are the most general concepts used by people in their evaluation of morality. Good is used to refer to the conduct and event that conform to the norms and principles of morality; evil is used to refer to the conduct and event that violate the norms and principles of morality. In different ages, different races and different social classes, there are different evaluation standards of morality. The standard of evil and good totally belongs in the domain of ethics, which will not be explored here. What we are going to introduce herein is the concepts of good and evil in the science of *Zhineng Qigong*. In order to further our knowledge and understanding of the concepts of good and evil in *qigong* science, let us begin with the concepts of good and evil in traditional *qigong*.

i) The Concepts of Good and Evil in Traditional *Qigong*

In traditional *qigong* in China, the concepts of good and evil are stated at the theoretical level essentially in the three systems of Confucianism, Daoism and Buddhism.

It is believed in orthodox Confucianism that the content contained in goodness, righteousness, ritual and wisdom is the standard for the evaluation of good and evil. What conforms to goodness, righteousness, ritual and wisdom is good; what goes against them is evil. Among goodness, righteousness, ritual and wisdom, goodness is essential. What is goodness then? According to Confucius, "If you want to establish yourself in society, you should let others establish themselves in society; if you want to attain your goals in life, you should let others attain their goals", "Do not impose upon others what you do not want to happen to yourself", and this is actually what "He who adheres to goodness loves people" means. That is to say, if we ourselves want to have good life, we should also hope and help others to have good life. Therefore, goodness according to the ancient Chinese people has another meaning of "helping to grow". The reason why goodness is considered as being good is because it conforms to the feature of "The grand *De* (moral force) of heaven and earth is to grow." Righteousness is to do what should be done under the guidance of good thoughts, while ritual is the regulation of all kinds of interpersonal relationships.

In Confucianism, there are different opinions as to the origin of goodness, righteousness, ritual and wisdom, the standards that are used to judge good and evil. For example, it was believed by Mencius that good which belongs to goodness, righteousness, ritual and wisdom was brought about from before birth: "The heart of compassion everyone has; the heart of shame everyone has; the heart of respect everyone has; the heart of making judgment about right or wrong everyone has. The heart of compassion is goodness; the heart of shame is righteousness, the heart of respect is ritual, and the heart of making judgment about right or wrong is wisdom. Goodness, righteousness, ritual and wisdom are not imposed on me from without. They are what I innately possess."(*Gao Zi* in *Mencius*)

Xun Zi, on the other hand, believed that goodness, righteousness, ritual and wisdom were acquired through education after birth. He said, "Humans are evil in nature. Their goodness is achieved by their efforts to do things well. If they are to follow their nature and yield to their sentiments, there will inevitably be contending and violation of their social identities and ethical principles, which leads to violence. Therefore, there must be education about laws and teaching from teachers and guidance of ritual and righteousness, then they will exercise forbearance, accord with culture and ethics and return to order and peace. "(*Xing E* in *Xun Zi*)

In the Daoist System, natural conduct that benefits humans and other existences is considered as being good. It is said in *Lao Zi*, "The utmost kindness is like water. Water is good at benefiting all existences and vies with none. So, being located at where people detest, it borders on *Dao*. Living in the kind place (lying low while avoiding being high), harboring kind profoundness (being void, calm, deep and unfathomable), giving benevolent goodness (benefiting all existences under heaven without wishing for reward), speaking with kind integrity (no empty talk or lies), guiding with kind governance (putting in order the high and low and getting rid of the foul and filthy), serving with kind capacity (being very skillful and flexible in dealing with people and things) and moving at good time (taking timely actions)—as it does not contend, there is no wrongdoing or grudge."

In the Daoist religion of later ages, there are relatively specific doctrines of "ten virtues and ten evils". The ten virtues are: a) Shoulder filial duties and show filial piety to parents; b) be loyal to Daoist teachers; c) be merciful to all existences; d) be patient and tolerant to others; e) candidly dissuade others from evil deeds; f) sacrifice one's own benefits to save other people; g) free captive animals and care for creatures; h) plant trees and build bridges; i) promote what is beneficial to people, eliminate what is harmful to them, and educate unenlightened ones; j) permanently revere the "fragrant flowers" (truth in humane studies and ethics) and always chant rules prescribed by worthy men and sages. The ten evils are: a) profane and filthy language; b) boasting and telling lies; c) abusive language; d) using language to sow discord; e) greediness; f) holding grudge; g) ignorance; h) killing; i) stealing; j) adultery.

In the explanations of goodness in Buddhism, there is goodness in the broad sense and goodness in the narrow sense. Goodness in the broad sense is "What accords with the doctrines in Buddhism is good, what goes against them is evil", the so-called "compliance with truth is good while violation of it is evil," or called "following the right concept is good while going against it is evil". Goodness in the narrow sense refers to the kind heart in *Faxiang* School (Dharmalakshana or Yogachara School). It includes integrity, repenting sins, being ashamed of one's misdoings, no greediness, no holding grudge, no ignorance, making endeavors to progress, being relaxed and calm (and using this feeling and state of mind in life and in doing good things), to be never lost in excessive indulgence, to be dedicated to alms giving and doing no harm. There are also ten virtues and ten evils in Buddhism. The ten evils are: a) Killing; b) stealing (also called to take what is not given); c) committing adultery; d) telling lies and boast; e) saying things that sow discord; f) saying rude and abusive language; g) using profane and filthy language; h) being greedy; i) holding grudge; j) having wicked views. The ten virtues contrast to the ten evils. They are: a) Do not kill; b) do not steal; c) do not commit adultery; d) do not tell lies or boast; e) do not speak words that sow discord; f) do not say rude or abusive words; g) do not say profane and filthy words; h) do not be greedy; i) do not bear grudge; j) do not have wicked views. In addition, in Buddhism it is also believed that there are the two aspects of "stopping" and "acting" in the ten virtues. "Stopping" means to stop committing the evil deeds (both of one's own or someone else's) of the past; "acting" is to cultivate great *De* (virtue) and bring help and peace to benefit all.

ii) The Concepts of Good and Evil in *Zhineng Qigong* Science

The concepts of good and evil in *Zhineng Qigong* Science are established on the basis of the theory of *hunyuan* whole entity and the basis that human beings' holistic life activities unify with Nature.

The standard in *Zhineng Qigong* Science of judging good and evil is whether or not a human being's conduct (including language, behavior and mental activity) is helpful for this human being's life activity and whether or not it is helpful for the progressing and development of the human species-being (or nature). The conduct beneficial to these two points is good and that against these two points is evil. As humans are social and communal beings in society and in community, the principles mentioned above cannot be used metaphysically to judge whether a person's conduct is good or evil; instead, it needs to be analyzed specifically, comprehensively and from the standpoint of whole entity. For example, when an individual is bringing harm to the community and to the progress of human beings in order to satisfy his or her own interests, such conduct is evil. When it is like this, though getting rid of this evil is not helpful to this human being's personal life activity, it is helpful to the whole community and to the progress of human beings. Therefore getting rid of evil is doing kind deeds. The standard of judging beautifulness and ugliness or righteousness and unrighteousness is also like this.

It is true that in a class society, different classes have different standards by which they judge what is good and evil. It should be known that this is a phenomenon which shows that human beings have not yet established a truly advanced civilization in their long historical process of development. Just think it over: Will this phenomenon still exist once great harmony is realized in the world?

Zhineng Qigong illustrates the connotation of goodness at the altitude of life science: Goodness is the conduct that is beneficial for the perfection of the body and heart and for the freedom and liberation of life, and it is the same whether it is in the conduct in oneself or in the conduct toward others. According to *Zhineng Qigong*, there are three levels of goodness. One is goodness in conduct, for example, to help people in danger and get rid of distress, to do things of benefit to society and mankind. Next is goodness in the mind, to be good and kind in each and every mental activity and wish others to be blessed with good luck. Then it is pure true goodness. Although people do not think of themselves as being good or kind, they never overstep the boundaries of kindness in whatever they say, do or think. Great goodness achievers have not only reached the level of having all the three virtues in themselves, but the scope of what they can benefit is extremely vast, with no selection as to who is to receive the benefit and who is not. To those who are kind, the great goodness achievers show their kindness; to those who are unkind, they also show their kindness. There is none that they refuse to help or bless. They emphasize kindness in the heart, yet they do not neglect kindness in conduct. They lay even more emphasis on teaching people approaches and theories of goodness in order to help everybody to rid themselves of the secular state and enter the realm of Grand Self. Achievers at this level of goodness show a pure image of kindness and have a pure mentality of kindness while there is no such thinking as "I am being kind" in their mind. Such a realm is actually Great *De* (virtue). Hence there is the saying of "*De* is goodness" according to the ancient Chinese to express the image of the pure flow of heavenly truth in grand goodness. In fact, it is the natural display of the holistic function of consciousness *hunyuan* or consciousness integrated in oneness (the unity of *Yi Yuan Ti* with *hunyuanqi*).

ii. The Concept of Conscience
i) What Is Conscience

Conscience is also acknowledged in Marxist ethics and is considered as the most deeply concealed regulator of human behavior and conduct. It is the conscientious awareness of the moral responsibility in one's own conduct and the moral responsibility that one should shoulder for others. It is the manifestation of the consciousness of self and the sense of self in one's own morality. As to the formation of conscience, it has been clearly pointed out in modern ethics that conscience is formed in a person's life after birth, and the opinion of Marx is quoted, "Conscience is determined by a person's knowledge and by all his or her way of life." Nevertheless, the real nature of conscience has not been made clear. For example, why can conscience frequently have effects? Why can conscience acutely express the feeling of moral satisfaction and dissatisfaction? If these questions are not answered, it is almost impossible to reveal the laws of conscience or to make conscience better exert its functions of supervision and guidance in human beings' advancement.

ii) The Nature of Conscience and Its Relationship with Morality

According to the theory of *hunyuan* consciousness, conscience is part of self consciousness. It is the self consciousness that maintains the continuance of morality activity. It has been pointed out in the theory of consciousness that self consciousness is the holistic function that allows human holistic life activity to go on smoothly, while human morality (which mainly refers to social *Dao De*) is the advanced life activity peculiar to humans. It is the fundamental characteristic of human social attribute. Human morality is the law that must be followed when the universe and Nature have developed to the stage of humans who have consciousness. Otherwise, humans as a specific species cannot exist. It is so with the abstract attribute of humanity as a species, and it is likewise so with specific and realistic human individuals.

Morality is the factor that accelerates the advancement of human beings through consciousness activities in the process when people associate and connect themselves with other people around. This is an active factor. With this factor repeatedly acting on *Yi Yuan Ti*, on the one hand, specific responsive patterns are formed which affect human consciousness activities and life activities; on the other hand, these patterns will enter *Yi Yuan Ti*'s Reference System and become part of self consciousness and set patterns with which people deal with interpersonal relationships.

When a person deals with relationships with people around, these set patterns in consciousness will unconsciously and automatically come into effect and cause correspondent changes at the level of life. This is a kind of reasonable and natural change. Such changes will display complementary and corresponding holistic mergence and change with external existences. If a person acts according to such prescribed information, the set pattern response in consciousness will be realized, which will be further fed back to one's consciousness. Thus the entire process of consciousness activity (from the mental commencement to the accomplishment of the instruction) is smoothly expressed and in one's self consciousness will be the feeling of content. On the contrary, if one changes his or her original intention and cannot act according to the information that has already been started, or in other words, the patterned motion process in consciousness pauses or is stopped, which makes the original wish unfulfilled, one cannot possibly get the

287

consummate feedback. Since these response processes have become the natural process of patterned response, or rather, they have become actually what should or must be done in the habitual tendencies of a person's self consciousness, a special feeling of discontent will occur in this person's self consciousness. This feeling of content or discontent is the result of whether or not the function of conscience itself is fulfilled.

It is true that the standard of patterned response of human conscience is neither from time immemorial or unchangeable nor can it be easily changed. This is because patterned response has become settled in the Reference System to some extent. It is closely connected with a human being's entire life.

II. Requirements of *Dao De* in the Theory of *Hunyuan* Whole Entity

The present age is an age in which human beings sail from the life realm of necessity to the life realm of freedom, an age that humans realize the free, conscientious species-being (or nature). Human beings will build a world of great harmony in which humans and society, humans and nature, humans and themselves are unified. This is an epoch-making progress in human beings' development history. In an era like this, *Zhineng Qigong* Science will not only make contributions to humankind in tapping and applying super intelligence, but will also conscientiously apply morality laws and fully exert the functions of morality in human beings' evolution. Therefore, in addition to the requirement that *Zhineng Qigong* practitioners observe discipline, abide by laws and follow social codes of public morality, we put forward the novel requirement of morality in accordance with the theory of *hunyuan* whole entity, that is, "Harmony, Joy, Naturalness and Dignity." It is both the requirement of individual humans and of collectivity.

i. Harmony

The harmony expressed here refers to overall order. It is the state shown in healthy existences: Every part is in motion according to specific laws and it is the special manifestation of beauty. It is the guiding principle of joy, naturalness and dignity and includes the harmony between humans and heaven and between humans and society.

i) Harmony between Humans and Heaven

Human-and-heaven harmony refers to the state that the unified whole entity of humans and Nature carries out orderly motion according to laws beneficial to human beings' development. It is pointed out in the theory of *hunyuan* whole entity that humans and Nature are a whole entity and that humans are part of Nature. To take a step further, Nature is the inorganic body of humans and humans are the expression of the consciousness activity of Nature. Ever since the appearance of human beings, nature has become a humanized natural world since human beings continuously change nature according to their wishes. Since human beings still do not know nor have they grasped the laws of the whole entity of humans and nature, conflict between humans and nature has occurred in the reformation of nature, and, as a result, humans have destroyed nature and endangered themselves. With the development of science and *qigong* science, the mystery in the relationship between humans and nature will be gradually revealed, and the laws and channels of the exchange of mass, energy and information between humans and nature will be grasped by mankind. So a harmonious whole entity of humans and nature in favor of the development and progress of human beings will be established. This whole entity is not the restoration of what nature once was but rather the humanized nature established according to the laws of the development of the human species' nature.

ii) Harmony between Humans and Society

Society is human beings' community organization as well as its facilities formed in the process of humans' interaction with Nature. Society guarantees human beings' living and is the bond with which the connection between humans and nature is realized. This is because the connection between humans is in essence the connection between humans and nature, and the connection between humans and the inorganic natural world is also accomplished through the medium of society. In view of this, in order to establish a harmonious relationship between humans and heaven (nature) that is beneficial for the development of the human species, harmony between people and society (family, community and nation are the display of the different levels of society) must be established. It includes the harmony of communities and the harmony between humans and objects (facilities).

a. Harmony of Communities

Human community (collectivity) is the premise for the realization of the nature of human beings. It is only owing to historical reasons—the low level of productive forces and the foolishness and ignorance accompanied with it, that humans kill each other for snatching life necessities and have degraded themselves in a brutal state for a rather long time. With the progress of human civilization, humans gradually realize that harmony of communities is the guarantee of human beings' progress. For the present stage, this should be done through harmony in thought, in action, in interpersonal relationship and in organization. A harmonious social community is the unity of entirety and part, of conscientiousness and discipline, and of equality and authority. This authority refers not to a person, but to unified thought and discipline. All this must be unified to the fundamental point of being beneficial for the realization of human species-being. The collectivist notion put forward by *Zhineng Qigong* about "integrating benefiting oneself into benefiting others" is the specific embodiment of this fundamental point. The harmonious relationship in the family thereby established is thus a spiral upward in comparison with the ethics stated by the ancient Chinese ("Father benevolent, sons showing filial piety;

elderly brother friendly, younger brothers showing respect; husband and wife as respectful to each other as if treating distinguished guests"). The friendly and mutually helpful social relationship and the senses of responsibility and honor thereby established have entered a more conscientious realm compared with chivalrous conduct of being "loyal and faithful, benevolent and righteous, honest and respectful" described by the ancient Chinese.

b. Harmony between Humans and Objects

In humans' own advancement process, human beings continuously humanize natural things and enable them to enter the realm of society and human life; particularly today when science is highly developed, humanized natural things in society have become human social organs. All this closely integrates humans and society and humans and nature. With human beings' continuous exploration and their deeper understanding of their own life features, a harmonious environment more beneficial to humans will be established.

It is not difficult to achieve the goal of harmony because, though there is different specific content in the cultivation of harmony, the general goal is all for the establishment of an overall positive order. This can be done from establishing a sense of responsibility as owners who love collectivity conscientiously, by consciously adopting oneself into collectivity, by conforming one's own conduct to the requirements (disciplines) of collectivity and by conquering individuality and small cliquism.

ii. Joy

Joyousness is the display of being full of vitality and is the result of harmony. As far as individual humans are concerned, it refers to gladness, joyfulness as well as cheerfulness and wholehearted delight. It is the manifestation of the harmonious flow of *qi* and blood and the clearness and brightness of *Yi Yuan Ti*. It is the display of the vigor of life force. As far as collectivity is concerned, it refers to a good, lively atmosphere, such as being unified and friendly, being helpful to and cooperative with each other in life, and being enthusiastic and responsible, diligent and assiduous at work. It is the outward sign of the healthiness of community and organization, the outcome of both joy in the individual and willpower of the collectivity. Joyousness in collectivity is also the outcome of benevolence and affection to each other. A benevolent heart can generate strength in oneself, and affection is to wish those whom you love to step toward perfection. In this sense, fraternity is the foundation of joy.

It is not difficult to achieve the goal of joy. If everyone in the community can act according to what Comrade Zhou Enlai, the first premier of the People's Republic of China, said "Respect and love each other; learn from and help each other; tolerate and forgive each other; console and encourage each other" and first begins doing this from oneself while adhering scrupulously to the principles of "Do not impose on others what you do not want to happen to yourself" and "Do good to others and enjoy helping them", the collectivity or community will undoubtedly display a joyous situation full of vitality.

iii. Naturalness

Although naturalness is the manifestation of the truthfulness in things or existences, here it essentially refers to the law that one's life activity is helpful to oneself and to the community. It not only conforms to the requirements of harmony but is also a stipulation of joy. As far as an individual human being is concerned, naturalness is not casualness, but the realm reached in a person's self-cultivation, which is shown in the naturalness in handling matters (such as being neither affected nor pretentious, neither domineering nor arrogant, neither haughty nor humble) and in the naturalness in language and manner and at work. It is an important factor that guarantees interpersonal harmony.

Besides, naturalness is also an important restriction in the harmony between humans and heaven (nature) and between humans and society. This harmony must be established on the natural laws helpful for human beings' development. It is not all up to nature, nor is it to return to primitive nature.

iv. Dignity

Dignity refers to the carrying of oneself with uprightness, dignity and ease. It is the external natural sign of being abundant in *shen* and *qi* and pure and profound in *Dao* and *De*. It is another outcome of self-cultivation and a special exhibition of beauty. It is interlinked with "Fullness is beauty" (*Jin Xin* in *Mencius*) pointed out in ancient times by Mencius.

It should be pointed out that dignity in one's carriage results from plenty of upright *qi* in one's self-cultivation. It is not simply the adorning of looks and appearance. The cultivation of dignity must begin with conduct. When treating others, one should be loyal and sincere, tolerant and benevolent, having an open heart and be candid. In language, one should speak from the heart and be as good as one's word. In behavior, one should be above-board and upright even when there is nobody else around. So long as one can do accordingly, one's dignity will manifest by itself and one's *Dao De* will grow increasingly stronger day by day.

In sum, "Harmony, Joy, Naturalness and Dignity" as stated above is not the average content of social *Dao De*, nor is it the average content of educating people how to think and work. Instead, it is the aggregation of humanism (morality related to the character of the human species-being), naturalism (human is nature humanized) as well as collectivism (which is the kernel of the thought of communism). It can be regarded as the special *Dao De* requirement in *qigong* science.

III. Influence of *Dao De* on the Life Activity of the Human Body

The fact that *Dao De* can affect human life activity seems to be difficult to understand at first hearing. However, a review of the theory of *hunyuan* whole entity will make it easy to comprehend. For example, it has been pointed out in Formation of Human *Hunyuanqi* (Section IV) in Human *Hunyuanqi* (Chapter IV) that *Yi Yuan Ti* carries out mergence and change with human body *hunyuanqi* through the three channels of the *hunyuanqi* of nerve cells, the *hunyuanqi* of the secretions of nerve cells and the direct transmission of *Yi Yuan Ti*. This is both the mechanism of the formation process of adults' *hunyuanqi* and the mechanism of how *Yi Yuan Ti* takes command of human life activities. Under the influence of this mechanism, *Yi Yuan Ti* has a commanding function in human physiological and pathological changes, just as it has been stated in Functions of Consciousness (Section IV) in The Theory of Consciousness (Chapter V) in this book.

In the Theory of Consciousness (Chapter V), the mechanism of the formation of emotions has been elaborated upon; in this chapter, we also illustrate the relationship between morality and emotions and between morality and self. This provides theoretical foundations for further recognizing, sensing and understanding the influence of consciousness activities (including morality consciousness) upon life activities. The influence of morality on the life activity of the human body is shown in the fact that morality causes changes in emotions through the medium of self and brings about changes in human body *hunyuanqi* through emotions, and that morality consciousness influences the generation and growth as well as the quality of human body *hunyuanqi* through one's self. Therefore, it has an enormous impact on *qigong* practice.

i. The Relationship between Morality and Emotions

When a certain external stimulation acts upon a person, there usually occurs in this person a certain change in emotions. The change in emotions is restricted by the notions of love and hate as well as acceptance and rejection, and the standard of these notions is determined by the mode of value judgment and such morality consciousness as what is right or wrong and what is good or evil. From this it can be seen that the occurrence of emotions is directly related to a person's state of morality (*Dao De*). For example, people with different levels of morality treat the same matter with different attitudes of love, hate, acceptance and rejection, and these attitudes are usually accompanied with different emotions that occur in the course of the matter. All this influences people's behavior and conduct. It is true that we cannot generalize all problems related to emotions, tastes and aspirations as moral problems, yet the occurrence of emotions is closely connected with morality. When reaching a certain limitation in intensity, emotions can directly become moral problems, and emotions have an enormous impact on the *qi* change in the human body.

We know that an emotion is part of the activity of consciousness. It is a special feeling and expression that occurs when *Yi Yuan Ti* combines with the change in human body *hunyuanqi*. It is the outcome that self consciousness has lost its feature of self-awareness and self mastery and rather clings to what has happened. It is a sign that a person has lost his or her own conscious self mastery and mental brightness. Once an emotion occurs, if one's self consciousness does not identify with it, the change that it brings about in human body *hunyuanqi* will be small; otherwise, the change will be intense. Intense change in emotions can cause abnormality in human body *hunyuanqi* and makes a person show a morbid state. Generally speaking, healthy emotions such as being active, aspirational, striving for the better and best, forging forward, and being optimistic and joyful can enhance a person's physical and spiritual health; on the contrary, negative emotions such as depression, pessimism, despair and distress can reduce the life force. As to unhealthy emotions like anger, sorrow, worry and fear, they can make a person fall ill.

There are many statements in the theory of traditional Chinese medicine about this. For example, in *Ju Tong Lun* of *Su Wen*, "Anger makes *qi* go upward, gladness makes *qi* flow smoothly slow, sadness makes *qi* exhausted, fear makes *qi* go downward... fright makes *qi* in disorder... thinking too much makes *qi* gather toward one point." Why can these emotions cause these changes in *qi*? What result will these changes bring about? Explanatory illustration is given later in the same book. "Anger makes *qi* reverse. Raging anger may even make one spit blood or have diarrhea, thus *qi* has flown upward. Gladness makes *qi* harmonious and aspirations accomplished, and *rong qi* and *wei qi* move without slightest hindrance. Thus *qi* becomes smoothly slow in its flow. Sadness makes the *qi* channels between the heart and the lungs tight, the lungs spread and the pulmonary lobes move upward, *shangjiao* (the upper energizer) obstructed, *rong qi* and *wei qi* not spread, and the hot *qi* stay in the body. Hence *qi* is exhausted. Fear makes one's essence lost. The loss of essence makes *shangjiao* close. The closing of *shangjiao* makes *qi* flow backward. The backward flow makes *xiajiao* (the lower energizer) bloated. Thus *qi* is stopped in its movement... Fright makes the heart (referring to the mind) have nowhere to reside, the spirit nowhere to return, the thinking in mind nowhere to settle. Thus *qi* is disordered... Thinking makes one have something at heart and the spirit has a place to return. The upright *qi* stays but ceases to move (when thinking too much). Thus *qi* become 'knotted.'"

The morbid state in the life activity of the human body caused by these emotions is usually shown in the system of the "five *zang* internal organs" according to traditional Chinese medicine. As to this, it is said in *Ben Shen Pian* of *Ling Shu*, "Those who are in states of fear, alertness or excessive thinking too much harm their spirit. With the spirit harmed, one feels dread and the life essence continuously flows out of the body. Those who are too sorrowful get exhausted in *qi* and lose vivacity. In those who are too glad and much delighted, their spirit dissipates and is not stored. In those who worry too much, *qi* is obstructed from flowing. Those in a rage are confused and ungoverned. For those in fear, their spirit is dissipated and their heart (mind) lingers around."

"With the heart: To be fearful, alerted or thinking too much harms one's spirit. With the spirit harmed, one feels dread and losses self-control; one loses weight and becomes gaunt. The hair loses its luster and the countenance withers away. One suffering from this dies in winter. With the spleen: Being unsolvably worried harms the mind. With the mind harmed, one becomes depressed and distracted and the four limbs become limp. The hair loses its luster and the countenance withers away. One who suffers from this dies in spring. With the liver: Being so sad as to touch the heart of hearts harms the soul. With the soul harmed, one becomes wild, forgetful and not accurate. Being like this is not upright. The person's *yin* withers, the tendons contract and the ribs are weak. The hair loses its luster and the countenance withers away. One who suffers from this dies in autumn. With the lungs: Extreme delight and gladness harm the breadth of spirit. With the breadth of spirit harmed, one becomes wild. The one who becomes wild has nobody in the mind. The skin becomes dried up. The hair loses its luster and the countenance withers away. One who suffers from this dies in summer. With the kidneys: Being in a rage that refuses to stop harms aspiration. With the aspiration harmed, one is liable to forget what one has said. One cannot bow or face upward or stretch and bend at the waist. The hair loses its luster and the countenance withers away. One who suffers from this dies in the third month of the summer season. Being in fear and unable to have it dissolved harms the essence of life. With the life essence harmed, one feels sore, weak, numb and withering in one's bones, and the essence of life then frequently goes down."

What has been stated above is generally emotional and aspirational changes caused by stimulation from the external world (which mainly consists of social factors), the so-called "mind unified to change *qi*", that is, abnormality in *qi* change causes disease. In addition, *qi* change in the internal organs of the human body can also bring about change in emotions. In daily life, it can often be seen that those who are unhealthy or weak in the body are usually unstable in emotion and are apt to be influenced by the outside world. It is said in *Ling Shu*, "If *qi* in the liver is deficient, one is liable to be afraid; if *qi* in the liver is excessive, one is apt to get angry." "If *qi* in the heart is deficient, one is apt to get sad; if *qi* in the heart is excessive, one tends to laugh without stop." The change of *qi* in the five *zang* internal organs and the six *fu* internal organs in the human body belongs in the domain of the human body's natural *Dao De*.

ii. The Relationship between Morality and Human Body *Hunyuanqi*

In Section I of this chapter, we have illustrated that, in the formation of *Dao De*, the human body experiences natural *Dao De*, social natural *Dao De* and social *Dao De*. It can be seen that there are different kinds of *Dao De* in the human body, with each of them existent in the human body to a different extent, while their different features all play an extremely important role in the continuance of the life motion of the human body. This is because *Dao De* is the entity and function of the whole entities of existences as well as the whole entities of humans. It is closely related to and integrated with human *hunyuanqi*.

The human body's natural *Dao De* originates from the state of the *hunyuanqi* of the zygote before the body of the zygote comes into existence. The state of this *hunyuanqi* determines that the zygote has both the general features of natural *Dao De* and the special features of the life activity of the zygote that is to develop into a human being. Formless and imageless though *hunyuanqi* is, it determines a person's life activities. After the zygote comes into form, the division and combination of cells continuously go on in the process of the zygote's development. In the division and combination in every cell, there exists the process of somethingness changing into nothingness and nothingness turning into somethingness, with *hunyuanqi* continuously produced at the same time, which determines and also nourishes human life activities.

Natural *Dao De* in the period of human fetus is in fact the entity and function of the fetus' *hunyuanqi*. *Dao De* and *hunyuanqi* are identical, and the fetus is the embodiment of *hunyuanqi* in shape of a human body. Yet after the fetus is born, consciousness activities and the notion of *Dao De* are gradually established and they have different impacts on life activities. In the process of the human body's life activities, cells are in the process of continuous division and generation. There is *hunyuanqi* when cells are divided; there is also *hunyuanqi* when cells are formed. This change is going on in the human body all the time. During this process, *hunyuanqi* is embodied in human natural *Dao* and *De* and is shown as the life force. This *hunyuanqi* changes according to human material tissues and structures, and also according to people's consciousness activities or spiritual activities. This change is shown in *ju san* (condensing and dispersing, including gathering and expanding) of the *hunyuanqi* inside the human body and also in *ju san* with the *qi* outside the human body in the natural world.

We know that in adults there exists not only the activity of consciousness in the general sense, but also evident social *Dao De* consciousness, which is one of the major factors that lead *Yi Yuan Ti* into partiality and adherence to partial reflections and opinions. The influence of *Yi Yuan Ti* upon human body *hunyuanqi* has already been comprehensively illustrated in Formation of Human Body *Hunyuanqi* (Section IV) in Human *Hunyuanqi* (Chapter IV). What is to be pointed out here is that Partial Identity *Yi Yuan Ti* will make human body *hunyuanqi* have a partial feature, thus reducing the similarity between the *hunyuanqi* of the human body and the *hunyuanqi* in Nature (particularly the primordial *hunyuanqi*). This will undoubtedly affect the human body's use of the *hunyuanqi* of the natural world.

If you do not think so, please have a look at the following example. For infants and babies in the stage of social natural *Dao De*, no obvious partiality has formed in their *Dao De* state, nor has obvious opposition between the subject and the object formed in their self consciousness. That is to say, the self enclosure feature has not yet formed and the identical feature between *hunyuanqi* in the human body and that in the natural world is relatively great. Because of this, *hunyuanqi* inside the human body and that outside it can smoothly carry out their exchange and *hunhua* of *kai he* (opening and closing) and *chu ru* (going out and coming in), which makes babies exuberant in their life force and guarantees the swift development of their spirit and flesh. However, as the baby's body is still relatively inchoate and its development is still immature, particularly so in the nervous system, it has not displayed praiseworthy abilities of doing work, but it already has great natural abilities.

It is pointed out in *Lao Zi*, "The profoundness of *De* (moral force) can be analogized to that in babies newly born. Poisonous insects do not sting them, wild beasts do not attack them, and fierce birds do not snatch them. Though weak in bone and soft in tendon, their grips are firm. Though they know nothing about sexual behavior, baby boys' penises can become erect due to utmost fullness in the essence of their bodies. They may cry all day long while showing no sign of hoarseness because of their greatest harmony." This paragraph is both a description of the newborn baby and the description of very high level in the cultivation of *Dao De*. As to the behavior of the poisonous insects, of the wild beasts and the big birds, it seems to be hard to understand. Yet reports in recent years from abroad about the wolf child and the tiger child, etc., show the possibility of such cases.

With increase in age and growth of the human body and with the transition from natural *Dao De* into social *Dao De*, various desires in humans grow strong and the notion of private possession is established. Consciousness like this works on the human body's *hunyuanqi* with an enclosure effect and restricts the *hunyuanqi* of the human body. It makes the human body have fewer channels to contact the *hunyuanqi* in the natural world and the exchange of *hunyuanqi* between humans and nature decreases, therefore human life force is affected. Besides, different kinds of desires and emotions in humans also consume a lot of *hunyuanqi* and affect the health of the human body.

In the normal course of reasoning, in adults who have developed into maturity in all aspects, their various bodily functions should be greatly enhanced. Yet in reality, their adaptability in many aspects and their ability to defend themselves from the stimulation of bad external environment are decreased. For example, babies who wear open-seat pants play outdoors in winter and they do not feel cold. Usually they will not be frostbitten or stricken with illness while the grow-ups cannot endure this (here not in the sense of ritual but in the sense of bearing the coldness). Another example is that some people are selfish in their social *Dao De*. They see nothing but their personal interests and know only to take care of themselves while never bothering themselves with even a tiny thought of others. They cannot endure even a tiny loss in their personal interests in society. At home their parents and spouse must all obey them and center on them. This kind of person usually suffers from bad health, for in their spiritual state, they enclose themselves with the prestigious "me", which obstructs the association between humans and Nature and the connection between humans and society. The *hunyuanqi* from the outside world cannot come into them while the *hunyuanqi* inside their bodies is exhausted as soon as emotions occur, so they cannot receive supplements from the outside world. Therefore, this kind of person often falls ill and even a slight illness in them cannot be easily cured.

It is true that the cultivation of *Dao De* does create a condition and lays a foundation for humans to absorb more *hunyuanqi* from Nature. But this cannot be equated to that cultivating *Dao De* alone is everything. One reason is that a high level of *Dao De* cultivation is not attained at once. In the stage when being for the public and being for the private are struggling in the mind in the process of *Dao De* cultivation, a person tends to be somewhat troubled instead of enjoying peace of mind. Another reason is that the degree of healthiness that a person can reach is determined by many factors, e.g., the power of will, the degree of one's initiative to become stronger as well as change in the life environment. This will be talked about in the theory of the enhancement of life (Chapter VII).

iii. The Relationship between Morality and *Qigong* Practice

From the comparison between babies and adults mentioned above, we can observe that the feature of partiality and adherence to partial reflections and opinions in humans has an immeasurable hindering effect upon the communication and *hunhua* of human *hunyuanqi* with Nature *hunyuanqi*. Therefore, to improve human species-being or nature through *qigong* cultivation and practice, the level of human morality or *Dao De* must be improved so that humans can gradually evolve from partial social *Dao De* into social free *Dao De*. This is because the less the feature of partiality exists in a person's *Dao De* or morality, the less bias and partiality exist in this person's self and the less its fixed effect on the mode of this person's Reference System is. This is rather essential for receiving Nature *hunyuanqi* because various modes in the Reference System directly affect the combination of the information in *Yi Yuan Ti*, and the information constructed according to the mode of a partial identity Reference System is also partial. This partiality feature screens the *hunyuanqi* outside the human body, making it difficult to enter the human body; and what's more, partial information can enshroud *Yi Yuan Ti*'s initiative feature and brightness and restricts its activities. It is only when the commencement force of the partiality in the Reference System (which concerns the force of the partiality of one's familiar intelligence self) is reduced that the influence of the Reference System upon *Yi Yuan Ti* can be greatly decreased and the void, bright and clear entity and attribute of *Yi Yuan Ti* that is full of initiative can be brought into full play. Hence it is easy for *Yi Yuan Ti*, this *hunyuanqi* simple in structure, to enter human *hunyuanqi* or *hunyuanqi* in the external world and form the state of "spirit entering *qi*". Meanwhile, the *hunyuanqi* in Nature, as the partial screen is gone, can easily enter the human body and be used by the human being.

People in ancient times in China also attached great importance to this. It is said in *Guan Zi*, "Therefore, this *qi* (which refers to *Dao*) cannot be stopped with force but can be settled with moral force (*De*); it cannot be called with voice but can be greeted with mind. Guard it with respect, lose it not, and this is called the accomplishment of virtue (*De*)." And "The growth of human life must rely on happiness. Worry loses the principle of *Dao* and anger loses propriety. Worry, sorrow, glee and anger make *Dao* have nowhere to rest. Pacify love and desire and rectify the chaos you meet. Neither lead nor aid it and blessings will naturally return. *Dao* comes by itself, and you can solve problems with it. Being peaceful, you get it; being irritated, you lose it. The *qi* with initiative is in the heart, and it comes in and goes out alternately. So fine is it that it has no within; so grand is it that it has no without. The reason why you lose it lies in restlessness. If the heart can reside in serenity, *Dao* will settle by itself."

This quotation mainly talks about the idea that *Dao* (which is equivalent to what we call *hunyuanqi*) is ubiquitous and can nourish the human body and heart when it enters the body. However, *Dao* "cannot be stopped with force" but can only "be settled with moral force". Therefore, when people can kind-heartedly treat themselves with composure and also respect and love others, their heart is adjusted and *Dao* (*hunyuanqi*) in nature can be used by them. The reason why people cannot get *Dao qi* is because of the disturbance of their various emotions (worry, sadness, glee and anger). As for this, it is even more clearly and comprehensively illustrated in *Geng Sang Chu* of *Zhuang Zi*:

"To become noble, wealthy, distinguished, authoritative, to have fame and to have fortune— these six things make people confused in consciousness and change in aspiration. Appearance, behavior, color, reason, diction and (postnatal) mental activity—these six things betray the heart. Detest, likes, glee, anger, sorrow, gladness—these six things encumber virtue. To get away, to get close, to take, to give, to adhere to the idea of having knowledge and to adhere to the idea of having capability—these six things block *Dao*. If the four sixes do not hold sway in the chest, one will be upright. Being upright, one will be calm. Being calm, one will be bright. Being bright, one will be void in the mind. Being void in the mind, one can be in a state of no action yet achieve all with no exception."

The ancient Chinese people called what we describe today as *qigong* practice the cultivation of *Dao* and the accomplishment in such cultivation attaining *Dao*. It is in fact that through *qigong* practice and the improvement of the level of *Dao De* or morality, one is not only able to perceive the entity and attribute of *Dao*, but can also retain the *Dao qi* (*hunyuanqi*) of the natural world for one's own use so as to attain greater freedom.

Section III *Dao De*, Self and Human Nature

The identical features between humans and Nature have been elaborated upon and the truth that humans and all other existences originate from the same *qi* has been pointed out in The Theory of Whole Entity (Chapter III). However, humans are after all the most intelligent beings of all existences and the most advanced creatures on the earth. What is the difference between humans and all other existences then? As early as the period of the Warring States, explicit illustration about this was already given by Xun Qing: "Water and fire have *qi* but not life. Grass and trees have life but not perception. Birds and beasts have perception but not righteousness. Humans have *qi* and life and perception and righteousness, so they are the most precious under heaven."(*Wang Zhi* in *Xun Zi*) This is to say that humans have not only the characteristics of all existences in nature (both animate and inanimate), but also righteousness in the domain of ethics and morality. This agrees with the philosophy of dialectical materialism in which it is believed that humans are both humans of nature and humans of society. Marx once explicitly pointed out, "The human essence is no abstraction inherent in each single individual. In fact, it is the sum of all social relationships," and "it is free conscious activity that constitutes the species-character" of the human species. The former quotation emphatically illustrates human social characteristic, and the latter emphatically states the natural (or biological) feature of human beings. In Synopsis of Morality (*Dao De*) (Section I) in this chapter, we again state that humans have natural *Dao De* and social *Dao De*. So far it is not difficult for people to see that the life activities of specific individuals are relatively complicated: Both natural attribute and social attribute, both natural *Dao De* and social *Dao De* exist. Facing such complicated humans, people cannot help but ask: What is human nature? How does a human being make such complicated aspects as the natural and social attributes and the natural and social *Dao De* unify? Psychologists have made careful studies into this question but they have not obtained a perfect explanation yet. Philosophers of dualism believe that spirit and flesh are two independent existences. We believe that the unity of a human being is realized through self in the individual. Now let us begin our illustration with *self*.

I. A Brief Introduction to *Self*
i. What Is *Self*

Although the term *self* is not unfamiliar to us, e.g., "self appreciation", "self intoxication", "self criticism", "self introspection", "self encouragement" and so on, rather few people can give an answer to how to recognize self and what the nature of self is. Rather few people have done research into it. Investigation shows that further studies about self have been made only in Buddhism and psychology. In order to understand self more deeply, let me first introduce the knowledge of self in psychology and Buddhism.

i) The Recognition of Self by Psychologists

It is pointed out in *Dictionary of Psychology* by Mr. Zhang Chunxing that self refers to the substantial entity of the existence of oneself of which one or an individual is aware. It includes bodily and psychological characteristics as well as the psychological activities and various processes engendered by these characteristics. From the degree of how much an individual or one is aware of oneself, there is also the difference of *subject self* and *object self* in a person's *self*. The *subject self* is the self that takes actions, the observer, namely, the self in the leading position; the *object self* is passive, the self being observed, and is the recipient. Generally speaking, an individual's consciousness of the subject self is relatively weak while the consciousness of the object self is relatively strong. The target of research in psychology is also in the aspect of object self. In the ancient Chinese saying "Every day I (*wu*) examine myself (*wu shen*) many times", the former *wu* is the subject self, while the latter *wu* is the object self.

Freud, who is famous for psychoanalysis, divided self into three levels. One is *id*, which refers to instinctive impulse and desire; one is *superego*, which refers to the ideal self; the other is *ego*, which represents the realistic part of a person's character and is a regulator between id and superego. The functions of ego are: a) to satisfy basic needs so as to maintain the individual's existence; b) to regulate the primitive impulse so as to meet the requirement of reality; c) to control the primitive impulse that is not accepted by superego and to maintain the balance and harmony between id, superego and ego.

ii) The Recognition of Self in Buddhism

In Buddhism, it is believed that there is not such a self as believed by average people. Average people feel that there is a self existent in their body and having the commanding function. They take their own human body as something that truly exists and this is called "the human self". They take their spirit and flesh as something that truly exists and this is called "I myself". All this is considered in Buddhism as the result of confusion and delusion that disagrees with Buddhist theories. Thus many theories and approaches have been put forward as to how to break away with "I myself". For example, it is said in *Zhi Guan*, "To observe with wisdom, there is actually no me. Where is my *self*? The head, the feet, the limbs, the joints—I look at them one by one and can not find me myself. Where are humans and all other creatures? Humans and all other creatures come from the mechanism of what they once have done and gather falsely and emptily together. They are born of multitudinous predestined karmas and have no inside commander as if dwelling in empty pavilions."

It is stated in *Yuan Ren Lun*, "The look of body and bones, the heart of worries and thinking—they come from where there is no beginning because of the force of predestined karmas. They are born and dead with each and every mental activity and seem to be immutable and permanent.

This is what the mortal and ignorant do not know and they regard these as their selves."

iii) The Concept of Self in the Theory of *Hunyuan* Whole Entity

According to *Zhineng Qigong* Science, humans are an objective existence with consciousness and therefore self in human beings does exist. But the recognition of self in *Zhineng Qigong* differs from that in psychology.

According to *Zhineng Qigong* Science, self in a human being is such a kind of genuine existence: When the activity in *Yi Yuan Ti* has established its connection with the life of the individual human being and can maintain its independence in its interaction with existences in the external world, this state of life of the human body is one's self. The reflection of this state in *Yi Yuan Ti* is self consciousness. Self consciousness is the synthesis of *Yi Yuan Ti*'s functions, which are based on the human body's life activities, and *Yi Yuan Ti*'s Reference System. It is the controller of human body's life activities (both macrocosmic and microcosmic) and can know the objective world, the individual human being himself or herself, *Yi Yuan Ti*'s own motion as well as the subject *Yi Yuan Ti*. It can merge and change with all kinds of *hunyuan* entities in the universe. What is more important is that self can carry out independent motions (such as thinking, imagination and aspiration) in a certain sphere.

If analyzed rigorously according to the theory of *Yi Yuan Ti*, *Yi Yuan Ti* as well as its functions of receiving, sending and processing information is one's true self, while the Reference System in *Yi Yuan Ti* is not one's self. This is because the Reference System is but the tool or mode used in the process when *Yi Yuan Ti* exercises its functions (and we know that the content in an individual's Reference System can also be changed, let alone the Reference System in human beings' entire development process). What we mentioned about *mistakening the false owner for the true internal master* in The Theory of Consciousness (Chapter V) refers actually to this. On the other hand, however, we cannot deny the status of the Reference System as self because, so long as the functions of *Yi Yuan Ti* are truly exercised and can be realized, *Yi Yuan Yi* must connect with specific things or existences and carry out a certain degree of mergence and change. Hence it must follow a certain mode. The content separated from any mode is unimaginable. Therefore, the functional activities of *Yi Yuan Ti* as one's self and the activities of the Reference System are usually hard to divide.

It is true that the mode of the Reference System indeed has, to a certain extent, an effect similar to "frequency locking" on *Yi Yuan Ti*'s functions as that in synergetics, which confines the functions of *Yi Yuan Ti* to a relatively narrow scope, thus making people show stubborn Atmengraha (or adherence to the normal-intelligent self). We think that it is a must to break through the restrictions of the Reference System and the approach is to enlarge the domain of the Reference System as much as possible. This is not only to broaden the extent of familiar intelligence, but to broaden the scope of intelligence into the realm of super intelligence. When super intelligence becomes dominant in the Reference System, the self of humans will naturally enter into a new realm.

ii. The Relationship between Self and Life of an Individual

Although we have illustrated at different levels the human whole entity features as well as the functions of the human spirit in the human body in The Theory of Whole Entity (Chapter III), Human *Hunyuanqi* (Chapter IV) and The Theory of Consciousness (Chapter V), we have not clearly explained the interrelationship between the self in charge of an individual's life, i.e., the embodiment of willpower, and the individual's life activity. After we have become clear about the connotations of self, it is relatively easy to explain their relationship. What is the relationship between the two of them, then?

i) An Individual's Life Activity Is the Foundation for the Existence of Self, which Is Shown in:

a. *Yi Yuan Ti* comes into being when the nervous system has developed to a certain stage. This is to say, without the life activity at the level of cells in the individual, the formation of *Yi Yuan Ti* is impossible. On the other hand, *Yi Yuan Ti* also results from the interaction between the life of an individual and social humans. Otherwise, the nervous system cannot accomplish its development process.

b. *Yi Yuan Ti*'s Reference System is established in the process in which each tissue and organ of the human body (particularly the sensory organs) exercise their own functions and continuously receive and reflect information from the external world (including information about the human body itself). That is to say, without the macrocosmic life activity of each tissue and organ (which refers to the receiving and the input of information), the formation of the Reference System is impossible. We should keep in mind that in the Reference System that is formed after birth, there are both consciousness structure and mode about science and arts and morality consciousness mode that belongs in social consciousness.

c. Self is established in the life activity of the individual. When the Reference System is preliminarily established in *Yi Yuan Ti*, with the reinforcement of the association between *Yi Yuan Ti* and the individual's life activity, difference between the flesh upon which life of the individual relies and the natural environment gradually appears in *Yi Yuan Ti*'s Reference System, which further separates the individual from the environment. Although such clear recognition has not formed in the baby's consciousness, change as this is already going on naturally in the Mind Oneness Entity (*Yi Yuan Ti*). Once the individual's own life activity is divided from the background of the natural environment in *Yi Yuan Ti*'s Reference System, *Yi Yuan Ti*'s Reference System will exercise its commanding function upon the individual's life activity. Hence self is preliminarily formed. Generally speaking, in the entire life process of an individual, change in a certain extent will happen in certain details and a person's self will divide into ideal self, realistic self, natural self and social self and so on. Yet all this is established on the basic mode of the Reference System. Thus it is relatively difficult to cause fundamental change in the Reference System. The basic mode of the Reference System will have profound effect upon the character of the individual, on the individual's consciousness of self, notion of self, self-respect and notion of value. However, under a certain circumstance, self can be changed. Cultivation of consciousness in *qigong* practice is put forward according to this point.

ii) An Individual's Life Activity Is the Means by which the Activity of Self Is Realized

During the process when self is being formed as well as after its formation, self sends information through *Yi Yuan Ti* to the life of the individual and conveys multitudinous *hunhua* content in *Yi Yuan Ti* to the life process of tissues in each part of the human body. Particularly after the formation of human consciousness activity, it becomes evident that the life activity of the individual goes on according to the instruction of consciousness activity, and this is the fundamental difference between humans and animals. Just as what Marx said,

"Animals directly identify themselves with their life activities. They do not separate themselves from their life activities. An animal itself is just this life activity. Humans, on the other hand, make their own life activities the target of their will and consciousness. A human being's life activity is carried out with consciousness. This is not the prescribed feature to integrate humans directly with their life activities. Life activities carried out with consciousness directly distinguish humans' life activities from animals' life activities. It is because of this that humans are a species existence, or in other words, it is because humans are a species existence that they are existences having consciousness, that is to say, their own life is to them their target. It is only on account of this point that their activities are free activities."(*The Complete Works of Marx and Engels*, Vol. 42, p. 96)

As to the commanding effect of consciousness on life, it has already been elaborated upon in The Theory of Consciousness (Chapter V) and will not be repeated here.

It should be pointed out that, as the mode in *Yi Yuan Ti*'s Reference System grows gradually more complicated, human consciousness activities also gradually improve and the position of self consciousness gradually becomes explicit and fixed in consciousness. This represents great progress in the development and change of *Yi Yuan Ti* and self because it makes the entire human species take a step forward toward human civilization. However, from another aspect, the reinforcement of the Reference System forms the imprisonment of self and confines the free feature of *Yi Yuan Ti* to the frame of (familiar intelligence) self and the Reference System.

From the analysis above it can be seen that self, on the one hand, commands life level activity in the individual through *Yi Yuan Ti* and, on the other hand, directs the individual human to communicate with society through the mode of morality in the Reference System. In summary, the natural attribute and the social attribute of a human being are organically combined into an integrated organic entity by means of self in the realm of consciousness.

iii. Ideal Self

Ideal self is the realm of self in which the human character of being true, good and beautiful is achieved through cultivation in all respects. It is also the ideal realm achieved through the cultivation of *Dao De*. In *Dao De* of *Zhineng Qigong* Science, it is the self established upon social free *Dao De*. In this state of self, truthfulness, goodness and beautifulness are unified.

We know that *true* is the original state as far as objective existences are concerned. For human beings, it is their correct reflection and recognition of objective existences, including reality of society and of natural things. When the true state is described, rational logic concepts are usually needed as the medium and this is science today. When a truthful psychological state is applied in treating people, it is the display of honesty. Yet there are two kinds of situation in the so-called true state described here: One is the truthfulness that conforms with or is beneficial to life activity, namely, the truth as found in "the heart of a newborn baby" which is generally thoroughly connected with kindness. The other is the direct expression of unhealthy emotions in the activity of one's inner world, which, though also an honest expression, yet may not be kind. This is because kindness refers to the consciousness and conduct in the individual that are of benefit to the development of society and other people (including oneself). In *qigong* science, kindness refers to the consciousness and conduct that conform to the laws of the human-and-heaven whole entity and therefore are beneficial to the whole entity in which human and heaven (nature) are one. This requires not only correct recognition of the laws of the human-and-heaven whole entity, but also the application of these laws in practice. Such practice is both the truth that correctly reflects the objective world and the good that is of benefit to humans and society. The true and the good are here unified. Social free *Dao De* is just a good exhibition of this state.

Generally speaking, truthfulness and goodness are closely related to beautifulness because the good that is shown from the original natural true state must be beautiful. Beauty is the manifestation of harmony and can give people a feeling of joy. According to the theory of *hunyuan* whole entity, the true, the good and the beautiful in humans are the display of the whole entity feature of self in different aspects, and *beauty* is the harmonious display of the whole entity of self in form. Since harmony is full of healthy information and can arouse human life force, beauty or beautifulness can bring a joyous feeling to people and make them feel vivacious and energetic. *Kindness* or good is the exposition of the outward good effect of the whole entity of self. As such outward effect is the embodiment of the motivation and the conduct conforming to the feature of the golden mean and is helpful for the formation of *qi* of great harmony in humans, goodness can encourage people with an inspirational power and makes people have a feeling of friendliness and cordiality. *Truthfulness* is the manifestation of self in nature. It results from the fact that *Yi Yuan Ti* has broken through the confinement of the partial identity Reference System and all of *Yi Yuan Ti*'s functions have fully displayed, or in other words, truthfulness is the result that self has become thoroughly emancipated.

Generally speaking, self is under the restriction, control and directed guidance of the partial identity Reference System and cannot fully exercise its genuine self-mastery function, which consequently displays the partiality feature in average people. Once the partial identity Reference System is overcome, *Yi Yuan Ti*'s holistic functions will take the leading position. Then, the language and behavior of humans will be not only kind but also beautiful. Beauty in language, in behavior and in the heart as frequently referred to by people in daily life will have become truly realized. Self will have reached the free realm of not overstepping the boundaries of right while not consciously abiding by them. All this is the inevitable manifestation of self having attained social free *Dao De* and is the unity of the true, good and beautiful that is elevated to a spiral upward compared with self in the stage of infanthood and babyhood.

Let me mention a point in passing here, that is, self exists not independent of society. We have previously mentioned that self is an objective existence, and here we also say that the ideal self "has reached the free realm of not overstepping the boundaries of right while not consciously abiding by them". This, at first sight, seems to be similar to education in existentialism which holds the subjectivity of "self-accomplishment" as its starting point. In fact, they are completely different. In existentialism, it is emphasized that humans, when choosing their own nature, are absolutely free and that humans are free from the influence of social environment and class status, which is absolutely impossible for people who are under the influence of the partial identity Reference System. Moreover, it is also believed in existentialism that natural environment and society are antagonistic to humans. All this is not only entirely different from the ideal self we talk about but also poles apart from the theory of *hunyuan* whole entity. As for the freedom that we talk about, it is not only the result of the recognition and the mastery of the laws of life motion of the holistic human being, but also the new realm that people enter when super intelligence is developed and tapped in them and when they have broken away from the partial identity Reference System. This freedom is the freedom of activity that human beings can enjoy to the utmost extent in the natural world. It is the fundamental display of the liberated human species-being or nature.

II. Current State of the Species' Nature of the Humans Race
i. The Character of Human Species-Being as well as Its Alienation and Distortion

As preciously talked about, human beings began to have consciousness and language after they separated themselves from the kingdom of animals. Very simple though the consciousness was in the beginning stage, it already became the commander of the life activity of human flesh and was able to freely issue orders to the body, making it follow the commandment of consciousness. For example, consciousness gave the instruction of labor and the body accordingly carried out work and labor; consciousness gave the instruction to eat and the body accordingly took food; consciousness gave the instruction to have fun and play games and the body had fun and played games accordingly; consciousness gave the instruction to contact people and the body accordingly carried out the instruction to make contact with people, and so on and so forth. In brief, whatever instructions consciousness gave, the body carried out corresponding life activities. The life activity of the human flesh has become the means to accomplish the life activity of consciousness.

Although the generation of human consciousness activity results from *Yi Yuan Ti*'s synthesis of the information in the internal and external environment of the human body, the process from the formation of mental activity to sending of the instruction is free and meanwhile perceivable by oneself. This is what Marx said, "...free conscious activity constitutes the species-character of man." This species character of humanity seems to have been shown to a certain extent in primitive commune society. But that was a display at a low level, the low-level free and conscious state in the period when various functions of humans were not fully displayed.

With the complication of labor and the improvement in productive forces, not only do tissues and organs in each part of human beings evidently improve in their functions, but great changes also take place in their shapes and forms. For example, the sense of hearing of human ears has acquired the ability to distinguish a variety of speech sounds and music; the visual ability of human eyes has acquired the capacity to distinguish complex colors and shapes; the senses of taste and smell have acquired the abilities to distinguish different kinds of tastes and smells; the hands have acquired the ability to perform exquisitely complicated skills; human consciousness has acquired highly efficient thinking ability and so on. In the normal course of reasoning, with the improvement in humans' various kinds of abilities, the character of the human species should be further manifested. It is not the case in reality, however.

When productive forces improved to a certain degree, the system of private ownership came into being. This change is a negation of the primitive commune and imprinted in human *Yi Yuan Ti* a mark of private ownership. Private ownership is to possess, to own the target object. We know that various human senses are the display of the essential features of different sensory organs. They are the reflection in *Yi Yuan Ti* of the mutual *hunhua* of human *hunyuanqi* with the *hunyuanqi* of existences in the external world. Such processes all reflect the free and conscious feature of human beings and are actually a colorful integrated *hunhua* process.

However, when *Yi Yuan Ti* is filled with the desire to own or possess privately, human behavior begins its directed activity to take into possession the target objects, which restricts the free feature of *Yi Yuan Ti*'s holistic functions: On the one hand, *Yi Yuan Ti* is no longer free in sending instructions but has to submit to the need of the desire to possess. At this time, labor, this life activity which is the special activity of humans, is no longer the means to satisfy the activity of consciousness (or life) of the individual, but the means to make a living instead. On the other hand, human senses are no longer the complete feeling of holistic time and space, but the senses restricted to the special features of related sensory organs. The content of senses turns from their original rich and colorful state onto the simplistic, destitute track, which results in the senses of our familiar intelligence.

What is more important to point out is that, when the desire to possess (the private desire) has taken up *Yi Yuan Ti*'s Reference System, human senses and actions are enshrouded with a veil of possessing, which deprives human senses and actions of their original human meanings. For example, when a person comes into possession of something, this something has also taken possession of this person. The information about good *hunhua* effects between humans and objects loses the objectiveness and is replaced merely with the possessive desire. This makes the person who owns this something more frequently haunted by the consideration of how to take good care of this something that he or she likes, and humans enter the realm that is encumbered with things and are often troubled with worries. On the other hand, in order to pursue perfect spiritual reliance, humans create gods and endow them with incomparable powers and then beg blessings from them. The free character with true-self awareness in the nature of the human species is thus alienated and distorted in the physical and spiritual realms.

Although human beings are loud in the slogans they put forward after capitalism has been developed—freedom, equality and fraternity, the fundamental reason that causes the deviation and distortion of human nature—private ownership and theology, instead of having been eradicated, has developed further. Therefore, the distorted human species' nature has not been "restored", but has become, in a certain degree and to a certain extent, further intensified. The increasing desire of chasing after things, the barren state of human spirit, the phenomenon of homosexuality and the communal suicide of the Peoples Temple in America in the 1970s are obvious evidence.

ii. Human Beings Today Still Do not Know the Character of Human Species-Being

Human beings today have acquired rather deep knowledge about Nature, e.g., they have conducted profound research into and have obtained profound knowledge about the microcosmic world counted in nanometers and the macrocosmic world counted in light years, and spaceships have not only landed on the moon, but are also flying into the immensity of outer-space. However, as far as the human body's life activity is concerned, only the basic structure (the double spiral structure) as well as its regeneration process is known so far and still very little is known about its holistic functions. It is so with the functions of a cell, and even less is known about the holistic functions of the entire human body. As for the proposition of the character of human species-being, it is rarely touched upon, except that Marx once illustrated it from the level of philosophy of natural science. In reality, researchers still study human body's life activity at the basic level.

i) Some scholars study the human body's life activity from the perspective of physics and have established biophysics, which includes biodynamics, electrobiology, biomagnetism, and also biological optics and bioacoustics recently put forward by some researchers. As great achievements have been attained in biophysics, some scholars have the inclination to conclude that human nature belongs to the level of physics. We think that a certain amount of physical motion indeed exists in the human body, but these physical changes in the life process of the human body differ greatly from the physical changes in the natural world. For example, biological electricity, biological magnetism, biological light and biological sound, etc., in the human body differ from the electricity, magnetism, light and sound in the physical world in general in their features and also in their mechanisms of how they are generated. Such physical features in the human body have their own biological features. The biophysical changes in the human body are just the fundamental activity in the human body's life activity. It is still quite far from the essential character of human nature.

ii) Some scholars study the human body's life activity from the perspective of chemistry. They have established biochemistry and then molecular biology, molecular cytology, chemical embryology as well as molecular genetics that is combined with genetics, and so on. As these fields of research have already studied life phenomenon to the depth of the molecular level and have revealed quite a lot of laws of change in the process of life activity, some scholars come to the conclusion that the nature of human life is biochemical motion. We think that biochemical motion indeed exists in the motional process of the human body, but it also exists in other organisms. The biochemical motion in some species of higher animals even resembles the biochemical motion in humans. Can we therefore equate humans with animals because of this? It is evident that we can't since people today commonly believe that humans have fundamental differences from animals. Biochemical changes are just the chemical characteristics of organisms. They are not the life motion itself.

iii) Some scholars study human life activity from the perspective of psychology and have established theories of different models of activities, e.g., the physiopsychological model, psychodynamic model (also called the model of psychoanalysis), behaviorist model, cognitive model and humanistic model. Although in the study of psychology they all emphasize that "psychology is a scientific and humanistic exploration", in fact, the basic content of what psychologists study is merely the various phenomena accompanying the process of psychological activities, including changes at the microcosmic level and the macrocosmic level. This is just the objectivity instead of the subjectivity of self in the realm of consciousness, whereas human species-being lies critically in the subjectivity of self. Of course, the objectivity should not be neglected either. Therefore, studying phenomena accompanying the process of psychological activities still cannot give the answer to the character of the human species-being.

In sum, human beings' science today has just recognized some, but not all, physical and chemical phenomena in the human body's life activities as well as the form and structure at the level of biology. Even in humanistic psychology, only simple analysis has been made at the level of familiar intelligence about human psychology (consciousness), and it has illustrated that the fundamental nature that distinguishes humans from animals, aside from their free will, is mainly humans' driving force of self-realization. This proposition bears some similarity to human species-being as described by Marx. However, humanistic psychologists have neither made detailed analysis of human species-being at the high level of philosophy of science as Marx did, nor have they demonstrated this nature from the standpoint of science, not to mention the revelation of the essence and laws of human nature from the altitude of super intelligence. What exactly are the nature and the laws of the character of human species-being, then?

III. Morality Is the Internal Grounds for the Species-Being of Humanity

When we have understood the nature of self, it is relatively easy for us to analyze what was said by Marx "free conscious activity constitutes the species-character of man" and "Human nature is human social connections in the real sense."(*The Complete Works of Marx and Engels*, Vol. 42, p. 25) Human nature is in fact all kinds of human life motions generated after human social connections are internalized into the Reference System. The species-character of such activities in humans are free and conscientious activities when self has been established, and all this is directly related to morality (*Dao De*). In order to further illustrate the relationship between morality and the human species' nature, analysis is to be made from two aspects.

i. *Dao De* Consciousness Is the Deeper Level Activity in Consciousness Activity

We have summarized content of consciousness activity in normal-intelligence people into the four aspects of consciousness of science, consciousness of arts, consciousness of morality and consciousness of *qigong* in The Theory of Consciousness (Chapter V). If analyzed further, the content of consciousness in normal-intelligence people can be divided into philosophical, political, religious, scientific, artistic and *Dao De* (morality) consciousness and so on. The activity of morality consciousness is in the deepest stratum in these consciousness activities. This is because:

i) *Dao De* Consciousness Is the Relatively Early and Deep Imprint in *Yi Yuan Ti*'s Reference System

It has been pointed out in The Theory of Consciousness (Chapter V) that *Yi Yuan Ti* in the later stage of human fetus and the early stage after the baby is born belongs basically in the domain of Pure Original *Yi Yuan Ti*. That is to say, *Yi Yuan Ti* over the two periods is still only at the level of instincts. The newborn individual at this moment is only a natural human being. We know that, in the evolutionary history of the human race, the evolution from a higher animal with biological impulses into a human being has experienced a very long time of labor, which enables the life functions and the forms and structures of tissues to continuously evolve and helps form consciousness and language and then morality (*Dao De*). The entire development process for a newborn individual to grow from a natural human being into a social human being, however, takes only a short period of several years. The newborn individual accomplishes the development of life functions and forms of tissues under the guidance of consciousness development.

In the formation process of consciousness, the content of morality is the earliest to appear, or in other words, *Dao De* is a relatively early and deep mark impressed in *Yi Yuan Ti*'s Reference System. As it has been stated in the formation process of *Dao De*, when a baby is able to show gladness to its caregiver who gives it either breast-milk or other kinds of milk, this shows not only that the baby has combined its earliest functional ability of being able to distinguish (which is the primary foundation of *Dao De* or morality) with its emotions of liking and disliking, but also that the earliest established in the baby's realm of consciousness is the consciousness activity of dealing with the relationship with people. And this is also the beginning for the newborn baby to establish its self.

From this it can be seen that the baby's behavior in the domain of emotions mentioned above means that one's self begins to take command of one's own life activity and that self begins to establish relationship with people around. That is to say, self is established on the basis of the life activity and morality of the individual from the very beginning. Consciousness and language of the individual are formed in the individual's development process thereafter. Although morality is a relatively early imprint marked in the Reference System of *Yi Yuan Ti* in the process when the individual's consciousness is being formed, complete maturity of morality is achieved relatively slowly. This is because human beings' morality is still in an advancing and changing process, whereas language and related consciousness activity are relatively mature.

ii) The Position and Function of *Dao De* in Consciousness Activity

With the newborn individual's development, command of consciousness over the individual's life activity (e.g., command over body movement) becomes gradually habitual and forms a natural process. As such commanding activity can already satisfy self's needs in life, this function then stays at the level of exercising control over macrocosmic life activities. The relationship between consciousness and morality activity is quite different, however. Humans live in complicated society, and they must follow certain norms when dealing with relationships between people. They must make *Dao De* judgment, give *Dao De* instruction and carry out *Dao*

De conduct at all times in an environment that is in continuous change. Therefore, morality (*Dao De*) is gradually reinforced in consciousness and gradually becomes the decider of people's motivations and aims, the supervisor and judge of conduct and the fulfiller of the value of life.

As it is well known, conduct results from the control under consciousness, while the norms of conduct that people must follow in making social contact belong in *Dao De*. In other words, all people's social connections, though the content may differ in tens of thousands of ways, must be restricted by the notion of *Dao De*. Morality consciousness activity is the fundamental point and the driving force of general consciousness activities. Meanwhile, it also provides them with explicit prescriptive guidance. Morality consciousness belongs to the deep level consciousness activity of "The heart in the heart" as the ancient Chinese people said. The structure and mode of *Dao De* consciousness in the fundamental stratum of the Reference System take an extremely important position in all the activities of consciousness. As human life activities go on against the grand background of social activities and each and every word and action of humans needs the involvement of morality consciousness, morality exerts a rather important effect on self that is in control of one's life activity. There are no exceptions whether it is in the formation process of self or in the process when self performs its duties.

ii. The Function of *Dao De* in the Evolution of the Human Species' Nature

It is true that morality is not born of thin air. Instead, it is determined by the mode of production and the way of life and reacts rather strongly upon them at the same time. Morality always advances simultaneously with human beings' material and spiritual civilization. When reformation takes place in the relations of production, the notion of morality corresponding to the new relations of production will become contradictory to the old relations of production. When the new notion of morality becomes the norm in guiding people's social contact, society advances, civilization of human beings also advances, and the nature of the human species accordingly strides toward the free and conscientious realm. Human beings' entire development history is in fact an evolutionary history of human nature. Changes in politics, economy, philosophy, religion, science, morality, etc., in the long historical process of human beings' evolution from primitive humans to what humans are today not only are all the signs of the evolution of human nature but also all affect the evolution of human nature. The effect of morality on this evolutionary process is particularly outstanding. This will become evident at a glance if we take a look at the characteristics of human beings in different historical stages.

Primitive humans at the age of barbarism, though they formed only uncivilized consciousness in that early stage of human beings, already surpassed the contradictory relationship existent between animal groups and animal "families" (the association established between male and female animals). They overcame the destructive feature brought by male animals to the animal group in the mating season when they repelled other male animals. They began to show the special species' nature of human beings—social connections: Humans must satisfy not only their physiological needs, but also their consciousness needs at the same time. Therefore, they must combine into society.

People who wanted to maintain their subsistence under the conditions of very low level productive forces and extremely unfavorable environments had to labor together to absorb life materials and share achievements of labor of what they could merely obtain, and this is the primitive commune. The relatively simple and innocent consciousness state at that time was a reflection of such kind of objective existence. The notion of *Dao De* as part of consciousness activity was for the collectivity of the clan and a state of being all for the public was displayed. Even sexual relationship between men and women was communal marriage which showed the feature of "for the public" at that time. Such *Dao De* reflected the character of the species nature of primitive humans who were still uncivilized and it belonged in the domain of natural social *Dao De*. Different functional abilities of the human body were also in an undeveloped state. According to *qigong* science, this state was "body and spirit in one" at a low level. Christopher Columbus discovered that the Indians (in clan communes) in South America had the characteristics of robust body and vigorous actions and this is a convincing proof. *Yi Yuan Ti* of human beings during this period was in the primitive self *Yi Yuan Ti* stage.

With the improvement in the forces of production, private property in the system of private ownership gradually appeared and the exploiting and exploited classes began to form in society. Although slave society deprived most people of the freedom they once had in the period of primitive commune, it is yet a kind of progress compared with the barbaric age when captives were killed. What is more important is that it created a condition for other people to engage themselves in mental labor. This is human beings' even greater progress. From then on, human consciousness activity became gradually complicated and alienated, and the free and conscientious species consciousness of humans started to suffer from the compulsion of living and the check of the desire to possess. For example, slaves were possessed by slave owners and were forced to work and they lost their freedom; slave owners were, on the other hand, driven by the possessive desire and their consciousness activity began to incline towards the realm of knowledge. This led to the second social division of labor in human beings' development history, with which came class differences, hence the fundamental change in the notion of morality, and the notions of private ownership, hierarchy, family and marriage (monogamy), etc. were preliminarily established. All this reinforced further development of the nature of the human species. The tapping and development of wisdom, for instance, though it was at the price of restraining super intelligence, did at that time make human beings take a progressive step forward toward the division of manual and mental labor, toward the enhancement of the functions of consciousness and better command of spirit over life activity. The morality and ethical notions in the integrated ritual system in the Zhou Dynasty played a particularly immeasurable accelerating role in the enhancement of the species' nature of the Chinese nation. The Contending of a Hundred Schools in the Spring and Autumn Period and the Warring States was the Golden Age in ancient Chinese civilization, and this has innumerable subtle links with these notions in the integrated ritual system.

An economy of scattered and small-scale farming by individual owners in feudal society is the liberation of slaves to a certain extent. Although farmers were heavily exploited by the class of landlords, they had the freedom to cultivate and labor on rented land and family became the basic unit of society. This increased laborers' enthusiasm to work. China had long been a feudal society, which is the historical reason for the formation of the outstandingly diligent, sincere and honest, loyal and tolerant character of the Chinese nation. The feudal ethics peculiar to China permeated into every corner of society. This not only had a decisive impact upon the formation of intense patriotism, but also facilitated the flourishing of intelligence and skills and the tapping and development of wisdom. The notion of private ownership controlled every class of society and egoism and hedonism continuously grew. Although this loaded the broad masses with a heavier burden, it promoted the development of arts, and the perceptive functions of various sensory organs of human beings also gradually enhanced and developed with the interaction with the beauty of nature and the beauty created by labor. All this facilitated the development of feudal culture. This feature is rather outstanding in the history of China.

The inequality in feudal ethics and morality (e.g., The Three Cardinal Guides: Ruler guides subject, father guides son, husband guides wife; The Five Constant Virtues: benevolence, righteousness, propriety, wisdom and fidelity) as well as the social consciousness hereby formed not only further intensified discrepancy between different social ranks in feudal society in Chinese history but also rid individuals of their physical and spiritual balance to "civilize the spirit and toughen up the body". This, I am afraid, is the chief reason for the formation of the "sick men of East Asia" in the late 19th and early 20th centuries.

Capitalist society is the last pattern of society of private-ownership society and it will create conditions for the complete liberation of human beings. Although in its preliminary stage it had pushed to a pinnacle the corruption of the class of feudal landlords (e.g., the reactionary act of destroying the surplus products overproduced, which was the momentary recovery of the decadence and degeneration of the landlord class), although the state of consciousness of entire human beings has developed toward extreme individualism and the partial feature of *Yi Yuan Ti* is also gradually nearing its summit, things must turn into their opposites when they reach the extreme. Human beings will break through the confinement of clan and family and enter into more extensive social connections, thus pushing the nature of the human species into a new stage.

With the development of capitalism, when the feature of capital appreciation has completely controlled capitalists and capital has truly become their soul, the feature of the feudal landlord class is gradually overcome, thereby modern bourgeois civilization is established: Materials and wealth in society become unprecedentedly abundant and cultural quality of people becomes unprecedentedly improved; social consciousness such as equality, fraternity and freedom, narrow and fake though it is from the very beginning, has after all liberated people from fossilized feudal thoughts. All this gives free rein to the ample development of familiar intelligence, the intelligence that is brought about by the notion and system of private ownership, starts from oneself, the individual human (the substantial entity) and develops to reinforce and satisfy individual needs. Although capitalist commodity production has driven the alienation of labor to the highest peak, the nature of the human species has never stopped cleaving paths for its own advancement in the entire process of human beings' development. The scope of social connection in capitalist society is unprecedentedly extended and the nature of social connection becomes deeper as never before. For example, large-scale socialized production has turned labor into collective labor and the production of labor is provided for the masses in society to consume, which is, in a certain sense, the reappearance of the mode of production in primitive communes. In light of the fundamental problem in society—the system of ownership is private and therefore people's notions are overwhelmed with private possession, the nature of the human species still cannot break away from the confinement of selfishness to obtain fundamental liberation.

Once this fundamental problem of the system and the notion of private ownership is solved, fundamental changes will happen in human morality consciousness. Humans will change from the small self of being for oneself into the big self of being for society. They will then become the grand self closely connected with Nature. The activity of human consciousness will gradually overcome the partiality feature of familiar intelligence based on reinforcing and satisfying individual needs. Super intelligence will be comprehensively tapped and developed. Human *Yi Yuan Ti* will gradually enter the altitudes of Complete *Yi Yuan Ti* and Developed Integrate *Yi Yuan Ti*. Human morality will also completely enter the realm of social free *Dao De*.

In the development history of the human beings in the past, the functioning of both *Dao De* consciousness and all human social activities on human beings' advancement is a slow automatic process. Therefore, it has taken a rather long time to develop from primitive humans into humans at the present moment. *Qigong* science today reveals this objective law of the function of *Dao De* in the evolution of the human species' nature and it will change the evolution of the human species' nature from the natural development state into the conscious development level. This will fundamentally change the evolutionary process of the nature of the human species.

A comprehensive survey of the development history of human beings tells us that the development goes forward along three main routes. One is the physical and mental development of individual humans, which includes techniques of labor (the use of tools and the mode of labor) and wisdom, the form and function of human tissues and organs, and the fact that *Yi Yuan Ti* develops from Pure Original *Yi Yuan Ti* through Self *Yi Yuan Ti* into Partial Identity *Yi Yuan Ti*. Nowadays, each isolated subject in science seems to be going into synthesis and form holistic connections under the effect and influence of borderline science and especially the outlook and methodology of the philosophy of dialectical materialism. This will undoubtedly help to change the partial and biased feature of Partial Identity *Yi Yuan Ti* in human beings. The second route of development is that the production unit of humans (which is as well the unit of material production in society) develops in the direction of being increasingly simple: from primitive clans, families of the same ancestor, families with several generations living together, to the families of husband and wife. This change indicates the evolution in the relationship between male and female in human beings. Family both as the unit of producing humans in the way similar to that machines produce products and as the unit of material production in society gradually steps into the relationship of true love. At present in capitalist society, a tendency of discarding the family pattern of husband and wife and stepping toward celibacy appears. This results from the distorted development of people's consciousness brought about by fetishism of materials in the system of capitalist private ownership. It will be solved with the realization of human beings' free species-being of knowing their true self. The third route of development is that the mode of connection among humans as social beings is developing in the direction of being increasingly great in scope: clans→tribes→countries controlled by the system of slavery→countries controlled by the feudal system→capitalist-system countries internationally connected. Human beings will finally enter the World of Great Harmony (*Da Tong Shi Jie*) without national boundaries.

Change in the notion of *Dao De* in humans is restricted by the development and change of the aforementioned three aspects and, at the same time, facilitates the development of these three aspects. It should be known that the advancement of human beings must be the change in the fundamental stratum of people's connections. Yet as far as each human individual is concerned, it is concentrated in the change in the stratum of self in the individual's *Yi Yuan Ti*. This is what Marx said that nation and family and so on can all be reflected in human nature. All this will rely on self to adjust the development at the level of life (which belongs to the domain of natural *Dao De*) and the level of consciousness (which belongs to the domain of social *Dao De*).

IV. Emancipation of the Character of the Human Species-Being

As it has been mentioned above, the character of the human species-being, on the one hand, is distorted and alienated in the long historical river of human beings' development, and, on the other hand, continuously develops forward with the evolution of human life functions and is shown through various functions of the human body. It is just as what Marx said, "The enjoyment of the human eyes differs from that of the primitive non-human eyes. The enjoyment of human ears differs from that of the primitive ears." He also said, "Therefore the senses of social humans differ from the senses of nonsocial humans. It is precisely because of the richness in the objective unfolding of the human nature, the richness in the subject of human and human perceptuality (e.g., the ears that have the sense of music, the eyes that can feel the beauty in form) that, in sum, those senses that can become human enjoyment, the senses of essential power that proves one is human, are partially developed and partially produced. This is because not only the five senses but also the so-called spiritual sense and practical sense (will, love, etc.), in brief, human senses and the human character of the senses are produced through the existence of their object, through the humanized natural world. Formation of senses of human sense organs is the outcome of the entire previous world's history."(*The Complete Works of Marx and Engels*, Vol. 42, pp. 125-126)

Up to now, though the fundamental character of the human species' nature of being "free and conscientious activity" is distorted and deviated, the essential functions of the character of the human species' nature displayed in the life activity of the human body have been enhanced for we do not know how many times compared with those in the primitive humans. The freedom that human beings enjoy in the natural world is even worlds apart from that of primitive humans. Human beings, having developed up to now, seem to have mounted the throne of human civilization and begin to feel somewhat complacent and proud, thinking that they have gone through the ages of barbarism, savagery, civilization and now they have come to the age of pleasure seeking. As a matter of fact, this is a misunderstanding because human beings up to now still do not truly know their own nature, to say nothing of their attainment of thorough emancipation of the human species-being. Once human beings have realized the character of their own species-being, they will undoubtedly conscientiously pursue the thorough emancipation of their own species-being and a new epoch in human beings' development history will undoubtedly be inaugurated.

How can human beings obtain this emancipation? Marx said, "Sublation of private property is a thorough emancipation of all human senses and characteristics. But the reason why the sublation is the emancipation is because these senses and characteristics have become those of human whether it is in subject or in object. The eyes have become human eyes when their object has become a social, human object, created by human and destined for human... Senses relate themselves to the thing for the sake of the thing, but the thing itself is an objective human relation to itself and to human, ... Need and enjoyment have thus lost their egoistic character, and nature has lost its mere utility by the fact that its utilization has become human utilization." And "Active sublation of private property, that is to say, perceptive possession of human nature, of human life, of human as the target and of human product for the sake of human and by means of human should not be merely understood as direct and partial enjoyment, should not be merely understood as occupying and possessing. A human, in an all-dimensional manner, namely, as an integrate whole being, possesses the human's own entire nature. The relationship of a human being with any other human being in the world—seeing, hearing, smelling, tasting, feeling, thinking, intuition, sense, wish, activity and love, in all, all the organs of the individuality of this person possess the target through this person' relationship with it."(*The Complete Works of Marx and Engels*, Vol. 42, pp. 123-125)

This passage means that the system and the notion of private ownership make the character of the human species' nature of being "free and conscious activity" attach to the possessive desire and become distorted. Therefore, when privately owned property is sublated (which refers to the sublation of not only objects in reality, but also the consciousness and notion of private ownership), the essential powers of human nature—the functions of different tissues and organs, will be freed from the confinement of the consciousness of private ownership and restore their free and self-awareness functions which act as the essential powers of humans. This is because society will have already had sufficient material conditions to satisfy the demand of people's living and people no longer need to weigh in their mind how to take into their possession those materials, but directly experience the *hunhua* that takes place when they establish their relationship or make contact with materials. This is both passive and active hence a kind of human's self enjoyment.

Although Marx's exposition about the emancipation of the human species-being is mainly illustrated from the aspect of social science, it has profoundly pointed out that, in order to attain the thorough emancipation of the human species-being, certain social conditions are required, namely, the living conditions with abundant materials created by highly developed science and technology have reached the high level of the World of Great Harmony (*Da Tong Shi Jie*) as was described by the ancient Chinese —"There is such an abundance of material things that they are left around and people need not take them away to keep them for themselves." Then people will be freed from the shackles of private possession and various human functions will truly become the functions of humans. This is of course the emancipation of the character of the human species-being, but it is only the emancipation of the part of essential powers of human beings' familiar intelligence. It is still not the thorough and complete emancipation of the entire human species-being.

It is only when human beings have overcome the partiality feature of their Partial Identity *Yi Yuan Ti* and have attained their emancipation in the realm of society and the realm of economy that human beings can possess their own essential powers of human nature in all aspects. It is only then that the thorough emancipation of humans as holistic humans, as humans in a united whole entity with nature and society can be realized, that it is the command of the entire essential powers of human nature (including the power of familiar intelligence and the essential power of super intelligence state), and it is the outcome that all human functions have been unfolded. Humans and Nature are unified on the basis of great harmony and it is the exhibition of human morality having developed to an extremely elevated realm—social free *Dao De*. Here human morality, self and the human species-being are unified at an advanced level.

In all traditional *qigong* theories, consciousness (including morality) has been considered as the function of the "heart" and as a special existence that exists independently, while the holistic connection of consciousness with nature and society has been neglected. Therefore, they have not given correct exposition about consciousness (including morality), nor have they correctly dealt with consciousness cultivation, this fundamental question in *qigong* practice.

Chapter VII The Theory of the Enhancement of Life and the Concept of *Hunyuan* Medical Treatment

The theory of *hunyuan* whole entity established in *Zhineng Qigong* Science is a human-oriented theory aiming at the complete emancipation of human species-being. *Qigong* practitioners take the theory of *hunyuan* whole entity as the theoretical means to enhance their own lives and to obtain their complete emancipation in the realm of life, while the theory of *hunyuan* whole entity uses *qigong* practice as the material means to realize the emancipation of the character of human species-being. The complete emancipation of the character of human species-being begins with the enhancement of life, as is shown in: a) As far as a realistic individual human being is concerned, it consists of firstly making the individual human divorced of the unhealthy state and enter into the healthy state, and then change from the normal healthy state into the enhanced life state; b) as far as a newborn individual is concerned, it is to make the individual become an enhanced life, and this requires optimal conditions in bearing and rearing. Both of these parts belong in the domain of the theory of the enhancement of life. But for the convenience of explanation, we call the former the concept of *hunyuan* medical treatment and the latter the theory of the enhancement of life.

Section I The Theory of the Enhancement of Life

I. A Conceptual Introduction to the Theory of the Enhancement of Life
i. The Meaning of the Theory of the Enhancement of Life

The mentioning of the enhancement of life is easy to be associated with eugenics. Yet the enhancement of life in *hunyuan* whole entity has its special content. Eugenics now talked about in society refers to better bearing, that is, to be able to give birth to a healthy child or healthy children. Its content is essentially the prevention of hereditary disease, and its means is chiefly through consultation about hereditary information, diagnosis of the womb and the fetus before the birth of the child, healthcare during the period of pregnancy as well as induced abortion by choice, etc., in order to prevent an individual with hereditary defect from formation and from coming into this world. The theory of the enhancement of life that we state here is not merely for this purpose, but to improve the level of intelligence of average people to a new level. Its ultimate requirement is to make humans become the humans with ideal self, with social free *Dao De* (morality) and with advanced level of both normal and super intelligence whose human species-being has attained complete emancipation. "To surpass the secular and enter the realm of the sage" is the comment and encouragement in traditional *qigong* to few practitioners who aspired after the cultivation and practice of *qigong*. Yet for thousands of years people who have "surpassed the secular and entered the realm of the sage" are actually "as rare as feathers on the phoenix and the horn on the Chinese unicorn" as the Chinese saying goes. The theory of the enhancement of life in the science of *Zhineng Qigong* is to start with everyone and to enable the entire group of humans to enter the realm of enhanced life.

ii. Different Claims about the Enhancement of Life

As to how to achieve the goal of enhanced life, in modern science, some lay emphasis on heredity, some lay emphasis on training after birth, both of which are reasonable to a certain extent with their good grounds yet both have their respective partiality. *Zhineng qigong* roots its arguments in the theory of *hunyuan* whole entity. These arguments will be stated respectively as follows:

i) The Theory of Heredity

According to the theory of heredity, though there are as many as a hundred thousand billion (10^{14}) cells in the human body which have over a hundred forms and functions and form different tissues and organs that perform different duties, they all originate from one common cell—the zygote if we seek their origin, and the zygote is the combination of the sperm that carries all the information about the father and the ovum that carries all the information about the mother. The substance that carries all the information about the parents has been discovered by modern science. It is a kind of biomacromolecule called deoxyribonucleic acid (DNA) which contains a great many hereditary genes (chromosomes in cells are formed with DNA helix that coil together) and can pass the information about the parents on to the next generation. It has now really been found that in the newborn individuals many of their parents' characteristics are retained, e.g., the height, the color of the skin, the color of the eyes, and so on. Many diseases are also hereditary. Investigation shows that the hereditary diseases known so far are as many as over 4,000 and they are the important content guarded against by eugenics. Moreover, it is believed in the theory of heredity that intelligence and wisdom can also be inherited. It is because of this that eugenics in the theory of heredity has put forward the argument of "ridding the backward and retaining the better". Under the instruction of thinking as this, some people in the West have put forward the incorrect claim of superior race and some have put forward their fantasy of reproducing "genius", e.g., some people advocate the formation of excellent newborn individuals by collecting the sperms and the ova of some outstanding people and carrying out external fertilization.

As regards the theory of heredity, we agree with its claims and means to guard against hereditary disease yet disagree with its argument that heredity determines everything. This is because hereditary genes mainly determine the heredity of human biological features. Even so, there will still be the occurrence of recessive genes. In the newborn individual, there is also the possibility of genovariation. On the other hand, education and cultivation after birth have an enormous impact on the formation of a person's talent and virtue which cannot be determined by heredity alone.

The argument in the theory of experience stated below is put forward from the opposite side of the theory of heredity.

ii) The Theory of Experience

In the theory of experience, training after birth is emphasized. It believes that all kinds of human intelligence, talents, wisdom and skills result from training and learning after birth. In ancient China, there already existed the sayings of paying attention to "the education of human fetus" as well as "education and cultivation are weightier than family status (family inheritance)." What is stated in the theory of experience, strictly speaking, is not the content at the same level of eugenics in the theory of heredity. The theory of experience does not admit the existence of hereditary features. It mainly talks about the sociological features of humans. All this is apparently closely related to postnatal learning. Here is a famous quote by John Watson, the behavioral psychologist: "Give me a dozen healthy infants, well-formed, and my own specified world to bring them up in and I'll guarantee to take any one at random and train him to become any type of specialist I might select -- doctor, lawyer, artist, merchant-chief and, yes, even beggar-man and thief, regardless of his talents, penchants, tendencies, abilities, vocations, and race of his ancestors." (*Behaviorism*, 1930)

We think that the theory of experience is correct in emphasizing the great significance of postnatal environment in the tendency of development of a human being. However, the argument that environment determines all as John Watson put forward is not completely correct. On the one hand, the babies John Watson talked about must be healthy ones, and this itself contains the premise of hereditary factors. In other words, these babies are the natural display of human beings having evolved into the present stage. They are the embodiers of the achievement retained through heredity after human beings have gone through evolution for over millions of years. This shows from one aspect that John Watson cannot train the primitive humans or anthropoids to display the various talents and capabilities of humans today. On the other hand, the training according to the theory that environment determines everything claimed by John Watson is only part of the content in the entire life journey of a human being. We have stated in Human Body *Hunyuanqi* (Chapter IV) that there are different stages in the development process of a human being and each stage has its environment, all of which has a certain effect upon the formation of the holistic feature of the human being. It is not difficult to see that the training in the environment claimed by John Watson is only part of all the environment in the entire development process of a human being. Therefore his point of view is partial.

iii) The Theory of *Hunhua* Enhancement of Life

This is the point of view in the theory of the enhancement of life in *Zhineng Qigong* Science. The theory of *hunhua* enhancement of life, with its starting-point in the theory of *hunyuan* whole entity, considers a human being as a *hunyuan* (merged-into-oneness) whole entity of time and space. This *hunyuan* time-and-space whole entity develops from the zygote's *hunyuanqi* to the *hunyuanqi* of the adult and experiences different stages of development, which enable the multitudinous potential information in the zygote to merge and develop with the surrounding environment and form the realistic human whole entity. The starting point of this process is hereditary information. According to the theory of *hunyuan* whole entity, hereditary disease in parents does affect the next generation; even disease that is not hereditary, when it has become serious to a certain degree (e.g., when it affects the *hunhua* process of the *hunyuan* entities of the parents), can also affect the health of the next generation. Solving these problems belongs to medical treatment that enables people to depart from the morbid state, which will not be talked about here.

According to the theory of *hunhua* enhancement of life, the enhanced state of a newborn individual results from *hunhua* of the hereditary factors with the environmental factors. It is not that hereditary factors determine everything. If you do not agree, please think it over: Is there any newborn individual who can inherit all the features of the parents? Facts prove that it is impossible. Even if the sperm and the ovum retain all the hereditary features of the parents, some information still become recessive and cannot be displayed into reality when the sperm and the ovum are combined. Not only in the stage of fertilization can some of the information be depressed, the different hereditary factors (normal or abnormal) can be depressed in every development stage of a human being due to the influence from the environment. With this clear in the mind, it is not hard to understand that it is impossible to find two persons exactly the same in this world. It has been proven through investigation in modern medical science that, even in the twins developed from one same ovum, hereditary disease may happen in only one of them. This further shows the correctness of the theory of *hunhua* enhancement of life from another aspect.

What is emphasized in the theory of *hunhua* enhancement of life is not only how to enable healthy parents to give birth to enhanced posterities, but also the enhancement of the newborn individuals into better humans with more accomplished talents. This is in fact a question of the enhancement of the entire process of human body *hunyuanqi*'s development. It is partial to only emphasize any one link in this process. The enhancement must be the enhancement of all the links in this process, which includes the enhancement of fertilization, of formation of the embryo, of nourishing the embryo, of education of the fetus, of delivery of the baby, of raising the infant and of education. In order to achieve this goal we must be clear about the development laws and features of human individuals so that we will not equate humans to animals. It is only when the features of humans are comprehensively and profoundly recognized that the enhancement of life can be carried out well-targeted.

II. Features of the Development of Human Individuals
i. Humans Have the Longest Development Period

Human beings are the most advanced creatures on the earth and their functions and structures are also the most complicated and consummate. Therefore, the time that it takes for an individual human to develop from a monoplast into a human being intact in form, perfect in function and sound in intelligence and wisdom is also the longest. According to modern psychology, a human being's psychological development is not mature until the age of 16. According to the theory of *hunyuan* whole entity, a normal newborn baby's tissues and organs are mature preliminarily, yet it does not have the ability to live independently; at the age of around 6, functions of every tissue are preliminarily settled, while the ability of logical thinking has not formed in the activity of consciousness; at the age of about 16, psychological activity in a person is reaching maturity yet *Dao De* (morality) consciousness is still in a restlessness stage; at about 25, the partial self is formed. A person of familiar intelligence generally reaches the summit of development when arriving at this stage. The life afterward is to engage oneself in all kinds of study and work under the guidance of the partial identity self. Although the human being will still make progress in career and will gradually attain achievements, which further fixes the partial identity self, all these are but changes in the domain of the partial identity self. *Zhineng Qigong* practitioners, however, are required to continuously advance on the basis of the partial identity self and get rid of bias and partiality. They are not only to tap and develop super intelligence, but should reach the level of Complete *Yi Yuan Ti* in their consciousness. In morality or *Dao De*, they should improve to the level of social free *Dao De*. In psychology, they will form truly free conscientious self with true-*self* awareness. This is what they need to spend decades of effort and even their whole life-time energy to accomplish.

ii. The Development of Human Individuals Is a *Hunhua* Process under the Guidance of the Holistic Time-and-Space Structure

It is easy for people to understand the "lengthiness" in human Individuals' development. However, a human individual develops from a zygote, a monoplast, into a complicated human being—this is not easy to imagine, though the double spiral shaped molecular structure of DNA with its hereditary function has already been found in molecular biology to explain the self multiplication of cells, which therefore solidifies the position of the theory of heredity in the evolution of organisms in science. Biologists have also found the Homeobox, a DNA segment, which appears at first as the essential component in homeotic genes. It lines up according to a certain order and is involved in the regulation of morphogenesis in organisms. The Homeobox in the genes of every animal's body is the same, and it is already known that the Homeobox is part of the genetic switch that turns on cascades of other genes. Up to this point, it seems that people can explain the development process from a zygote the monoplast to a complicated holistic human being according to the nature and function of genes. In fact it is not really true because the essential problem has not yet been solved.

For example: Every cell in the diploid and tetraploid (even the octaploid in lower animals) in the blastomeres of the zygote is basically the same, or that there is no difference between them (as has been proven in biology through the culture of one cell in them which develops into a newborn individual animal of this species). Why, after the octaploid is formed, the cells that are divided into existence show their differences and then further develop into different tissues and organs? It is evident that this cannot be clearly explained merely with the function of the Homeobox segment as the genetic switch, for cells divided into the same generation form different tendency in development just because of their different positions, e.g., cells in the different positions of the three germinal layers form different tissues and organs; cells in the primordia of different organs develop according to the needs of the organs. What is particularly hard for people to understand is that implantation of the nucleus of a cell in the body into an ovum with its nucleus removed can make a new individual grow into being and born into this world. All this shows that the development of organisms, particularly of advanced organisms, is not determined by DNA alone.

It is pointed out in the theory of *hunyuan* whole entity that the growth and development process of any organism is a *hunhua* process in which the *hunyuanqi* that this organism obtains through heredity merges and changes with the *hunyuanqi* in the external world. The directive development of the organism is the result of the guidance of the holistic time-and-space structure of the organism's *hunyuanqi*. The *hunyuan* entity of the organism merges and changes with the *hunyuanqi* that is in the same structure and is complementary to it in the surrounding environment and forms a biological *hunyuan* entity in a new stage. This *hunyuan* entity in a new stage then merges and changes with the *hunyuanqi* that is complementary to it and is in the same structure in the environment and forms a newer *hunyuan* entity. The mergence and change repetitively goes on like this. A step is taken forward with each time of mergence and change. The former step is the cause of the next step, while the next step is the result of the former step and meanwhile the cause of the further next step of mergence and change.

For example, in the time of the diploid, tetrapoid and octaploid blastomeres of the zygote, the internal environment and the external environment of each cell during cleavage are the same and the *hunhua* goes on in each cell is the same, so the feature and function shown in each cell are also the same. Ever since then, the external environment of each cell begins to differ, e.g., some cells have more contact with the external environment, some cells have more contact with the cells of the same category as themselves. The *hunyuanqi* in the surrounding environment of each cell becomes different, which causes difference in the mergence and change in the cells. Thus the tendency in the division of cells begins to differ and form the yolk sac, the allantois and the blastodisc and so on. After this, the blastodisc divides into the entoderm and the ectoderm and then the three germinal layers. The three germinal layers then divide into different tissues and organs ...

From this it can be seen that the development of the ovigerm and embryo results from the *hunhua* of the *hunyuanqi* of the zygote and the *hunyuanqi* around it. If there is only the genetic information in the zygote and there is not the ovum's cytoplasm, this *hunyuanqi* as a special environment, the development of the ovigerm cannot be accomplished, to say nothing of the development of the embryo.

The *hunhua* process mentioned above includes such two parts of content as the *hunhua* between the sperm and the nucleus of the ovum and the *hunhua* between the fertilized zygocyte (zygote) and the ovum's cytoplasm. The directive development is the result of the interaction between the holistic time-and-space structure (*hunyuanqi*) of the zygocyte and the *hunyuanqi* of the ovum's cytoplasm which is complementary to and structurally identical with the zygocyte's *hunyuanqi*. After the formation of the human fetus, the holistic life structure of the whole entity of the fetus and the placenta comes into being. On the one hand, it is the *hunhua* of the fetus with the *hunyuanqi* of the placenta; on the other hand, it is the *hunhua* of the placenta with the *hunyuanqi* of the mother's womb. This complicated development process is all carried out according to the laws of biological heredity. Therefore the fetus or the newborn baby that experiences this development process is but a natural human being.

The development from when the baby is born to adulthood, though it still follows the features of biological genetics in a lot of content, shows the fundamental difference in this stage between humans and animals, that is, the formation of consciousness (or called the development of consciousness, though people are not accustomed to a term like this). This is the fundamental display of the character of human species-being—sociality. Although there exists holistic guidance in both the development of consciousness and the above-mentioned development process of the natural human being, the two kinds of holistic guidance differ in essence. This is shown in: The beginning reason of a natural human being is the zygote which already has all the information structure to develop into a human being. It can develop into a human being so long as there is the structurally-identical and complementary *hunyuanqi* in the environment. A social human being's beginning reason, however, is *Yi Yuan Ti* that is established on the basis of cranial nerve cells. In *Yi Yuan Ti* is the special substance that is even in texture without any difference and only has the ability to receive (reflect) and process information about the objective world. It cannot form consciousness until social consciousness is internalized into *Yi Yuan Ti*. This is another kind of *hunhua* fundamentally different from structurally-identical complementary *hunhua* or structurally-identical regeneration *hunhua*. In this *hunhua*, the guidance of social consciousness plays a decisive role, without which not only the development of consciousness cannot be accomplished, but the development of cranial nerve cells after birth will also be affected, as has been proven in the case of the wolf child.

Human consciousness activity is the most advanced and complicated mode of motion in the universe. It is formed in the external environment established by the elder generations. When a human being's consciousness is established, it will also affect and reform the external environment and helps to create the guidance condition for the new generation.

iii. The Transition from a Natural Human Being into a Social Human Being Is a Process in which Consciousness Activity Is Established to Guide Life Activity

Although human beings are the most advanced creatures on the earth, individual humans do not have the functional abilities to live independently as normal animals do when they are born into this world. You may wonder why it is like this. According to the theory of *hunyuan* whole entity, this is the inevitable result when the biological world evolves into the stage of humans. In the evolutionary history of the human race, humans evolved from primates. Although primates already had relatively developed brain, they had not formed genuine conceptual consciousness activity and their life activity agreed with their nerve activity, whereas human beings' life activity is the means to realize their consciousness activity. This is to say, the activity of consciousness plays a decisive leading role in all human life activities. We know that human consciousness activity is gradually formed in the social life after the fetus is born into this world, so there must be a process in which the consciousness activity combines with the life activity. Nature has created all existences, and natural evolution has formed in them natural reasonability. The formation and the entire development process of individual humans are just the embodiment of this reasonability. A mature fetus has accomplished the morphological development of a natural human being, yet it has to gradually learn how to use its functions in the practical life after its birth. The process of learning to use these functions (including the functions of life activity) is taught and trained by the adults with the aid of language and the example of actions. That is to say, the learning of how to use its own life functions is carried out under the instruction of language as soon as the newborn baby begins to learn how to use them. It is well-known that instinctive activity that belongs in unconditional reflex, i.e., the life activity comes with birth, takes up a rather small proportion in the life activities of human beings. In other words, the major content of human life activities is acquired after birth. The process that human beings learn to use their own life functions is the process in which consciousness activity is formed and combined with life activity.

Psychology and the philosophy of dialectical materialism tell us that human consciousness is not born of thin air but is all well-grounded in things in objective existence. The brain and *Yi Yuan Ti* of the newborn baby after birth already have the internal material foundation for the formation of consciousness activity. Not only is conditional reflex established step by step, but information about adults' consciousness activity and about their language is also received, which helps form babies' preconsciousness and prespeech utterances (the sounds that babies make in the early stage when they learn to speak). All this brings about the reinforcement of nerve functions and the display of various functions and makes life activity and consciousness activity closely combine. Let me analyze this with examples. Whatever actions adults teach a baby, they inevitably aid their teaching with language. For example, when teaching the baby to stand, they not only support the baby to stand, but guide it with such words as "stand, stand". When teaching the baby to walk, they help the baby walk forward and say such words as "forward, forward". Different kinds of simple actions that babies learn in the early stage are all carried out under the instruction of adults' speech. When they have just learned how to do an action, babies will accordingly act as long as the adult says instructive words of this action. This fully shows that babies' life activity and consciousness activity not only accompany each other, but their life activity is accomplished under the guidance of their consciousness activity.

From the introduction given above it can be seen that consciousness has a guiding function on the macrocosmic motion of the human body's life activity. Even in the microcosmic field of life activity, consciousness can also have a rather great effect. This has already been elaborated in the Theory of Consciousness (Chapter V) and the Theory of Morality (*Dao De*) (Chapter VI), hence it will not be repeated here. What is to be emphatically pointed out here is the development of consciousness as well as its partial guidance over human body's life activity.

Since in the hereditary information about the human species is deposited all the information in the long evolutionary chain of the human species, human life functions should be all dimensional. To be specific, it means that the balance of human life activity can be maintained with both tangible and intangible substances. Various human functions can be manifested in both normal and super intelligence. However, people with whom babies have contact are generally all familiar intelligence people and the information babies receive is mostly familiar intelligence information. Therefore, the consciousness activity that forms in babies also falls into the domain of familiar intelligence. The result of the guidance of such consciousness activity is that human life activity steps onto the partial track of familiar intelligence. For example, when a person is hungry, this person will have to eat. This is already plain common sense. Must it be really like this? Our answer to it is: It is like this and also not like this. We say that it is like this because, according to the standard conclusion, a person who has not eaten anything for seven to ten days will suffer from irreversible malnutrition, and this has become an unchangeable final conclusion proven in modern medicine. We say it is not like this because the above-mentioned life phenomenon results from the partial guidance of the partial identity consciousness over the life activity of the human body. If the partiality in consciousness is overcome, normal life activity of the human body can still be maintained without eating any food. *Qigong Bi Gu* (eating no crops or eating nothing) is a convincing proof. In addition, though human super intelligence is suppressed and cannot be unfolded due to lack of appropriate guidance, experiments into paranormal abilities in the human body in recent years have proven that giving training of super extraordinary capabilities to familiar intelligence children can make many of them able to use the techniques of super intelligence, including but not limited to reading with the ears, penetrating eyesight, telescopic eyesight and teleportation. The practice in *qigong* science has further proven that adults can also tap into their potential intelligence and energy under appropriate guidance of consciousness.

III. Enhancement of Every Link of Life

According to the theory of *hunyuan* whole entity, the enhanced life is an enhanced holistic time-and-space structure. It is the result of the enhancement of each and every link or stage of the entire life process. It is very important whichever link is enhanced. If all the links are enhanced, it is total enhancement; if part of the links is enhanced, it is just partial enhancement. Of course, there is a question of quality and quantity in the enhancement of each and every link. The degree of enhancement in each link determines the overall level of the enhancement of holistic life activity. We know that, in the evolutionary history of the human beings, every step that human beings took forward was an enhancement process of "getting rid of the bad and retaining the good", a process which was, however, accomplished through "natural selection". Today, human beings have entered the rational age of science. We already have a deep knowledge about the harmful factors in genetics and have therefore begun a process of "human selection" in ridding ourselves of harmful elements. However, this selection in modern eugenics is just to get rid of the bad and retain the normal. It has not reached the level of creating better lives. The theory of *hunyuan* whole entity has revealed the laws of the formation of human body *hunyuanqi* and has laid a theoretical foundation for the creation of enhanced lives. The following gives a brief illustration about the principles of the enhancement of life from relevant links in the formation of human body *hunyuanqi*.

i. The Enhancement of the Zygote's *Hunyuanqi*

The enhancement of the zygote's *hunyuanqi* is determined not only by quality sperm and ovum but also by the environment in which they combine.

i) Quality Sperm and Ovum

Quality sperm and ovum mainly refer to:

a. They Have Abundant Amount of Information (This Essentially Refers to the Sperm)

According to the theory of *hunyuan* whole entity, the genetic information that a new individual life receives, though it consists of the inherent information about generations of ancestors, is most closely related to the individual's parents. Generally speaking, the amount of information in a person is greater in the stage of maturity than that in the immature stage and will increase as the degree of the person's maturity grows. Although the old Chinese saying that "The son and the daughter one has in one's old age are mostly intelligent and wise" is not always true, as heredity is only one link in the enhancement of life, the amount of genetic information passed on from parents does have something to do with the intelligence and talent of the offspring.

b. They Have Abundant Life Force (This Mainly Refers to the Ovum)

Generally speaking, the prerequisite for having plenty of life force in the sperm and the ovum is the parents' robust health and exuberant vitality.

The two points mentioned above determine the principles of the enhancement of life in this stage. First, it is not proper to fertilize and bear children when the parents have not reached the stage of maturity; secondly, it is not proper to fertilize and bear children when the man and the woman are in a poor state of health; thirdly, those who want to be parents should "nourish the essence and reserve the vigor" before fertilization and bearing of children. This is because it takes a certain period of time for the sperm to become mature and frequent ejaculation of semen will result in the release of immature sperm. The sperm that are not strong in life force will undoubtedly affect the enhancement of conception.

ii) The Environment in which the Sperm and the Ovum Combine

It is proven in modern medicine that a series of changes takes place before and during the process when the sperm and the ovum are combined. These changes include the increase in the energy of the sperm, the exposure to its specific antigen, relieving the enclosure of its acrosome reaction and so on. All this needs specific conditions in the environment. Otherwise, the aim of impregnation cannot be realized. This has already been grasped and successfully applied in artificial insemination in modern medicine. However, in the process of natural conception, it is not as simple and convenient as artificial conception in regard to how the optimal environment and condition for conceiving a child can be achieved. This is because only when great life force is aroused in both the man and the woman that the environment of conception can reach an enhanced state. There is systematic and profound illustration about this in traditional *qigong*, which can be found in *A Brief Introduction to Different Qigong Schools* in *A Comprehensive Statement of Traditional Qigong Knowledge*, so it is omitted here.

ii. The Enhancement in the Embryonic Stage
i) In the Embryonic Stage

The *hunyuanqi* in this period, as introduced in Human Body *Hunyuanqi* (Chapter IV), has an intensified feature of expanding outward and will not be taken back to the embryo until the nerve canal is formed. In this period, if the *qi* cannot be taken back into the embryo, a healthy fetus cannot be formed and this usually leads to miscarriage. In view of this, the relevant principles of enhancing life in this period are: a) Facilitate the outward opening of the embryo's *hunyuanqi* in the beginning stage and make it merge (and change) more sufficiently with the *hunyuanqi* in the outside world. Therefore, the creation of an appropriate environment is essential. Those who have sufficient time, energy and money to travel would be advised to arrange an appropriate tour in order to facilitate the absorption of *hunyuanqi* from the outside world. b) When the nerve canal is formed and is becoming closed, the mother-to-be should help the *hunyuanqi* of the embryo return to the embryo and the taking back of the *qi* must be complete and integrate. In order to better reinforce this opening and closing process, grasping relevant knowledge about embryo is essential. During this period, the mother-to-be should especially avoid rashly using medicines in case it affects the sending out and the taking in of the *qi*. To adopt the practice of sending out internal *qi* and taking in external *qi* in *Zhineng Qigong* is a more active method.

ii) In the Fetal Stage

When the fetus develops into the seventh month of pregnancy, its *Yi Yuan Ti* is already formed. Such a fetus already has the ability to receive different kinds of information and to form consciousness activity (as proven by the survival of premature babies born during the seventh month). However, real consciousness activity cannot form in it due to the special environment in which the fetus is located where the reinforcement and repetition of specific information is hardly available. In spite of this, the fetus' *Yi Yuan Ti* loses its fundamentally pure and blank feature immediately after its formation. Genetic features make the information about the life activity of the individual to be born naturally reflected in the fetus' *Yi Yuan Ti*, as forms the most fundamental background of *Yi Yuan Ti*'s Reference System. It will become the foundation for the establishment of consciousness activity and the guidance and commanding of life activity. This part of content is in fact what is referred to in Buddhism as Alaya-vijnana.

As the fetus' *hunyuanqi* in this period is still in the prenatal *hunyuan* whole entity state, the life-enhancing principles in this period are: a) Increase the fetus' *hunyuanqi* and the mother-to-be should take in more essential nutrition; b) the mother-to-be should frequently send good information to the fetus. The information sent to the fetus should be simple, explicit in content and repeated many times instead of being sent in passing. A better effect will be achieved if the mother-to-be can often send *qi* to the fetus with her mind. The means adopted during this period for enhancing life can facilitate the development of functions of different tissues and organs of the individual to be born. During this period, the pregnant woman should remain positive and joyful at heart and settled and serene in the mind. All this will have a good effect on the fetus' *Yi Yuan Ti*. Impatience, worry, sorrow, resentment and indignation, etc., on the contrary, have a certain bad effect on the fetus' *Yi Yuan Ti* and life activity.

iii. The Enhancement in the Stage of the Newborn Baby and Infancy

This stage does not only concern the formation of the background of *Yi Yuan Ti*'s Reference System, but also the formation of consciousness activity as well as its holistic connection with life activity. This is the starting point from the natural human being to the social human being. It is also the beginning point that the true character of the human species-being is formed and displayed. The essential content in this stage is that *Yi Yuan Ti* establishes connections with the internal and external environment of the human body, and speech, language, consciousness activities and various life activities are but the specific display of these connections. It should be noted that *Yi Yuan Ti*'s Reference System has not formed in this period, thus there is still no restriction from partiality and adherence to the partiality. This allows *Yi Yuan Ti* to be able to receive information from all aspects, including the information about mass and energy of specific things as well as the information about human consciousness not expressed with speech.

The principles of the enhancement of life in this period are to carry out enhanced rearing and to enable the baby to have diversified contact with the external world. That the adult touches the baby's skin with good information in his or her mind is an important factor to facilitate the baby's healthy development. One should pay attention to the nurture of the baby's sensori-motor thinking and thinking in terms of images. Those who are accomplished *qigong* practitioners or who have super intelligence should pay attention to inputting *qi* and super intelligence information into the baby. It is only through this that the baby's Reference System of *Yi Yuan Ti* can be impressed with the mark of all dimensional information so as to prevent the formation of the feature of partiality in the fundamental stratum.

The influence of adults' language, behavior and consciousness activity on the baby in this period is more evident and more specific by far than that in the fetal period, and this influence is the major factor for forming the baby's primary base of *Dao De* (morality). If we can often organize *qi* field and transmit *qi* to the baby and guide the baby's development with intense mental information, it will have an enormous reinforcing effect. If the baby is only given familiar intelligence information in this period, the baby's Reference System will be led onto the partiality track, with which the baby's life activity will be commanded. Such partiality has an ingrained effect on the baby's life activity afterwards and is hard to be noticed and even more difficult to be changed. This is the primal ignorance (also known as "ignorance without beginning") in Buddhism. It should be understood that the so-called "ignorance without begin-ning" is just that human beings in the evolutionary history of the human race have never realized the reason for their ignorance, for their differentiation and evaluation of all the dharma in the universe, and for their absurd adherence to the partial reflections and opinions. In Buddhism, it is believed that the world and human beings have no beginning and that human beings lost their nature from the time of no beginning, thus it causes life, death and samsara. This is the absurdity caused by not knowing human beings' evolutionary history.

iv. The Enhancement in the Stage of Childhood

This is a stage in which a human being's wisdom is comprehensively developed and also a stage in which the human body outstandingly grows. The principle of the enhancement of life in this period is to carry out enhanced education. This, on the one hand, includes nurturing good steadfast belief in children so as to strengthen and increase their willpower, and, on the other hand, involves comprehensively developing children's familiar intelligence and super intelligence. Meanwhile, the education of their morality must be reinforced. The relationship between being for the public, being open-minded and the taking in of *qi* should be particularly made clear to them. All this must be carried out on the basis of enhancing their physical and mental health.

To practice *Zhineng* Motion *Qigong* is a simple and effective method. As far as children are concerned, this not only should include the practice of the actions of *qigong*, but the application in all respects must also be carried out, such as organizing *qi* field, treating disease, pouring *qi* into their body and seeing *qi*. It is only by means of this that super intelligence can be tapped at the quickest possible speed, that typical super intelligence work order can be established and typical super intelligence habits can be cultivated, and that the uncomfortable exhausted feeling caused by using super intelligence in oneself can be prevented. If training and tapping of super intelligence can be added to the improvement of familiar intelligence education during childhood, favorable conditions will be created for children's comprehensive development in all aspects.

v. The Enhancement in the Stage of Youth

The stage of youth is a time when great changes take place in a person's life, among which the most fundamental change is the maturity of sexual function that makes human *hunyuanqi* complete with all the information about the life of the human body. In young people's lives, they begin to step out of their parents' family and set out on their journey in society. With outlook on life, aspirations and ideals surging in their mind, young people's life force gradually grows into a vigorous stage. The principles of the enhancement of life in this period are, first of all, to correctly deal with sexual impulse. This requires one, on the one hand, to break away with the mysterious sense of sex by learning the knowledge about it, and on the other hand, to have the correct rational knowledge of cherishing one's sexual function: People in their youth should fully realize that the male and female essence is the carrier in which all the information about life activity is congregated and that it differs from common cells and common body fluids. The viewpoint in modern medicine that to lose one's semen is nothing is not exactly correct. It is true that one need not to be over worried about the loss of sexual fluids. If one can better understand and deal with sex from the level of human body *hunyuanqi*, it will have an immeasurable effect on the enhancement of life in the period of youth.

In the stage of youth, tenacious willpower and industrious learning spirit in academic pursuit should be nurtured in the individual, as they are the foundation for career accomplishments in a person's life and also the guarantee of improving the person's cultivation in science, arts and moral consciousness. Young people should particularly be dedicated to studies into the theory of *qigong* science. It is only through this that they can consciously cultivate their body and heart, that they can instruct themselves to assiduously carry out *qigong* practice and that the power of super intelligence can be better unfolded.

vi. The Enhancement in the Prime of Life

The guiding ideology about enhancing life in this stage is to comprehensively perfect oneself and defer aging. The fundamental point about enhancing life in this stage is the all-sided reinforcement of self-cultivation in *qigong*. We know that the prime-of-life stage is a golden age for career achievements and that it is rather difficult for people to spare some time for *qigong* practice in order to enhance their own lives. I just want to take the opportunity to remind those who hold this point of view of one Chinese saying, "Sharpening the axe won't delay the cutting of the firewood." To spare some time for life enhancement can improve efficiency at work and prolong the years of abundant life force. Here I will just mention some simple principles. So long as the truth and the theory are understood and then *qigong* practice is integrated into daily life, the goal of life enhancement can be achieved, for *gongfu* in *qigong* is neither quietude nor action of the body, but being peacefully settled in the mind. To attain settled peaceful mind is to maintain a certain mode of *Yi Yuan Ti*'s activity and make it become a fixed pattern, e.g., the settled mode of the opening and closing of *qi*, so that the driving force in consciousness will be generated and one can in due course enter super intelligence from familiar intelligence. One can also frequently sense and perceive the formless *hunyuanqi* around one's body. This is what is said in *Yu Quan*, "The method of self-cultivation is not to guard the heart but to guard *qi*. *Qi* gathers at the beginning of *hunyuan*. Its birth comes before the heaven and earth." If one can consciously and repeatedly pour *qi* into one's body from above one's head, it will be even better. It is also said in *Yu Quan*, "The foundation of the practice lies in *qi*. In *qi* there are the five elements and their own minute collaterals. To guard this *qi*, one need not take it from the heart or the kidneys, but pour it into the middle *dantian* from *kunlun*, and this is the very prenatal ancestral *qi*. If one accumulates it from the five *qi*s, it is but the side branches that cannot most directly improve one's *gongfu*."

The method of *Peng Qi Guan Ding* (Holding *Qi* up and Pouring *Qi* down Method) has further specified this content stated above.

vii. The Enhancement in the Stage of Old Age

The stage of old age is a period leading to degradation and death. The guiding ideology for the enhancement of life in this stage is to nurture heavenly age, enjoy good health and prolong life. In order to achieve this, one must decrease meaningless consumption of one's bodily and mental energy. As people in old age are relatively stable in thoughts compared with young people and those in their prime of life, it is easy for them to be calm and concentrated in the mind. The various desires in them are also relatively few, and they generally have more time for *qigong* practice. Therefore, old age is a very good time for self-cultivation and *qigong* practice. The time of old age is also a stage of maturity when one's talents and wisdom have reached their summit. Therefore, people in their old age should be all the more active to bring into full play their talents and integrate benefiting themselves into benefiting other people. For the enhancement of life at the old-age stage, I just put forward the two methods of retaining spirit and nurturing form (the human body). The so-called retaining spirit means that one must understand that one's spirit has actually no form and no shape. As for how to retain one's spirit, one should merely think that *Yi Yuan Ti* is actually a mass of integrated *qi* that seems to be existent and meanwhile seems not, and that is enough. One should not have any other activity in the mind.

In *Yu Quan*, it is written, "If mental activity clings to an object, it disturbs one's spirit. If desire occurs in the mental activity, it leaks spirit. If an evil deed is spiritually or physically committed because of desire, it muddies the spirit. If retribution follows the misdoing, it sinks the spirit. When *shen* is muddied and sunk, retaining one's essence *qi* is out of the question despite one's intention to retain it." And also "If I can retain this *shen*, pure and exquisite, not only my body can remain and be protected, with goodness hidden in every evident application of it, but the grand doctrine of holy teaching I can follow and promote, and the weal and woe of people I can universally relieve. ... Therefore for a gentleman who is dedicated to the cultivation of life, the first and most important thing is to retain *shen*."

The so-called nurturing form is to nurture the human body that has form and adjust its *qi*. This is further explained in *Yu Quan*: "*Shen* relies on form for its life. This *shen* I cannot nurture or destroy before my birth. At the time after my birth, this *shen* has already entered my body, and then form and *shen* become interdependent. Therefore for the gentleman who cultivates the learning of profundity, it is indeed urgent to nurture *shen* yet the nurturing of the body should not be delayed either. However, the so-called nurturing of form is not that one must eat the cream of crops, that one must wear the delicately embroidered silks and satins, that one should gratify one's desires and take nutritious herbal medicine and food and then one's form is nurtured. So long as one is well-balanced between action and quietness in daily life, as flowers blooming in spring, as fish swimming in water, so long as one follows the heavenly seasons, adjusts one's earthly *qi*, deals appropriately with the relationship with others, respects life and worships heavenly gods, one's form will be distant from harm."

"For example, walking for too long harms tendons, yet walking slowly nurtures the tendons. Standing for too long hurts bones, yet standing in a straight, still and relaxed manner nurtures the bones. Sitting for too long harms muscles, yet sitting in the relaxed manner at a rest nurtures the muscles. Lying for too long harms *qi*, yet lying peacefully at ease nurtures the *qi*. Looking at something for too long harms blood, while looking inward into oneself nurtures the blood. Being in action for too long harms joints, while engaging in slight actions nurtures the joints. Bending the body for too long hurts *mai* (blood vessels), while bending it slowly nurtures *mai*. Stretching for too long harms the acupoints of *shu*, while stretching slightly nurtures these acupoints."

"Another set of examples: The *qi* in spring is harmonious, so it is good for one to be gentle in aspiration and free from worries. The *qi* in summer is bright, so it is good for one to breathe out with a little sound and get rid of depression. The *qi* in autumn is peaceful, so one should not work too hard nor be impetuous. The *qi* in winter is intense, so it is good for one to sit meditatively and speak less."

"One more set of examples: When the cold *qi* comes suddenly and fiercely, one should massage oneself to get rid of it. When the *qi* that moves too quickly comes suddenly and fiercely, one should clench one's fists tight to get rid of it. When the *qi* of intense summer heat comes suddenly and fiercely, one should slowly breathe in and out to get rid of it. When the warm *qi* of spring comes suddenly and fiercely, one should warmly massage oneself with one's hands to get rid of it. When the dry *qi* comes suddenly and fiercely, one should moisten oneself to get rid of it. Such kinds of *gongfu* are all what one should do to nurture one's body or form."

And "Now that one's body is nurtured, one's essence of *qi* will not be muddled by things outside the human body, nor can the *qi* from outside take it over. Thus *yuan shen* will be strengthened and the nature of *Dao* in oneself will be gradually enriched. Therefore, people who cultivate the learning of profundity should, secondly, nurture their body."

Zhineng Motion *Qigong* integrates the nurture of spirit and body in one and practitioners' efforts have proven that the impact such nurture has on enhancing life in people at old age is remarkable.

viii. The Enhancement in the Stage of Weakening and Death

This chiefly refers to the stage of approaching death. The task in this period is no longer the enhancement of life activity, but the enhancement of the process of death. When the average people are approaching death, they at first feel doubtful about it, thinking that they perhaps will not die, or show anxiety and worry about death. When they know that they are really dying, they show restlessness with anxiety or anger, which is succeeded with despair and the regret for what they have done wrong in their whole life. At last, they enter the peaceful state of death.

If a person has really paid attention to the enhancement of life when being alive, then the coming of death is the result of natural exhaustion of life force. This is inevitable on the one hand, and, on the other hand, death for this kind of people is peaceful happy death, without pain or bitterness. In regard to this, people around (including the family members) should create the environment and condition conducive to a peaceful and happy death for the person who is leaving this world. That the people around organize a *qi* field for this person and transmit *qi* to him or her with calmness in their mind and the wish for him or her to naturally develop will be the best means.

IV. The Ultimate Objective of the Enhancement of Life in *Zhineng Qigong* Science

The theory of the enhancement of life in *Zhineng Qigong* Science shares the aim of *Zhineng Qigong*, which is also to change human beings' natural instincts into conscious intelligence, to make human beings advance from the life realm of necessity to the realm of freedom, and help human culture develop fantastically into a more advanced stage.

For realistic individuals, its ultimate goal is to enable them to study and grasp the theory of life science through education so that they can take command over their natural nature with the theory of life science and their consciousness can freely and consciously guide their own human life motions. This requires not only the enhancement of familiar intelligence, but also the tapping and development of super intelligence. Normal life activity in the human body can be carried out either with the *hunyuanqi* of the concrete substance or by means of the *hunyuan* entity without form or mass. For society, the ultimate goal is to establish the modes of production and life in which super intelligence and familiar intelligence are unified. That will be the complete emancipation and realization of the nature of the human species. Humans will be truly happy and free humans and the world will be a truly equality and fraternity prevalent world.

Section II The Concept of *Hunyuan* Medical Treatment

The theory of *hunyuan* whole entity can be applied into different realms to establish different branches of *qigong* science. When applied in medical treatment, *qigong hunyuan* medical science will be established. This is a special kind of medical science different from both traditional Chinese medicine and Western medicine. It has its special viewpoints in physiology, pathology, diagnosis and methods of medical treatment. This is what we call the concept of *hunyuan* medical treatment. *Qigong hunyuan* medical science as a major branch in *Zhineng Qigong* science is rich in content. Here a brief introduction to its basic ideas will be given from the perspective of the theory of *hunyuan* whole entity.

I. A Conceptual Introduction to *Qigong* Medical Science and *Qigong Hunyuan* Medical Science

i. What Is *Qigong* Medical Science

Qigong medical science is a branch of learning that illustrates the diagnosis and treatment of disease by the means of *qigong*—no matter it is by means of *qigong* practice or by using external *qi*. There are many *qigong* schools in China. They differ widely in their theories, e.g., the theories of *yin yang*, *wu xing* (the five elements), *jing qi shen* (essence or body, *qi* and spirit), *ba gua* (the Eight Trigrams), *jing luo* (*qi* channels), *zang fu* (the five *zang* internal organs of the heart, liver, spleen, lungs and kidneys, and the six *fu* internal organs of the stomach, gallbladder, *sanjiao* or called the three energizers, bladder, large intestine and small intestine), etc.. These schools also have great differences in their methods and content of *qigong* practice. So the guiding ideologies and methods in treating disease also vary in different schools.

As traditional *qigong* in China and traditional Chinese medicine are subtly related to each other in many ways (e.g., they both emphasize *qi* and *jing gluo*, etc.) and they influence and permeate into each other in the treatment of disease, quite a few people regard *qigong* medical treatment as a method of medical treatment in traditional Chinese medicine. For example, it is said in *Bing Chuan Pian* in *Ling Shu*, "I browsed all kinds of prescriptions in private and found that there are such methods as guiding *qi* (*dao yin*), activating *qi* (*xing qi*), massage, cupping, acupuncture and moxibustion, taking medicine, ..." Guiding *qi* and activating *qi* mentioned here are types of *qigong* medical treatment. In addition, in a lot of traditional Chinese medical literature, details of *qigong* methods in treating different kinds of diseases are recorded. Nevertheless, *qigong* medical treatment should not be equated with traditional Chinese medicine because great differences lie between them. *Qigong* medical science is a medical system parallel with traditional Chinese medicine and Western medicine.

ii. Features of *Qigong* Medical Science

In China, ancient, modern, Chinese and Western medical sciences are congregated. There are now three big medical science systems in China, i.e., the systems of traditional Chinese medicine, Western medicine and *qigong* medicine having been mentioned above. These three big systems have their own respective features and should not be mixed with each other. Let me illustrate their features respectively as follows:

i) The System of Western Medicine

Although Western medicine deals with internal medicine, surgery, paediatrics, gynecology, ophthalmology and otorhinolaryngology and so on in various medical departments, using physical, chemical and biological means and so on in the method of examination and giving therapy with drugs, operations and physiotherapy in the method of medical treatment, their basic medicine is the same. Therefore their general medical ideology is identical. In Western medicine, no matter it is in diagnosis or in medical treatment, emphasis is laid on the disease while the whole entity of the human being is ignored. In diagnosis, it is emphasized that the focus of disease and the cause of it should be found out. In treatment, the emphasis is to get rid of the focus and the root cause of the disease. This is called by some people "treatment based on differentiation of diseases." The advantages of it are that the examinations are detailed and objective and can stand up to the test of repeated examinations, so the diagnosis is explicit, and that the means of treatment is well-targeted and researchers pay attention to seeking new methods and medicine of special efficacy and these methods are applicable for people on a large scale. And this is the reason why people call Western medicine a form of science.

Western medicine is not perfect, however. Whether it is in diagnosis or in medical treatment, it more often than not falls to a certain extent into mechanical materialism by paying more attention to the disease than to the human being and by stressing the external reason that causes the disease while neglecting the internal human functions. Even the methods aiming at enhancing human life functions such as vitamin therapy, hormone therapy and immunity therapy merely start from the external factors rather than from the internal factors in humans themselves.

ii) The System of Traditional Chinese Medicine

Traditional Chinese medicine starts from the fundamental guiding ideology of "heaven and human are one whole entity". Although it also has different departments in its clinical medicine, the different departments are actually an integrated entity. The historical doctors who were considered as the masters of traditional Chinese medicine and who also wrote important medical books and established medical theories could generally deal with all kinds of illnesses very well. This is because in traditional Chinese medicine the human body is regarded as an organic entirety, with *shen* (human spirit) as the commander, the five *zang* internal organs and the six *fu* internal organs as the nucleus, each part of which is connected by *jing luo* (*qi* channels) to maintain its normal functions. In traditional Chinese medicine, disease is considered as a changing process in which the life functions of the human body and the factors that cause the disease react upon each other. In this changing process, the unhealthy (the factor that causes the disease) and the upright (human life function) display a tendency that one decreases and the other increases. When the upright conquers the unhealthy, the disease is cured. If the unhealthy takes the upper hand, the patient will die. Although in traditional Chinese medicine diagnosis is also especially emphasized, it, unlike Western medicine, does not only pay attention to the focus of disease. Instead, it grasps the situation of disease and the opportunity to treat disease in the general trend of the fight between the evil and the upright. To be specific, it makes clear the nature of the disease through the different physical signs shown in the patient, including the symptom of the illness, the coating of the tongue, the pulse, the complexion, and the changes shown in the body of the patient. This is in fact a process of "diagnosing symptoms" in which the factors that cause the disease and the human life functions are integrated.

In medical treatment, traditional Chinese medicine emphasizes adjusting the balance of the human body's functions. Although there are such different means as medicine, acupuncture, moxibustion and massage, they all draw support from human upright *qi* to drive away the evil of disease and to restore the human body's normal functions. Its advantage is to make accurate concrete analysis according to varied concrete conditions, namely, "treatment based on diagnosing symptoms" as it is often so called, and in each and every case it is human upright *qi* orientated. Therefore it enables many patients who suffer from incurable diseases according to the diagnosis in Western medicine to pull through and survive.

Traditional Chinese medicine, which pays attention to both the internal and the external of humans and conforms to the logical content of dialectical materialism, seems to be perfect. However, it also has its disadvantages which are mainly the mysteriousness, subjectiveness and ambiguity in the means of diagnosis, such as feeling the pulse, viewing the complexion of the patient and listening to the patient's voice. These methods do not always lead to the same objective examination results with different doctors. Moreover, traditional Chinese medicine is too variable in the treatment of disease. Traditional Chinese medicine doctors, particularly the famous ones, are rather varied in the prescriptions they give and the methods they use and they do not stick to what is already conventional, and this makes traditional Chinese medicine difficult for learners to understand and grasp.

iii) The System of *Qigong* Medical Science

Although *qigong* medical treatment already has a history of thousands of years in China, it has not become an independent branch of learning in history. This is because, on the one hand, traditional *qigong* essentially aims at nurturing life instead of medical treatment, therefore most of the *qigong* practitioners in ancient times generally neglected medical treatment; on the other hand, a theoretical system of medical treatment peculiar to *qigong* medical treatment has not been established. For so many years, *qigong* medical treatment has merely been a component of traditional Chinese medicine and resides in the medical treatment system of TCM which is not its own home. Yet there are actually great differences between them. First of all, *qigong* medical treatment focuses on the human body's functions. It is believed in *qigong* medical treatment that people will not catch disease as long as their *qi* and blood flow smoothly and maintain a normal state. Therefore, *qigong* medical treatment treats disease entirely starting from the internal functions of the human body. Generally speaking, though there are also different methods for treating different diseases in traditional *qigong*, it is not like traditional Chinese medicine in which diagnosis and differentiation of the nature of the disease, such as *yin yang*, *biao li* (outside and inside), *han re* (coldness and hotness) and *xu shi* (deficiency and excess: deficiency of the healthy *qi* and excessiveness of the pathogenic *qi*), are particularly emphasized. What is emphasized in *qigong* medical treatment is the exertion of the function of human upright *qi*. "With upright *qi* inside, unhealthy *qi* cannot disturb" is a reflection of this thought.

The reason why the treatment methods of *qigong* have been recorded in detail in traditional Chinese medical literature is actually because *qigong* is integrated into the system of traditional Chinese medicine. *Qigong* is affiliated to "treatment based on diagnosing symptoms" in TCM. For example, the *qigong* methods recorded in *Zhu Bing Yuan Hou Lun* and *Za Bing Yuan Liu Xi Zhu* are all examples based on such classification. However, genuine *qigong* literature differs from this. For example, in *Song Shan Tai Wu Xian Sheng Qi Jing*, it is said, "The sages say that humans are in *qi* and *qi* is in humans. Humans do not separate from *qi* and *qi* does not separate from humans. Humans rely on *qi* to live and they die when *qi* totally dissipates. The reason for death and life is all in *qi*. So long as one adjusts and practices *yuan qi*, one cannot die even if one seeks death." If we say that this point of argument illustrates the guiding ideology and the principle of treating disease in *qigong*, then *Yuan Qi Lun* gives a concise, clear and specific illustration about the treatment of disease in *qigong* as follows:

"In regard to the flow of *yuan qi*, there is nowhere it cannot reach. As far as the human body is concerned, whether it is inside or outside it, in the place where the disease is located, if metal, wood, water, fire and earth are kept in the mind with their laws of growing and controlling one another being applied while you dwell your mind in the place of the illness, there is no illness that cannot be immediately cured. There is also this excellent knack. Though there are *he, si, hu, chui, xu, xi*, the six ways of breathing, it is not as effective as the two of the cold *qi* and the hot *qi* in healing all kinds of diseases. To unhurriedly swallow a mouthful of *qi*, keep it in the mouth for a while longer, and swallow it when it becomes warm—this *qi* is called warm *qi* which can be used to cure the symptoms of pathogenic coldness. If the *qi* has just been accumulated and is swallowed as soon as it becomes full in the mouth—this *qi* is called cold *qi* which can be used to cure the symptoms of pathogenic hotness. These methods are used with the mind at the time needed." It is also written: "In another case when someone falls ill, just let this person breathe out *qi*, to breathe out ten to thirty times and its effectiveness and efficaciousness will be known. The poison of alcohol and food will all be driven out with the breathing out of *qi*."

This is to say in *qigong yuan qi* is emphasized in treating disease and there is no need to lay so much emphasis on treating disease according to diagnosing symptoms as in traditional Chinese medicine. This has been more specifically stated in *Dan Ting Zhen Ren Chuan Dao Mi Ji*:

"Heaven brings all existences into being with *yin yang* and *wu xing* and all existences take form with the help of *qi*. Nothing in humans—actions, sorrow, joy, pain and other feelings, is without the effect of this *qi*. Therefore, lacking in this *qi* will cause disease and blocking this *qi* will cause disease. It is only when this *qi* is flowing all through the whole body that one can achieve health, peacefulness, bliss and blessedness. I tell you this as I know that what grows the human body is but this *qi* and that which makes one ill is but this *qi*. To cure one's disease, you need firstly to cure one's *qi*. Now doctors use the roots of grass and the barks of trees to get rid of people's pains and sufferings. This is but to use the postnatal *qi* to adjust and nurture the human body. And the five flavors of spiciness, sweetness, saltiness, bitterness and sourness of the roots of grass and the barks of trees are but the five *qi*s[1] of coldness, coolness, warmness, hotness and *ping qi* that is neither hot nor cold. These are the partial *qi* of heaven and earth, yet they can still eliminate disease. What great functions the *qi* of humans can have as human *qi* is connected to heaven and earth and contains the completeness of *qi*."

From this it can be seen that *qigong* treating disease in the real sense starts completely from and is focused on human *yuan qi*, which, therefore, makes *qigong* medical science have a fundamental difference from traditional Chinese medicine and Western medicine. What a great pity it is that this shining pearl of ancient Chinese culture is still buried in the ground, since *qigong* medical science is still not recognized in this world. It should be pointed out here that though there are records of treating illness with external *qi* in traditional *qigong*, e.g., "to arrange or distribute *qi* around oneself or in a particular place" in *Song Shan Tai Wu Xian Sheng Qi Jing*, what is elaborated and more emphasized is treating illness with *qigong* practice. Since *qigong* practice needs a certain amount of time, it is quite understandable that *qigong* medical science was not fully developed in old China when people's subsistence was not guaranteed.

iii. *Qigong Hunyuan* Medical Science

Qigong hunyuan medical science can also be directly called *hunyuan* medical science. It is an important component of *Zhineng Qigong* Science, with the theory of *hunyuan* whole entity as its guidance, with practicing *Zhineng Qigong* and treating disease with external *qi* as its means, and with getting rid of disease and prolonging and enhancing life as its objectives. It is developed upon the foundation of *qigong* being used in medical treatment.

Qigong hunyuan medical science differs not only from Western medicine and traditional Chinese medicine, but also from common *qigong* medical science. It emphasizes the unified feature between the human body's internal *qi* and Nature's external *qi* and focuses on enhancing the

Note:

The five *qi*s are a special term in traditional Chinese medicine. Here, hot and cold do not refer to the medicine itself, but the reaction and life condition in the human body after the medicine is taken.

motions of opening and closing, going out and coming in, gathering and expanding (including condensing and dispersing) as well as change and transformation of human body *hunyuanqi* in order to continuously refine, purify and sublimate the human body *hunyuanqi* until it finally enters the level of *lingtong hunyuanqi* (the miraculously functional *hunyuanqi* in which human consciousness that has initiative is thoroughly combined with primordial *hunyuanqi*). At this point, it is identical with the function and purpose of *Zhineng Qigong* practice.

The reason why it is called *hunyuan* (merged into one or oneness) medical science is because: a) Unlike Western medicine which can only cure postnatal disease but not prenatal hereditary disease, it can cure both prenatal and postnatal diseases so long as *hunyuanqi* is abundant and normal in the human body, and this can be called the *hunyuan* of healing both prenatal and postnatal diseases; b) unlike traditional Chinese medicine which can only cure the disease of the body and abnormal emotions, but not the disease of heart and mind, it integrates treating disease with the cultivation of temperament and character, and this can be called the *hunyuan* of healing both the illness of body and the illness of heart; c) unlike common *qigong* medical science which can cure only common diseases of the body and the mind but cannot create the lost (e.g., to grow bones and flesh from the impaired or lost state), dissolve the existent (e.g., to dispel spur and other abnormal things out of the human body) or detoxicate poison, there is no disease that it cannot cure and no poison that it cannot detoxify. Therefore it can be called *lingtong hunyuan* (the miraculously functional merged-in-oneness).

II. The Physiological Concept of *Hunyuan* Medical Science

Unlike traditional Chinese medicine and traditional *qigong* in which internal organs and *jingluo* are emphasized, what is emphasized in *hunyuan* medical science is *hunyuanqi* in which the three of essence, *qi* and spirit are in one entity. The human body with form is the manifestation of *qi* condensed and the ingenious spirit in which miraculous thinking takes place is the exposition of *qi* magically connected. It is said in *Han San Yu Lu*, "*Jing* (essence or body) and *shen* (spirit) are nothing else but *qi*. Therefore the Daoists talk merely about *qi* instead of *jing* and *shen*." Yet though *hunyuanqi* is emphasized in *hunyuan* medical science, it lays even more emphasis on the function of *shen yi* (spirit and mind). It is believed in *hunyuan* medical science that *shen yi* is the most exquisite and subtle and the most miraculously functional *qi* of human *hunyuanqi*. It has the feature of subjectivity and being at will and can command the life activity of the whole human body. It is a further development of what is said in *Yuan Qi Lun*: "With *shen* intact, *qi* is intact; with *qi* intact, the body is intact; with the body intact, the hundreds of acupoints are well adjusted within and the eight evils are driven and dispelled without". According to *hunyuan* medical science, the entire process of human life activity is a generating and changing process of human *hunyuanqi*. The generation and change of human *hunyuanqi* have been elaborated in Chapter IV in this book, so it will not be repeated here.

Although in the theory of *hunyuan* whole entity it is comprehensively illustrated that the maintenance of normal life activity of the human body needs exchange at the level of concrete substances, the level of energy and the level of information with the outside world, the ultimate aim of *Zhineng Qigong* practice is to gradually substitute the former with the latter. This is the reason why *hunyuan* medical science does not lay emphasis on the internal organs and *jingluo*, as the functions of the internal organs and *jingluo* are essentially for the maintenance of the life of people with familiar intelligence. If you do not think so, please consider the examples below: a) It has been discovered that some people can maintain their normal life without eating food for years; b) in *qigong Bi Gu* practice of eating and drinking nothing, people can also live without having food and drink; c) in traditional *qigong*, when the practitioners have reached a certain level in their practice (e.g., when breath, heartbeat and pulse stop), they can stop their blood circulation and their *qi*'s and *jingluo*'s motion (namely, "the sun and the moon stop their movement" as it is often so called). The first two examples stated above show that people can maintain their normal state of life without drawing support from the exchange of concrete substances; the last example shows that functions of *jingluo* on life have significance only in people's state of familiar intelligence.

What is the nature of *jingluo* according to *hunyuan* medical science?

According to *hunyuan* medical science, the phenomenon of *jingluo* stated in traditional Chinese medicine in fact has nothing to do with the existence of certain concrete channels. *Jingluo* differs from the blood vessels that convey blood. In *Ling Shu* it is said, "The *qi* channels (*qi mai*) are where *ying qi* is stopped from escaping to anywhere else" and "*Ying qi* flows inside *mai* while *wei qi* flows outside it". This is to say that even in traditional Chinese medical classics it is also believed that the twelve *jings* and the blood vessels are not the same thing. What actually does *jing* in the term of *jingluo* refer to? According to *hunyuan* medical science, *jing* is the gathered "streams of *qi*" formed with the *hunyuanqi* of related tissues in the human body. We know that there is *hunyuanqi* around different kinds of substantial or concrete tissues in the human body. Such *hunyuanqi* aggregates into "*qi* streams" that flow in certain directions under the effect of different factors in the process of human life activities. The *qi* in the *qi* streams has more evident advantage than the *qi* around not in the *qi* streams. These *qi* streams can cause certain changes in the biological tissues that they flow past, thus displaying certain features. They can not only be felt by *qigong* practitioners but can also be detected with apparatus and shown by means of certain indices.

As for the sense of *qi* flow in the "*qi* streams", it is caused by a) the influence of the flow of blood (strictly speaking, the *hunyuanqi* of the blood); b) the influence of the flow of lymph (strictly speaking, the *hunyuanqi* of the lymph); c) the influence of the transmission of the excitation signal in the nerves and d) the influence of holistic life activity. Yet it should be made clear here that the *qi* in the *qi* streams formed with the *hunyuanqi* of the tissues in each part of the human body does not actually flow. The flowing feeling or phenomenon brought about by the different influences mentioned above in fact results from the flow of the *hunyuanqi* of relevant substances involved in this effect. As the directions of these flowing substances differ, so is formed the *jingluo* or meridian transmission phenomenon of different directions in the same *qi* channels. This is what has already been proven with experiments on acupuncture or on *jingluo* sensitivity. The *jingluo qi* that "flows" in certain directions caused by the flow of blood belongs in the category of *ying qi* in traditional Chinese medicine, while the rest *qi* belongs in the category of *wei qi*.

How can we explain these statements that contradict the classic theories of traditional Chinese medicine? We know that traditional Chinese medical theories (particularly the theories of *jingluo* and *qi* transformation) were indeed established in the process of ancient *qigong* masters' practice. However, because of the low level in science in ancient times, knowledge about substance was restricted to the level of tangible substance. Therefore it was difficult for the practitioners in ancient times to perceive elaborate details and describe accurately the more subtle state of substances. Even in *Su Wen* and *Ling Shu* in *Huang Di Nei Jing* which is considered as the classic of traditional Chinese medicine, there are also a lot of statements that contradict each other. With careful analysis and comparison, it can be found that there is also quite a lot of content extremely similar to what we have talked about, e.g., some statements about the flow and motion of *wei qi*, the statement that *yang qi* flows from the head to the four limbs, the statement about the *qi* street and so on. Besides, there are also some statements in the classics of traditional Chinese medicine that are actually incorrect. The reason why the concept of *hunyuan* medical treatment in *hunyuan* whole entity differs from traditional Chinese medicine lies generally in this point.

According to the theory of *hunyuan* whole entity, people's normal life activities include three aspects: Firstly, normal spiritual activity in the domain of spirit and mind; secondly, normal life activity at the biological level in the domain of the body and *qi*; thirdly, normal connection between spiritual activity and life activity. This has been introduced in Human *Hunyuanqi* (Chapter IV) in this book, so here only relevant life activity will be briefly analyzed.

The theory of *hunyuan* whole entity states that the human body's life activity goes on mainly at three levels. First, the exchange of various substances (including mass, energy and information) carried out at the level where humans make contact with the outside world (which includes the skin, lungs, membranes of the stomach and intestines, and different sensory organs); second, the exchange of substances inside and outside the cells and also the change in the substances inside the cells at the level of cell; third, the conveyance of various substances inside the human body. These life activities that originally belong to the biological level show their special features in the human body. That is to say, these activities are restricted not only by biological genetics, but also by consciousness activities (consciousness activities in the broad sense). The purpose of *Zhineng Qigong* Science is to gradually reinforce the latter and enable humans to gradually attain control over their own life activities at will.

In addition to this, it is believed in the theory of *hunyuan* whole entity that normal continuance of human life activities must draw support from sufficient internal *qi* that integrates with external *qi*. Otherwise, the maintenance of normal life activities will be difficult. The cultivation and practice of *Zhineng Qigong* is intended to reinforce this process. Traditional *qigong* has also touched upon this point. In *Dao Shu*, it is written: "Among those who engage themselves in the cultivation of their body and moral character, there are some who are lacking in *yuan qi* and consequently not intact in spirit. It is only when their spirit is intact and their *qi* complete that their lives can be truly nurtured and they can bring about miraculous changes in their lives. Otherwise, if their internal *qi* cannot integrate with the external *qi*, it is hard to attain *Dao*. Why is it like this? The internal *qi* is the human *qi*; the external *qi* is the *qi* of heaven and earth. Heaven, earth and humans are all born from the true oneness *qi*. If the internal *qi* is lacking, the original body (before one's self-cultivation) is incomplete in essence, *qi* and spirit."

III. The Concept of *Hunyuan* Pathology

Readers may ask what makes people fall ill. According to traditional *qigong*, "*Yuan qi* in the human body is generated every day. It is only because people do not know how to protect and nurture it that *yuan qi* is depleted every day by two evils. What are the two evils? One is wind, coldness, hotness and wetness that are the evil of *qi*. The other is glee, anger, grief and delight that are the evil of sentiment. The two evils are the thieves of *yuan qi*. They invade and attack *yuan qi* day and night, thus making it become less and weak, until the coming of death." (*Xiu Dao Zhen Quan*)

According to *hunyuan* medical science, all kinds of diseases people suffer from are the result of abnormality in the motion of *hunyuanqi*—in its opening and closing, its going out and coming in, its condensing and dispersing, and its changing between somethingness and nothingness. However, the abnormality is displayed in different aspects of the body, *qi*, mind and spirit. As far as the cause of disease is concerned, there is the abnormality of *qi* caused by biological and chemical substances, the injury to the body caused by physical factors, the abnormality in the *qi* caused by emotions, the prenatal disease formed through hereditary factors, and so on and so forth. This will be explained respectively as follows:

i. Pathological Change in *Qi* Caused by Biological and Chemical Substances

We know that new changes continuously take place in human *hunyuanqi* and that the motion of mutual transformation between *qi* and form also goes on continuously. During this motional process, it is only when the various materials from the outside world have become assimilated and have undergone mergence and change that they can become human *hunyuanqi*. The concrete substances in these materials must be decomposed into fundamental bricks before they can be assimilated by the human body. If the substances from the outside world, after having come into the human body, cannot be decomposed into fundamental bricks that can be assimilated by the human body, or if the substances, though small in volume, cannot be assimilated by humans, they cannot form part of human *hunyuanqi* and will still retain their own features. In other words, the *hunyuanqi* of these substances is not identical with human *hunyuanqi*. When the amount of substance like this reaches a certain limit, normal motion of human *hunyuanqi* will be affected and human normal life activity will be hard to be carried out, thus showing a morbid state. Various biological factors (e.g., bacterium, virus, rickettsia, protozoon, spirochaeta and even parasite) and various chemical substances (including toxin produced by bacterium) that make people become ill belong in this category.

ii. Illness Caused by Injury

Illness caused by injury as described here refers to wounds caused not only by mechanical effect, but also by abnormal temperatures (high and low), electricity, radiation, etc.. The reasons why such factors cause illness seem to because these factors directly act on the human body and cause pathological change in it. In fact, it is not so, because in the *hunyuan* whole entity any change in form follows the change in *qi*. Any external stimulation is in fact also the effect of the *hunyuanqi* of external things. It first destroys the *qi* field of a particular part of the human body, makes the *qi* field directly lose its normal state, and then change takes place in the human body. Some stimulation comes so fast that it instantly acts on the human body and makes obstruction happen in the motion of human *hunyuanqi*, so part of the *qi* field of the human body becomes abnormal and finally brings about change in *qi* and form (body). If no abnormal change happens in the particular part of the *qi* field of the human body, more often than not it is hard for the human body to be harmed.

What should be particularly noted here is that change in human *qi* field is closely related to human consciousness activity. For example, some somnambulists jump off houses ten to twenty feet above the ground and remain safe and sound; it is not easy to injure a sound-sleeper or a drunkard who falls down to the ground from a very tall bed; what is more miraculous, nothing unfortunate happened to some children who fell down to the ground from the third and fourth floor. The author of this book has encountered four cases like this. Moreover, when a person encounters a situation that is really urgent, thinking whole-heartedly of "*hunyuanlingtong*" in the mind can mostly turn the bad luck into a good one. For example, a *Zhineng Qigong* trainer in Beijing who is 73 years of age and is customarily called Sister Cai was once knocked down by a big truck rushing toward her at a high speed when she was walking across the street. When the wheels were running over her body, "*hunyuanlingtong*" flashed instantly across her mind. As a result, Sister Cai's body, after having been run over by the wheels of the big truck, was fortunately unhurt. This is much more marvelous than the performance of "let a truck run over the body" in hard-form *qigong*.

Readers may wonder the reason why this can occur. According to the theory of *hunyuan* whole entity, consciousness can control and command the motion of *qi* and a great powerful *qi* field can be formed when a person attentively thinks of *hunyuanqi*. As the wheels of the truck were not very sharp and could not make an injury that immediately separated or cut the body, the *qi* field of the relevant part of the human body was not destroyed. Based on the premise of the body being intact, the function of the human body's *qi* field can be normally displayed and the human body can thus remain unhurt. Some *qigong* masters give the performance of stabbing a knife into their body, yet their body does not bleed and the wound in their body becomes immediately healed to the original unwounded state when the knife is pulled out. The reason for this is the same as that of the previous example of Sister Cai. Another example: Some *Zhineng Qigong* practitioners silently chanted "*hunyuanlingtong*" in their consciousness the moment they were accidentally scalded by boiling water. The result was that no blisters appeared, and the reason for this is the same.

In addition to this, the climate in the outside world which is called "wind, coldness, hotness, wetness, driness and fire" in traditional Chinese medicine can also rid the human body *hunyuanqi* of its normal state. The mechanism for this is mainly that these factors in the climate act on the *hunyuanqi* at the membrane level of the human body and make the motions of *hunyuanqi* (opening and closing, going out and coming in, condensing and dispersing, and change and transform) become abnormal and therefore make people ill.

If, in an adverse climate condition, people can composedly arrange a *qi* field around their body, usually they will not catch disease. Records exist about Chinese people in ancient times who, before they went to a family that were suffering from plague or to a plague-stricken area, arranged *qi* around their bodies and did not get infected after they went there. The reason for this is the same as what we have just talked about.

iii. Spiritual Factors that Make People Ill

There are two kinds of conditions here. One is that the change in emotion causes the change in *qi*, "mind unified to change *qi*" as it is often so called. Anger making *qi* move upward, gladness making *qi* flow slow, sadness making *qi* exhausted, fear making *qi* go down, fright making *qi* disordered, and (excessive) thinking making *qi* 'knot' as stated in traditional Chinese medicine belong in this type of pathological change. The other is that *Yi Yuan Ti* is not sufficiently bright so that it cannot correctly reflect and appropriately deal with various information from the outside world, which makes one's constitution prone to catch illness. This is usually related to a person's prenatal state.

Similar statements in regard to this can also be found in the classic of traditional Chinese medicine *Ling Shu*. For example, it is said in *Ben Cang Pian*, "There are some people who do not leave their chamber screened and sheltered and have no fear or fright or apprehensiveness. Yet they still cannot avoid being attacked by illness. Why is that? I need to know the reason. *Qi Bo* replied, "…The five *zang* internal organs are what humans rely upon to correspond to heaven and earth, to divide into *yin and yang*, to connect the four seasons and to imitate the five seasons[1]. There are the features of being big or small, high or low, firm or fragile, upright or slant in the five *zang* internal organs and small or big, long or short, thick or thin, bended or straight, slow or swift in the six *fu* internal organs. The twenty-five situations differ from each other, being either good or evil, prodigious or ominous." And "The ones whose five *zang* internal organs are all small rarely become ill; they are just often fidgety and anxious and seriously worried… Those whose five *zang* internal organs are all weak never divorce themselves of illnesses."

It needs to be pointed out here that we have no intention to use the quotations above as examples of prenatal disease in *Yi Yuan Ti*. They are merely quoted as an indication that a person's prenatal quality or state is the reason for catching disease.

In order to illustrate that prenatal insufficiency in spirit can make a person catch disease, let us cite a quotation from Daoist literature *Yu Quan*, "If *shen* (spirit) is turbid, in the essence of *qi* is hidden some reason of insufficiency. It is either lacking in wood, or metal, or water, or fire or earth. With the lack in the five elements, so long as one meets evil in the outside world, the inside will be distracted thus one becomes ill. The illness is in fact the illness of the essence *qi*, while as for curing it the essence *qi* cannot cure itself." This is to say that when one is insufficient in spirit, with some external stimulations, the balance of one's spirit will be disturbed, which leads to an imbalance in *qi*. This kind of disease is different from the abnormality in human *hunyuanqi* caused by external factors previously stated.

It must be also made clear here that there is another kind of morbid state of *Yi Yuan Ti*, that is, the influence of the Partial Identity Reference System of *Yi Yuan Ti* on life. Generally speaking, for familiar intelligence people, the content in the Partial Identity Reference System, be it science, arts, or the mode of *Dao De* (morality) in the Reference System, has both its positive aspect and negative aspect. After having reached a certain level in their *qigong* practice, contradictions in such activities going on in *qigong* practitioners' *Yi Yuan Ti* that are particularly related to their

Note:

The four seasons plus *changxia* which is the last 18 days in the sixth month according to traditional Chinese lunar calendar.

personalities and methods of thinking (e.g., benefiting oneself and benefiting others, conquering oneself and enduring others, etc.) concerning *Dao De* consciousness will become intensified. This is because, with the progress achieved in one's *gongfu* or *qigong* power, the brightness of *Yi Yuan Ti* is continuously increasing, the sensitivity to the reflections of things around is improving and the ability of response also intensifies accordingly. If the mode which is partial in the Reference System is not improved simultaneously, one will easily become emotionally imbalanced on account of trivial issues and be taken ill. Such disease can hardly be cured with the method of adjusting *qi* because *Yi Yuan Ti*'s ability of commanding *qi* has already been intensified with the improvement of one's *gongfu*. The adjustment of *qi* can only adjust a minor part of it. Even if *qi* becomes temporarily normal, it will still be of no help because *Yi Yuan Ti* often sends morbid information which continuously drags one into the morbid state. These diseases cannot be eradicated until *Yi Yuan Ti*'s degree of brightness is improved.

iv. Excessive Exhaustion of *Hunyuanqi*

Excessive fatigue is the main reason for the consumption and exhaustion of *hunyuanqi*.

Excessive exhaustion of the essence of reproduction not only directly consumes human *hunyuanqi*, but also affects the prenatal *qi* field of kidney *qi*, which will inevitably affect the assimilation of nutrition in the stomach and small intestine and also the *qi* change and *qi* transformation of the substances having been taken into the human body. This mechanism that causes disease is relatively easy to understand, so it will not be described here in great detail.

v. Hereditary Disease

This is the defect in the *hunyuanqi* of the newborn individual that is formed because of insufficiency in the parents' hereditary factors. Although modern science places great hopes on genetic cures for this kind of disease, so far no feasible method has been found. However, in the preliminary experiments into treating hereditary disease with *hunyuanqi*, encouraging developments have been manifested, e.g., there are already cases of hemophilia having been cured.

From the above it can be seen that, though the concept of pathology in *hunyuan* medical science is relatively simple, it has grasped the most essential point. Although disease in people is shown in different forms as the disease of the body, *qi*, mind and spirit, they all have in them *qi* as the medium and the spirit as the guidance. Once spirit and *qi* are brought under control, all problems will be easily solved. It is true that in the process of specific pathological change in the human body, there is change in *qi* that causes change in form, e.g., the condensing of *qi* that forms different kinds of morbid concrete substance, and there is also the imbalance in *qi* which causes the suffering of the attack of disease-causing biological substance. However, all this has not gone beyond the domain of change in *qi* and *shen*.

IV. The Concept of *Hunyuan* Medical Treatment

The medical treatment based on the theory of *hunyuan* whole entity differs not only from medical treatment in traditional Chinese medicine and Western medicine, but also from *qigong* medical treatment carried out in many *qigong* schools. When treating disease, *hunyuan* medical science lays emphasis not on the condition of the disease, but on the motion of human *hunyuanqi* and restores the *hunyuanqi* that has lost its normal motion to its normal motional state. To be brief and to use common expressions, it does not emphasize the treatment of disease by means of differentiating symptoms or differentiating diseases. *Hunyuan* medical treatment is a simple, convenient, easily applicable and highly effective method which can "deal with tens of thousand (of diseases) by using one means" of *hunyuanqi*. This argument seems to be against common sense, but in fact, there are similar statements in traditional *qigong* literature. For example, in *Fu Qi Liao Bing* in *Yun Ji Qi Qian*, it is said,

"If you at times feel tired and restless, guide *qi* and close it in the body to attack where the trouble lies. You must keep in mind that the head, face, nine apertures[1], five *zang* internal organs, four limbs, and even the end of the hair are all where *qi* is. You feel that *qi* flows in the body, beginning from the nose and reaching downward to the ten finger-tips. Then your spirit will become crystal clear and harmonious. You need not receive acupuncture and moxibustion or take medicine and you will be fine again. Whenever you want to get rid of any kind of disease by means of activating *qi*, you should make your mental activity where the disease is. If the head is aching, you keep your mind focused on the head; if the feet are aching, you keep your mind focused on the feet. Bring *qi* to attack the ache and it will disappear in two hours. Sometimes when you feel cold *qi* within, you can close your *qi* in to cause perspiration. When the sweat is out from the whole body, you will be cured."

That is to say that taking *qi* to cure disease (a treatment method used in traditional *qigong*) does not require the differentiation of the symptoms of the illness. So long as you can guide *qi* to where the disease is with your mind and make *qi* flow smoothly, the disease will be naturally cured.

In addition, treating disease in *hunyuan* medical science emphasizes the treatment of *shen* (spirit). On the one hand, *hunyuan* medical science involves calming down the patient's spirit with language, informing the patient of the power of *hunyuan* medical science and establishing in the patient the confidence to conquer the disease. This we call "conversation therapy", which

Note:

Nine apertures: Seven apertures of eyes, ears, nostrils and mouth, and pharynx and larynx.

is an important method to cure not only psychological disease (otherwise known as "disease of the heart"), but also the disease of the body. On the other hand, in *qigong* practice and in the adjustment of *qi*, *hunyuan* medical science pays attention to the adjustment of *shen*. If the patient's spirit can be adjusted, greater effect will be achieved with less effort. If the spirit of the patient is not adjusted, some diseases are not easily cured. There is also similar statement in *Han San Yu Lu*:

"The illness in the body can be treated, the disease in the mind is hard to cure. The devil outward can be conquered, the devil inward is hard to control. ... I feel pity for the great multitude and become a doctor to cure the disease in their heart and make their internal devil submitted. Once the disease in the heart is eradicated and the internal devil is eliminated, how can the devil outside invade? If a person's heart is not cured, the disease can enter into the acupoints of *gao* and *huang*. If the devil is enclosed in the body and gets stronger, the devil from the outside will also grow stronger."

The reason why the adjustment of *shen* can have such a great impact lies critically in *shen*'s commanding effect on *qi*. It is said in *Tai Shang Dao Yin Yang Sheng Jing*, "*Qi* is controlled by the heart (spirit). If the heart is evil, *qi* will be evil. If the heart is upright, *qi* will be upright. The motion or action of *qi* as well as glee, anger, sorrow and delight are from nowhere else but the heart. Activity in the heart is nothing but the motion of *qi*. *Qi* is felt and connected by the mind and the mind follows the heart. Therefore, if the *qi* is intact, the body will be intact; if the *qi* is exhausted, the spirit will be extinguished. If the spirit is extinguished, the body is dead."

All this sufficiently shows that in traditional *qigong* spirit is especially emphasized in treating illness.

According to *hunyuan* medical science, all diseases are curable. This is because, though there are different forms of diseases of the body, of *qi*, of mind and spirit, they all result from abnormal changes in the different levels of *hunyuanqi*. Now that normal *hunyuanqi* can integrate and change with the *hunyuanqi* from the outside world and becomes illness, it will undoubtedly be able to, in the case of disease, merge and combine with the *hunyuanqi* in the outside world again, transform, and be restored to the normal state. This is to say, any disease should and can be reversible. With regard to this, there is a similar statement in the traditional *qigong* literature *Tai Qing Yang Sheng* (Part II).

"The classic says that the human body has twelve big joints (which actually refer to big bones) and 360 small bones, with hole to hole, *qi* channel to *qi* channel, and with the new *qi* and the old *qi* flowing in them. The new *qi* either is abruptly blocked or moves smoothly along. The old *qi* either smoothly flows or is clogged up. The new *qi* and the old *qi* are either completely blocked or rush along together. The clogged up is the gathering of *yang qi* and it agglomerates into a lump. The abruptly blocked is the congesting of *yin qi* and it becomes the swell and ulcer. Now that *qi* can be accumulated, it can also be dissipated. Wherever the trouble is, *qi* can be guided there to dispel it, can be brought there to attack it, can be sent there with the mind to drive it away, can be nourished with the clear *qi* and can be replenished with the swallowing of saliva."

The quoted passage not only illustrates that diseases can be cured but also enumerates the methods to cure them. The methods stated here about curing disease seem to be complicated. To get to the core of it, it is none other than the mind, *qi* and guidance of *qi*, which is very similar to the propositions in *Zhineng Qigong*.

In curing disease with *hunyuan* medical science, though there are two ways of treating disease with *qigong* practice and with external *qi*, they both start from *hunyuanqi* and emphasize the conquering of disease by means of *shen yi* (spirit and mental activity) commanding *hunyuanqi*. The broadness in the extent of treatment and the swiftness in the effect of therapy are almost incredible. This has already been proven by the experience of millions of practitioners.

As for the power of *hunyuanqi*, particularly the power of *lingtong hunyuanqi*, it has been expounded in Human *Hunyuanqi* (Chapter IV). In order to make *Zhineng Qigong* practitioners further understand the power of *hunyuan* oneness *qi* (*hunyuanqi*), I would like to quote a para-graph from *Qi Jing* written by the Old Man Wuyuan Hunyi, "I know the subtle rule of Tathagata as well as *hunyuan* prenatal oneness *qi*. There is no place with its presence that is not illuminated by it. This light is not the common light, not the human light, nor gods' light, or Buddha's light. It is not the light that is seeable, not the cosmos light or the light of the six directions (above and below, left and right and front and back). It is the profound reason, the inerasable dauntless light of all. This is the light that can see and we can also say that it does not see. It can also illuminate. There is nowhere that this light cannot illuminate, and it can illumi-nate all. It is the light of the heart... This is called the heart light of *hunyuan* prenatal oneness *qi*."

This quoted passage illustrates that *hunyuanqi* is the prenatal light of the heart (i.e., the function manifested by *Yi Yuan Ti*) which can enter all realms—"There is no place with its presence that is not illuminated by it".

Frankly speaking, it is not that *hunyuan* medicine can cure all patients with any kind of disease. This is because whether a patient can be cured or not is closely related to the patient's spiritual state. If the spiritual state of the patient cannot be improved, a disease that is not serious (referring to the disease that has not burdened the patient's consciousness) can undoubtedly be quickly cured under the effect of the powerful *qi* field. That many patients who have their sebaceous cysts and fatty tumors instantly dispelled when the *qi* field is organized is the clear proof. But for patients who are seriously ill, particularly those who are heavily burdened in their spirit and have no confidence in getting cured, it is hard to achieve good effect. This is because such state of mind in the patients alone can induce illness. It is pointed out in *Xing Ming Gui Zhi*, "If the heart (which refers to the mind) is ill, the body will be ill. If the heart is not ill, the body will not be ill. Therefore the disease in the body results from the disease in the heart." As to what diseases can be cured and what cannot—the outcome can only be known through summing-up of practice. Before the treatment is carried out, it is hard to judge whether or not a disease can be cured.

Some people say that *qigong* can only cure the diseases which neither traditional Chinese medicine nor Western medicine can treat effectively. We think that this kind of opinion is not only groundless, but does not conform to logic thinking. Just think it over and we will know that it is incorrect. Considering that the diseases that neither traditional Chinese medicine nor Western medicine can cure (relatively speaking, the diseases that are more difficult to treat) can still be cured by *qigong*, how can it be that the diseases that both traditional Chinese medicine and Western medicine have the ability to cure make *qigong* medical science unable to exercise it magic power? In actual fact, the effects of *qigong* treatment on common ailments and frequently-occurring diseases are usually very remarkable as well.

What has been stated above is essentially related to the treatment of disease, which belongs to *hunyuan* medical science in the narrow sense. In the broad sense of *hunyuan* medical science, all the effects of *qigong* practice in *Zhineng Qigong* are exclusively the treatment of the partiality and biasness in the human body and mind. *Qigong* practice in the preliminary stage that aims at improving the health level of the human body and mind is for this; eradication of atma-graha (adherence to the familiar intelligence self and partial reflections and opinions) and sublimation of the Reference System that are higher level *qigong* practice also belong in treating the disease of the heart, and this is the fundamental task of *Zhineng Qigong* Science. Neither traditional Chinese medicine nor Western medicine can give solution to these problems. It has been explicitly pointed out in *Yang Zhen Ji* that "Famous doctors cure the disease in the human body while sages in the three sacred schools (Confucianism, Buddhism and Daoism) cure the disease of the human heart." The disease of the heart stated here includes not simply psychological problems, but also those states of mind that hinder people from "surpassing the secular, entering the sacred realm and achieving the goal of becoming the true man."

In *qigong* science, the adjustable and controllable activities in the mind are categorized as regular right mental activities that can gather energy and thus enable one to display a healthy spiritual state. Unadjustable and uncontrollable activities in the mind are irregular distracting mental activities that exhaust energy and therefore make one display a morbid state of spirit. *Qigong* practice is to get far away from this morbid state, to enter the healthy state and to make humans gradually reach the free realm in their own life activities.

In *hunyuan* medical science, effects of organizing *qi* field among group of people are emphasized in treating disease. Organizing *qi* field is emphasized whether it is in treating disease with external *qi* or in treating disease with *qigong* practice. As to how to use external *qi* and how to organized *qi* field to treat disease, this belongs to "Diagnostics of *Zhineng Qigong* Science".

Bibliography

Abhidharmakosa-sastra

Bai Yun Guan Zhi

Bao Hun Yuan Xian Shu in *Xian Shu Mi Ku*

Clinical Psychology by Zhang Chunxing, 1989

Da Xue (or: *The Great Learning*)

Dan Ting Zhen Ren Chuan Dao Mi Ji

Dao De Jing

Dao Mai Tu Jie

Dialectical Materialism and Historical Materialism

General Morphology by Ernst Haeckel

Gongyang Zhuan

Guan Zi

Han Fei Zi

Han San Yu Lu

Hua Shu

Huai Nan Zi

Huang Ting Wai Jing Jing Shi Zhu

Hun Yuan Sheng Ji

Introduction to Psychology by Atkinson and Hilgard, et al.

Jin Ye Huan Dan Nei Pian in *Dao Shu*

Le Yu Tan Yu Lu

Learning Approaches and Their Application in Education (translated from Chinese into English) by Walter B. Kolesnik

Li Ji

Lie Zi

Ling Shu (Jing) in *Huang Di Nei Jing*

(or: *Miraculous Pivot* of *The Yellow Emperor's Classic of Internal Medicine*)

Ling Xian

Mai Wang

Meng Zi (or: *Mencius*)

Nan Jing

Outline of Dialectic and Materialism by Mao Zedong, 1937

Peng Niao Fu

Psychology and Life by Philip Zimbardo

Qi Jing

Shi Ji

Shi Zi

Shuo Gua

Shuo Wen

Song Shan Tai Wu Xian Sheng Qi Jing

Su Wen in *Huang Di Nei Jing*

(or: *Plain Questions* of *The Yellow Emperor's Classic of Internal Medicine*)

Tai Ping Jing

Tai Qing Yang Sheng Xia Pian

Tai Shang Dao Yin Yang Sheng Jing

The Complete Works of Lenin Vol. 2, 7

The Complete Works of Marx and Engels Vol. 3, 20, 42, 46

The History and Status of General Systems Theory by Ludwig von Bertalanffy

Wen Shi Zhen Jing

Wu Liu Xian Zong

Xi Ci in *The Book of Changes*

Xin Chuan Shu Zheng Lu

Xing Ming Gui Zhi

Xiu Dao Zhen Quan

Xuan Men Bi Du

Xun Zi

Yang Zhen Ji

Yu Quan

Yuan Ren Lun

Yun Ji Qi Qian

Zheng Meng Qian Cheng

Zhi Guan

Zhi Ming Pian

Zhi Yan Zong

Zhong Yong (or: *The Doctrine of the Golden Mean*)

Zhou Yi Zheng Yi

Zhuang Zi